Ostracodology – Linking Bio- and Geosciences

Developments in Hydrobiology 197

Series editor
K. Martens

Ostracodology – Linking Bio- and Geosciences

Proceedings of the 15th International Symposium on Ostracoda, Berlin, 2005

Edited by

Renate Matzke-Karasz[1], Koen Martens[2] & Michael Schudack[3]

[1] *Department of Environmental and Geosciences, Palaeontology, and Geo-Bio-Center[LMU], Ludwig-Maximilians-Universität München, Richard-Wagner-Str. 10, 80333 Munich, Germany*

[2] *Koninklijk Belgisch Instituut voor Natuurwetenschappen, Afdeling Zoetwaterbiologie (Royal Belgian Institute of Natural Sciences, Freshwater Biology), Vautierstraat 29, 1000 Brussels; and University of Ghent, Biology, K.L. Ledeganckstraat 35, 9000 Gent, Belgium*

[3] *Institut für Geologische Wissenschaften, Fachrichtung Paläontologie, Freie Universität Berlin, Malteserstrasse 74-100, 12249 Berlin, Germany*

Reprinted from Hydrobiologia, Volume 585 (2007)

Sponsors:

RWE Dea

Micro Kern
Systeme für die Mikrotechnik

Raczek

LITERATURHOEKEREI WIESE

Micro Tech Lab
Hans Friz Weg 24 , A-8010 Graz , Austria
Fax: +43-316-324763

Freie Universität Berlin

Springer

Library of Congress Cataloging-in-Publication Data

A C.I.P. Catalogue record for this book is available from the Library of Congress.

ISBN-13: 978-1-4020-6417-3

Published by Springer,
P.O. Box 17, 3300 AA Dordrecht, The Netherlands

Cite this publication as Hydrobiologia vol. 585 (2007).

Cover illustration: *Eucypris virens* (Cypridoidea, Ostracoda) from Sicily, length about 1.3 mm.
Photo: Renate Matzke-Karasz

Printed on acid-free paper

Printed in the Netherlands

TABLE OF CONTENTS

EVOLUTIONARY SIGNIFICANCE OF OSTRACOD MORPHOLOGY

OSTRACOD REPRODUCTION AND ONTOGENY

Hydrobiologia (2007) 585:1–11
DOI 10.1007/s10750-007-0719-1

OSTRACODA (ISO15)

Ostracodology in time and space: looking back on fifteen International Symposia on Ostracoda, and the times in between

Renate Matzke-Karasz · Michael Schudack · Koen Martens

Abstract Fifteen International Symposia on Ostracoda (ISO) have been organized between 1963 and 2005, each of them followed by one or more proceedings volumes. In the present study, 906 titles presented at the occasion of one of these fifteen symposia were evaluated regarding their prevailing and their secondary theme or method, thus obtaining a retrospective on the history of ostracod research between ISO1 and ISO15. This review may help in moving ostracodology into the 21st century.

Keywords Ostracodology · International Symposium on Ostracoda · IRGO

Guest editors: R. Matzke-Karasz, K. Martens & M. Schudack
Ostracodology – Linking Bio- and Geosciences

Electronic supplementary material The online version of this article (doi:10.1007/s10750-007-0719-1) and accessible for authorized users.

R. Matzke-Karasz (✉)
Department of Environmental and Geosciences, Palaeontology and Geo-Bio-Center LMU, Ludwig-Maximilians-Universität München, Richard-Wagner-Str. 10, 80333 Munich, Germany
e-mail: r.matzke@lrz.uni-muenchen.de

M. Schudack
Institut für Geologische Wissenschaften, Fachrichtung Paläontologie, Freie Universität Berlin, Malteserstrasse 74-100, 12249 Berlin, Germany

K. Martens
Koninklijk Belgisch Instituut voor Natuurwetenschappen, Afdeling Zoetwaterbiologie, (Royal Belgian Institute of Natural Sciences, Freshwater Biology), Vautierstraat 29, 1000 Brussels, Belgium

K. Martens
Department of Biology, University of Ghent, K.L. Ledeganckstraat 35, 9000 Gent, Belgium

Introduction

Naples, Italy, 1963: Dr. Harbans Singh Puri, a pioneer in interdisciplinary ostracod research, organises the first International Symposium on Ostracoda (now generally abbreviated as ISO) at the historic Zoological Station of Naples, or *Stazione Zoologica Anton Dohrn*. True to the motto "Ostracods as Ecological and Palaeoecological Indicators", some 20 ambitious ostracodologists come together and, recognizing the need for more joint organization of ostracodology, found the Committee on Recent Ostracoda, which was later transformed into the International Research Group on Ostracoda.

Germany, Berlin, 2005: 42 years later, Prof. Dr. Michael Schudack, Prof. Dr. Steffen Mischke and Dr. Ulla Schudack invite the ostracodologists for ISO15 at the Freie Universität, and 155 highly motivated young and established researchers in the field of ostracodology come together to present their latest work in posters and talks and to start fruitful discussions, if not new collaborations.

Many research projects have been finished and countless papers on ostracods have been published in the time between Naples and Berlin. Among the publications, there were quite a few directing ostracodologists' attention (and not only theirs) to the unbounded scientific possibilities that applied ostracodology may open. Some of these are: (1) a permanent reflection of practical ostracodology on one hand, and (2) a clear invitation to use this outstanding group of microcrustacea as a tool in applied biology and palaeontology (e.g. De Decker & Forester, 1988; Martens & Horne 2000; Holmes & Chivas, 2002; Boomer et al., 2003).

In contrast, only a very few authors reviewed the achievements of ostracodologists in history. From the ISO proceedings, three reviews are noted which partially touched that topic. In 1976, Sohn reviewed the papers on ostracods of the previous decade, with reference to the geographical distribution of authors and to the age of ostracods being studied (Sohn, 1977). Six years later, Kesling gave an overview of ostracodology's state-of-the-art by means of new techniques and new concepts being used for research (Kaesling, 1983). A more historical study was published by Neale (1988), summarizing the development of the most important research areas in ostracodology through the centuries. However, this overview was done in a descriptive way, mainly focusing on the achievements of single ostracod-researchers who generated outstanding results in their fields. In contrast, the present study is based on numerical data, trying to show the paths that our research actually took between 1963 and 2005, the time covered by the ISO events. Several questions are addressed: Did ostracodology react to general trends and requirements in science and/or business? How fast did new techniques find their way into ostracodology? Which research methods are well represented in ostracodology, and which are not (yet)? On the occasion of the present proceedings volume (part 1) of ISO15, a retrospect on fifteen ISO proceedings may shed a light on the recent history of ostracod research, and may even help to see where the future will take us.

Materials and methods

For the presented study, the following proceedings volumes were analyzed (themes in parentheses):

ISO1, Naples 1963 (Puri, 1964); "Ostracods as Ecological and Paleocological Indicators".

ISO2, Hull 1967 (Neale, 1969); "The Taxonomy, Morphology and Ecology of Recent Ostracoda".

ISO3, Pau 1970 (Oertli, 1971); "Colloquium on the Paleoecology of Ostracodes".

ISO4, Newark 1972 (Swain, 1975); "Biology and Paleobiology of Ostracoda".

ISO5, Hamburg 1974 (Hartmann, 1976); "Evolution of Post-Paleozoic Ostracoda".

ISO6, Saalfelden 1976 (Löffler & Danielopol, 1977); "Aspects of Ecology and Zoogeography of Recent and Fossil Ostracoda".

ISO7, Belgrade 1979 (Krstić, 1979); "Taxonomy, Biostratigraphy and Distribution of Ostracodes".

ISO8, Houston 1982 (Maddocks, 1983); "Applications of Ostracoda".

ISO9, Shizuoka 1985 (Hanai et al., 1988); "Evolutionary Biology of Ostracoda".

ISO10, Aberystwyth 1988 (Whatley & Maybury, 1990); "Ostracoda and Global Events".

ISO11, Warrnambool 1991 (McKenzie & Jones, 1993); "Ostracoda in the Earth and Life Sciences".

ISO12, Prague 1994 (Riha, 1995); "Ostracoda and Biostratigraphy".

ISO13, Chatham 1997 (Boomer & Lord, 1999; Holmes & Horne, 1999; Horne & Martens, 2000); "Non-marine Ostracoda: Evolution and Environment; Marine Ostracoda and Global Change; Evolutionary Biology and Ecology of Ostracoda" and unpublished abstract volume.

ISO14, Shizuoka 2001 (abstract volume, unpublished); "Towards the new ostracodology in the 21st century" (proceedings volumes by Ikeya et al., (2005a) and Ikeya et al., (2005b) were not evaluated).

ISO15, Berlin 2005 (abstract volume, Kohring & Sames, 2005); "Ostracodology—linking bio- and geosciences".

ISO15 and ISO14 have been evaluated by their abstract volumes because of the much higher numbers of presented projects than is reflected by the subsequently published manuscripts in the proceedings. ISO13 has been evaluated by both the published proceedings volumes and the unpublished abstract volume of talks and posters. The proceedings volumes of ISO12 and ISO11 included poster abstracts, which were also included in the dataset. From the other ISO events, no abstract volumes were available, thus the data all rely on the published proceedings. However, these older volumes tended to contain the majority of presentations, as rejection of manuscripts only routinely started when proceedings were published as special issues of international journals.

Very few manuscripts or abstracts presented in the proceedings were not considered in the present study, such as non-ostracod papers, general reviews on ostracod research, or laudations. In total, 906 titles presented at ISO events were categorized.

The topics of all projects, presented either as abstracts or as complete publications, were evaluated and sorted with respect to their prevailing and their secondary theme or method. No differences were made between neontological and palaeontological research, e.g. 'biodiversity' and 'palaeobiodiversity' were not differentiated.

Within the eight main topics, namely taxonomy, morphology, genetics, reproduction, biodiversity, ecology, biogeography and biostratigraphy, several subcategories were formed according to the context in which (for example) a taxonomical or ecological study had been carried out. Thus, as an example, the collected data now allow a distinction between morphological studies carried out to gain new insights into phylogeny and evolution and those to relate ostracod body characters to ecological parameters of the habitat. Of course it was not always evident how to classify the projects, but, in general, every manuscript finally contained clear arguments to choose one of the eight main categories and a secondary category.

Results

All data obtained on the projects presented in the analyzed ISO proceedings and abstract books can be viewed in the Electronic supplementary material.

From these raw data, several diagrams were prepared to visualize (1) the differences in the composition of research projects presented at the different ISO events and (2) the development of ostracod research through time. Figure 1 shows an overview of the eight main categories of ostracod research presented on the International Symposia on Ostracoda, while Figs. 2–4 show some more details on the purposes of taxonomical, morphological, and ecological studies by breaking the data down into the different fields in which context these studies were executed. Finally, the development of research on ostracod taxonomy, morphology, (palaeo)biodiversity, (palaeo)ecology, (palaeo)biogeography and biostratigraphy and the relations between them are presented in Figs. 5 and 7.

Discussion

International Symposia on Ostracoda are interdisciplinary, international meetings of ostracod researchers, convened by invitation of different organizers in different countries. The basic idea is to provide a forum for presentations and discussions among all ostracod researchers. However, traveling to symposia costs money, and not all ostracod researchers can cover the costs for traveling around the world to meet their colleagues. Moreover, fellow ostracod researchers may not be able to leave their country for political reasons. As a result, the present analyses of the ISO outcomes may not fully represent ongoing ostracod research at that time. This weakness is of course a disadvantage for the present study—however, it is a much bigger hiatus in the meetings themselves, and the editors of the present proceedings look forward to the day when *every* ostracod researcher will be able to attend the coming international symposia on Ostracoda.

Each International Symposium on Ostracoda has had a main theme. Most of them are rather generally phrased, but some of these themes have of course influenced the composition of the presented work—a fact that can be noted for several, particularly the earlier symposia, such as ISO1 with its high ecological or ISO5 with its high morphological score. Nevertheless, in these symposia, many contributions were still not related to the given motto, as seen particularly in the poster abstracts of ISO12: while the manuscript contributions very much reflect the biostratigraphic bias of the given theme, the symposium abstracts included come from all possible fields. This reflects a selection of oral presentations conforming to the theme, followed by a publication in the proceedings volume. However, these pre-selections most likely did not influence the general trends of the development of contemporary ostracod

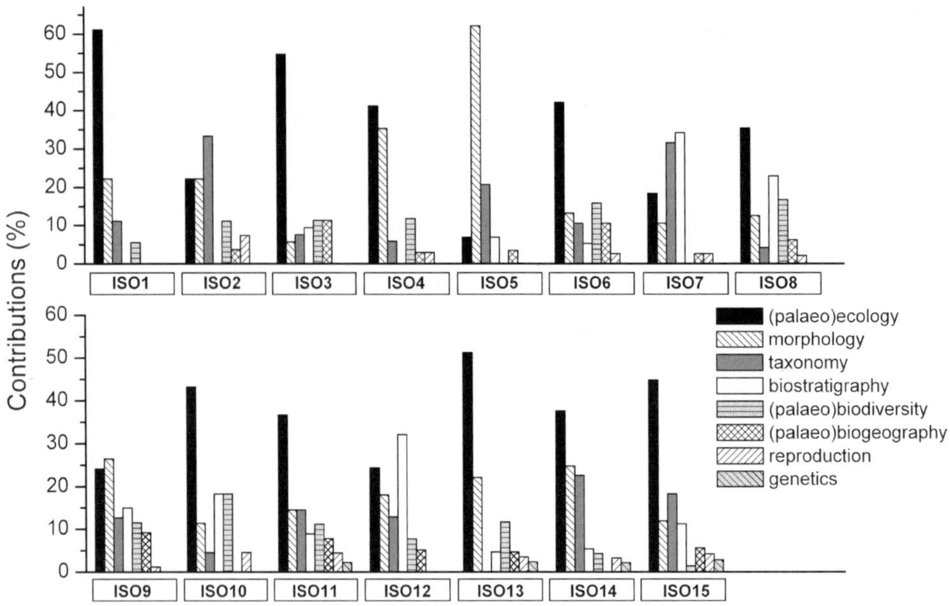

Fig. 1 Proportions of the eight main categories of ostracod research presented at the 15 International Symposia on Ostracods between 1963 and 2005

research, but are rather seen as a mirror of the respective research climate.

Taxonomy is the base of all ostracod research. Without well-defined alpha-taxonomy, ecological, genetical, biostratigraphical, biogeographical and biodiversity studies would have less relevance (and of course, vice-versa, these fields do contribute to a reliable taxonomy).

While ostracod taxonomy played the main role at ISO2 (33.3% of the presented topics; Figs. 1, 5), a

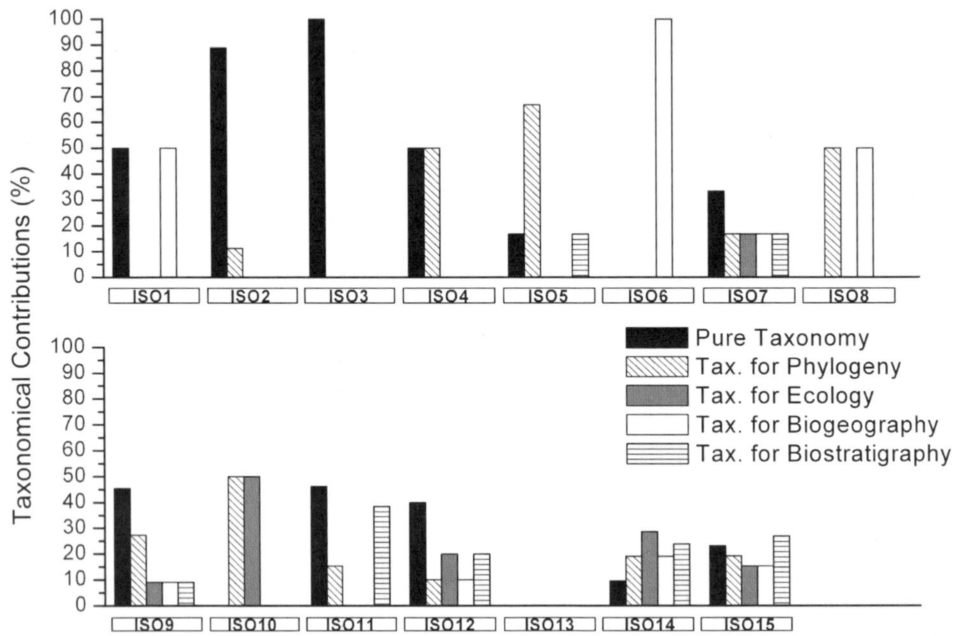

Fig. 2 Taxonomy at the different ISO events, broken down to five subcategories

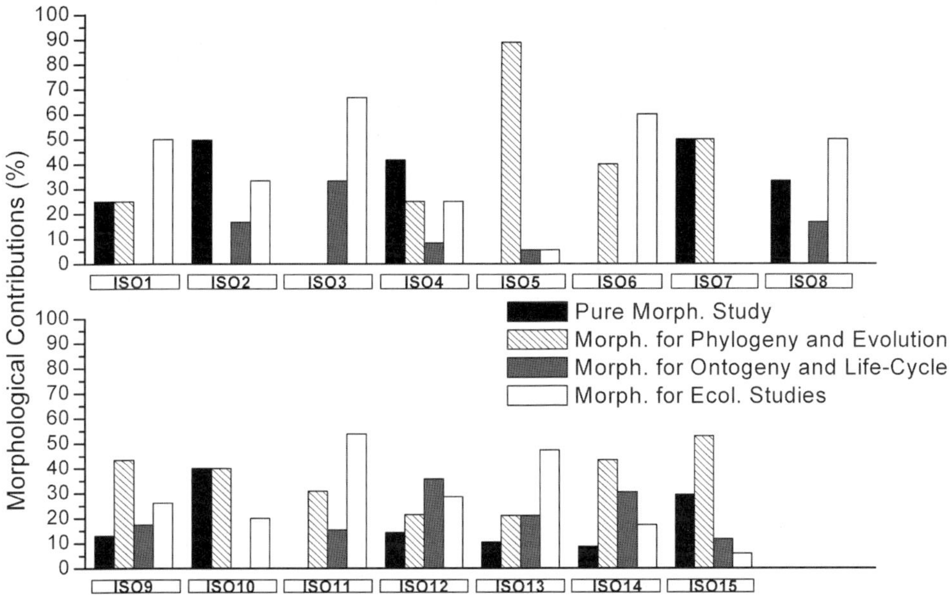

Fig. 3 Morphology at the different ISO events, broken down to four subcategories

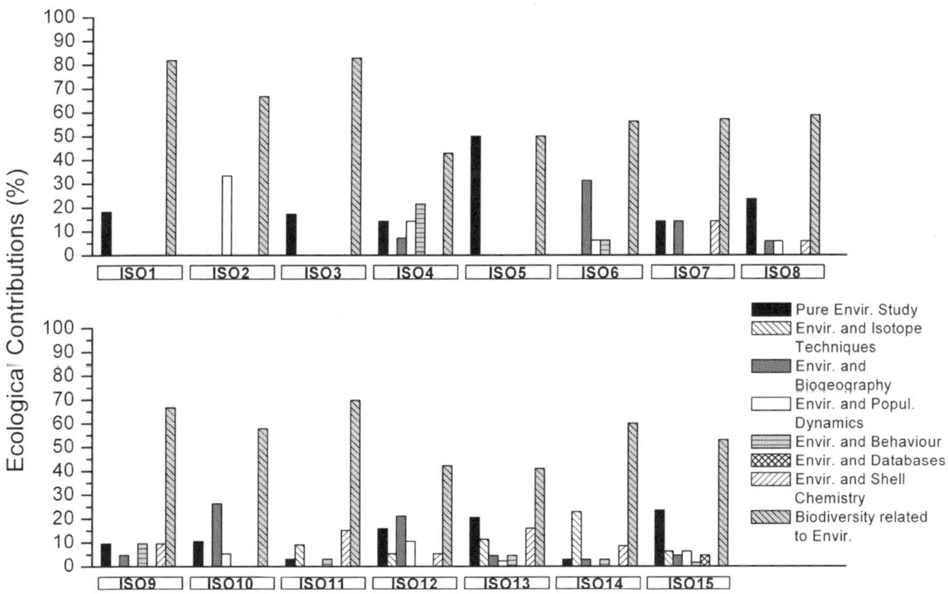

Fig. 4 Ecology at the different ISO events, broken down to eight subcategories

slow decrease is to be noted (apart from a peak at ISO7 which may be understood as an attempt to revive this area by including it in the main theme) until a stunning complete absence at ISO13 in 1997 (even in poster abstracts). When breaking down the data to details (Fig. 2), it becomes evident that within taxonomy itself there is a hidden second trend: at ISO1 to ISO3, most of the taxonomic studies were carried out as pure, fundamental taxonomy, while as from ISO4 this "exclusivity" mostly vanishes in favor of studies that feature taxonomy in the context of phylogeny, ecology, biogeography or biostratigraphy.

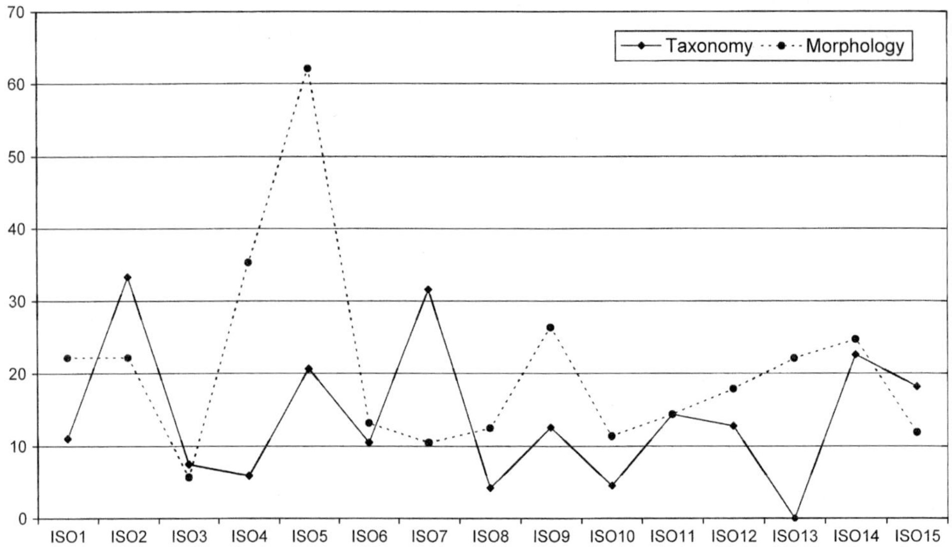

Fig. 5 Performances of taxonomy and morphology at the different ISO events. *y*-axis: proportions of taxonomical and morphological contributions in percent

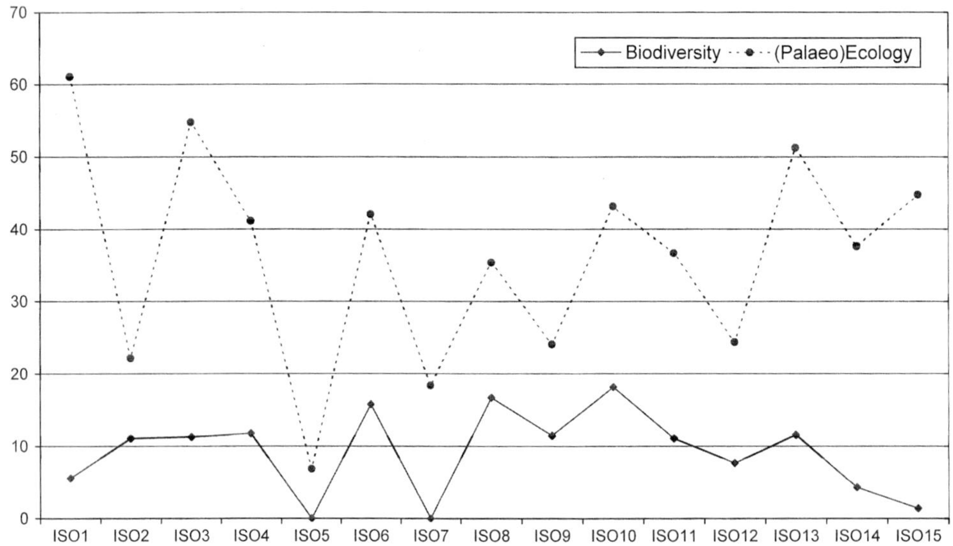

Fig. 6 Performances of biodiversity and (palaeo)ecology at the different ISO events. *y*-axis: proportions of biodiversity-related and ecological contributions in percent

However, at ISO9, ISO11 and ISO12, the percentage of taxonomic contributions seems to be balanced around 10–15% (at ISO12 quite evenly divided among fundamental and applied studies), to be followed by the complete breakdown of taxonomy at ISO13 (Fig. 6). Possibly, this effect can be explained by a change in funding policies which (at least in the European research area) largely neglected the necessity of fundamental taxonomic research. Taxonomy went out of fashion, and had to be included in other, more applied and/or fashionable, project proposals. However, countless horrifying reports from conservation biologists about extinction trends, as well as worldwide taxonomic actions (such as the Global Taxonomy Initiative, and others) initiated by ambitious biologists, seem to have set a limit to the extinction of

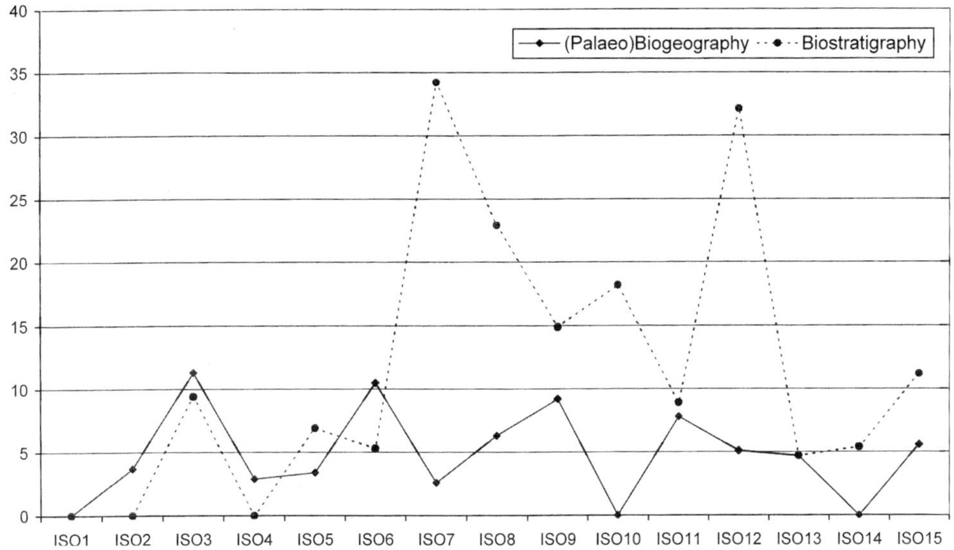

Fig. 7 Performances of (palaeo)biogeography and biostratigraphy at the different ISO events. *y*-axis: proportions of biogeographical and biostratigraphical contributions in percent

taxonomy. A budding new self-confidence of taxonomists is probably also reflected by the re-birth of ostracod taxonomy at ISO14 (22.6%) and ISO15 (18.2%), both presenting a good mixture of fundamental and applied taxonomical studies.

Ostracod taxonomy is traditionally built on *morphology*. Since hypotheses regarding evolution and phylogeny are mainly based on ostracod taxonomy, morphology also serves in these domains. Additionally, morphological knowledge of juvenile ostracods serves ontogenetic investigations as well as life-cycle-studies. Apart from a high peak at ISO5 (owing to its theme, morphology for phylogeny and evolution was highly pronounced), morphological investigations represent the more or less continuous background technique at all ISO events (Figs. 1, 5). Since environmental factors are often reflected by morphological traits, morphology is to a high degree an integral part of ecological studies, as shown in Fig. 3. Increasing interest in soft-part preservation in fossil ostracods has added a new aspect to the understanding of the group as a whole, and new technologies such as non-invasive methods might even give a major new impulse to this traditional research area in the near future. There is still a lot to be expected, the future is bright!

Ecology (including palaeoecology) is one of the major fields of investigation within ostracodology (Figs. 1, 6), and apart from a clear 'depression' at ISO5, it was always one of the main topics in ostracod symposia, if not the largest part of all ISO contributions (which was the case at 10 out of 15 ISO events). Additionally, at ISO1, ISO6 and ISO8, a high proportion of presented taxonomic studies were aimed at an improved understanding of ecological data (Fig. 2). Within the field of (palaeo)ecology, most studies deal with the relationship between (palaeo)diversity and various environmental factors, followed by pure environmental studies and those combining (palaeo)ecology and (palaeo)biogeography (Fig. 4). At ISO7 (1979), shell chemistry was introduced as an innovative technique for the use of ostracod shells as environmental proxies (peak at ISO13), followed by isotope analyses from ISO11 (1991, peak at ISO14). At ISO15, no project on pure shell chemistry was presented—it seems as if isotope analysis has currently succeeded as the most relevant technique if deductions from ostracod shell compositions on former environmental conditions are required. Behavioral studies related to environmental factors play only a minor role.

The outstanding performance of ecological research within ostracodology is not surprising: along with wealth and high living standards in the industrial countries, the last five decades have brought severe man-made impacts on our environments. To understand the ongoing global and regional changes, ecological survey techniques had to be improved and data were increasingly processed by more and more sophisticated computing technologies. Ostracodologists made and will make use of these developments to learn more about ostracod ecology, and—vice versa—will apply this knowledge to contribute to the understanding of both past and future ecological trends.

The study of ostracod *biodiversity* is rarely a stand-alone discipline. New insights into faunal assemblages of a given time and space are usually embedded within ecological, phylogenetical, biogeographical or biostratigraphical investigations. In particular the two fields 'ecology' and 'biostratigraphy' are often inextricably interwoven with faunistic analyses. Therefore, if ISO projects featured a biodiversity study within an ecological context, or a biodiversity study within a biostratigraphic context, respectively, these were categorized within the groups 'ecology' and ' biostratigraphy'.

The remaining analyzed biodiversity studies (Figs 1, 6 and data sheets) tell us two major things: (1) biodiversity studies executed to gain new insights into ostracod evolution and phylogeny are relatively rare; (2) in contrast, a much larger proportion of biodiversity studies combine faunal analyses with biogeographical issues, particularly at ISO10, which was dedicated to Global Events. However, from ISO10 onwards, the proportion of biodiversity–biogeography studies gradually decreased from 18.2 to 0.7% at ISO15. This development is both surprising and alarming, particularly if viewed together with the main theme *'biogeography'*, which also shows a more or less continuous decrease after ISO11 (Figs. 1, 7) with a complete absence at ISO14. Is it possible that such an important facet of ostracod phylogeny and evolution is no longer explored?

Biostratigraphy is certainly the oldest applied discipline in ostracodology. Ever since the first description of a fossil ostracod in 1813 (*Cypris faba* Desmarest), ostracod biostratigraphy has been a proven method for dating sediments and strata, not least in the economically important field of hydrocarbon exploration. Biostratigraphic contributions were relatively pronounced in the ISO7 proceedings from 1979 (34.2%), reflecting increased exploration as a consequence of the global "energy crisis", then gradually decreased down to 8.9% at ISO10 (Figs. 1, 7). At the occasion of ISO12, organizers may have wanted to pinpoint this negative trend by choosing the theme 'Ostracoda and biostratigraphy', but the following two symposia again featured only around 5% biostratigraphy within their proceedings.

However, since the oil industry has largely replaced biostratigraphy by other exploration techniques in recent decades (in order to decrease personnel costs), it now becomes obvious that biostratigraphy is still the most powerful tool for both immediate strata dating and palaeoenvironmental assessment. The new demand for well-trained micropalaeontologists meanwhile led to the establishment of popular international open short courses in applied micropalaeontology, e.g. at the University of Bonn.

This renewed interest in biostratigraphy certainly follows new requirements in the oil exploration industry. The relatively high performance of this working field at ISO15 (11.2%) is indicative of this trend.

Of the eight main categories of ostracod research, reproduction and genetics represent the smallest groups. *Molecular genetic research* first appeared at ISO11 in 1991 (Fig. 1) with two fundamental studies—surprisingly late when compared to the general development of this research field in the rest of biology. In the present proceedings volume, genetic contributions remained below 3% of the total number (four papers or 2.8% at ISO15). There may be several reasons for this. Firstly, molecular biological studies on ostracods are restricted to few research groups only. Ostracods may not be the most ideal model organisms for such studies, because many species remain difficult to culture in the lab, and their small size (amongst other reasons) makes DNA extraction less than equivocal. Secondly, molecular biologists working with ostracods often do not consider themselves ostracodologists per se, and attending ostracod conferences might not be their priority. Finally, molecular studies on any animal or plant groups are often seen as being more competitive than more classical disciplines, and therefore such papers might not be offered for publication in the symposium proceedings.

The general under-representation of *reproduction* biology within ostracod research (Fig. 1) is due to different reasons. Major studies on ostracod reproduction, namely on the morphological background of reproduction, were carried out before the first ISO, mainly in the first decades of the 20th century, by technically brilliant microscopists like Zenker, Müller, Bauer, Lowndes (e.g. Zenker, 1854; Müller, 1889; Lowndes, 1935; Bauer, 1940). With the rise of SEM and TEM techniques in the 1960s, several potentially ground-breaking studies on ostracod giant spermatozoa were published (e.g., Gupta, 1968; Reger & Florendo, 1969; Reger, 1970; Zissler, 1966, 1970), but unfortunately they were (1) not published within an ISO proceedings volume and (2) obviously not ground-breaking enough: apart from Wingstrand's monograph on ostracod spermatozoa (1988), they had no 'pollinating' effect on ostracod research. Additionally, it has to be mentioned that these studies were not carried out by 'genuine' ostracodologists entering the field of spermatology, but by spermatologists, who were only temporarily interested in ostracod spermatozoa (e.g. Zissler, Reger).

Within ISO proceedings, most studies on ostracod reproduction were related to morphological aspects, followed by investigations of the effects of reproduction on population dynamics. Karyological, ecological, behavioral and biogeographical aspects of ostracod reproduction have so far played a very minor role.

However, ostracods reproduce either sexually or asexually and asexuality (parthenogenesis) is limited to certain populations or even species, but never cyclic. Furthermore, the occurrence of parthenogenesis in certain species is correlated with geographical/climatological aspects. These biological peculiarities indeed bear a high potential for research in, for example, ostracod evolutionary ecology, palaeo-biogeography and genetics. The ongoing *SexAsex* EU Marie-Curie Research and Training Network on the geographic parthenogen *Eucypris virens* will certainly give this field a new impulse, so that a significant output is to be expected. Not least, the continuous improvement of affordable camera systems will provide increasing possibilities in particular to studies on ostracod reproductive behavior. Ostracod reproduction appears to be a research field in the starting blocks.

Conclusions

Within the spectrum of ISO contributions, *taxonomy* generally played a minor role and was mostly presented in a combination with other research fields, i.e. as applied taxonomy. From the 1970s, the percentage of taxonomic contributions decreased until a complete breakdown in 1998. Since then, a re-birth can be noted, perhaps as a sign for a budding new self-confidence in this area. *Morphology* may be termed the permanent background research at high level within ostracodology, implemented in literally every working field. Along with a growing interest in soft-part preservation in fossil ostracods, it will be more and more important to have a profound knowledge of soft-part morphologies of all ostracod groups.

Ecology is one of the major fields of investigation within ostracodology, and new techniques seem to be introduced quite easily and quickly. Shell chemistry studies had their peak in 1998, but were then gradually replaced by, or combined with, isotope analyses. Behavioral studies played a minor role, whereas most ecological studies made use of *biodiversity* analyses to use assemblages as environmental proxies, and often include biogeographical implications. Apart from this application, biodiversity analyses mainly functioned as a tool for biostratigraphical and biogeographical projects—only very rarely were phylogenetic studies based on biodiversity analyses. *Biogeography* as such has not performed well recently. In spite of useful new tools like GPS and affordable database software, not the slightest growth of this field is to be noted within the contributions of the last ISO events. In order to contribute to developments in general (palaeo)biogeographical research, work on ostracod biogeography urgently needs to be revived. In contrast, *biostratigraphy* seems to have overcome its low mark, which was clearly at ISO13. Ever since, renewed interest in this area is to be noticed. *Molecular genetic* research found its way into ostracodology rather late and after its introduction at ISO11 in 1991 no significant flourishing can be seen in the present analysis. Several possible reasons can be cited for this. Finally, research on ostracod *reproduction* has had its special chronology. The main morphological data on reproductive organs and cells were collected by researchers before the first ISO in 1963. Among the ISO contributions, the percentage of studies on reproduction were generally low, featuring morphological peculiarities, a few chromosome investigations, sex-ratio studies and only one manuscript (in total) in either an ecological or a biogeographical context. All spermatology research of the 1970s was performed by non-ostracodologists outside any ISO event. Since ostracod (a)sexuality is extraordinary for several reasons and even challenges evolutionary hypotheses by featuring putative ancient asexuals, this research area is expected to continue to flourish in the future.

Although the present study suffers from a number of shortcomings (cited above), we consider this retrospective of some value, because it might help us to decide which directions to take in our coming research projects in the 21st century.

Gemeine Köpfe lesen in der Geschichte die Vergangenheit, kluge die Zukunft—Simple minds read the past in history, clever minds read the future.

Acknowledgements We wish to thank all people who organized previous International Symposia on Ostracods, and who put a lot of work into the Symposia Proceedings. We also wish to thank all authors who published in previous ISO Proceedings, thus providing an overview on the current state of the art of ostracodology.

References

Bauer, H., 1940. Über die Chromosomen der bisexuellen und der parthenogenetischen Rasse des Ostracoden *Heterocypris incongruens* Ramd. Chromosoma 1: 621–637.

Boomer, I., D. J. Horne & I. Slipper, 2003. The use of ostracodes in paleoenvironmental studies, or what can you do with an ostracod shell? Paleontological Society Papers 9: 153–180.

Boomer, I. & A. Lord (eds), 1999. Marine Ostracoda and Global Change. Marine Micropaleontology 37: 227–294.

De Deckker, P. & R. M. Forester, 1988. The use of ostracods to reconstruct palaeoenvironmental records. In De Deckker, P., J.-P. Colin & J.-P. Peypouquet (eds), Ostracoda in the Earth Sciences. Elsevier, Amsterdam.

Gupta, B., 1968. Aspects of the motility in the non-flagellate spermatozoa of freshwater ostracods. In Miller, P. L. (ed.), Aspects of motility. Symposia of the Society for Experimental Biology 22: 117–129.

Hartmann, G. (ed.), 1976. Evolution of Post-Paleozoic Ostracoda. Abhandlungen und Verhandlungen des naturwissenschaftlichen Vereins in Hamburg (Neue Folge) 18/19: 1–336.

Hanai, T., N. Ikeya & K. Ishizaki (eds), 1988. Evolutionary ecology of Ostracoda. Its Fundamentals and Applications. Elsevier, Amsterdam.

Holmes, J. & A. Chivas (eds), 2002. The Ostracoda: Applications in Quaternary Research. Geophysical Monograph 131: 227–247.

Holmes, J. A. & D. J. Horne (eds), 1999. Non-marine Ostracoda: Evolution and Environment. Palaeogeography, Palaeoclimatology, Palaeoecology 148: 1–290.

Horne, D. J. & K. Martens (eds), 2000. Evolutionary biology and ecology of Ostracoda. Hydrobiologia 419: 1–197.

Ikeya, N., A. Tsukagoshi & T. M. Cronin (eds), 2005a. Earth environments and dynamics of Ostracoda. Palaeogeography, Palaeoclimatology, Palaeoecology 225: 1–334.

Ikeya, N, A. Tsukagoshi & D. J. Horne (eds), 2005b. Evolution and diversity of Ostracoda. Hydrobiologia 538: 1–256.

Kesling, R. V., 1983. Ostracodology in 1982. In Maddocks, R. (ed.), Applications of Ostracoda. Department of Geosciences, University Houston, 1–7.

Kohring, R & B. Sames (eds), 2005. 15th International Symposium on Ostracoda, September 12–15: 2005, Program and Abstracts. Berliner Paläobiologische Abhandlungen 6: 1–160.

Krstić, N. (ed.), 1979. Taxonomy, biostratigraphy and distribution of Ostracodes. Zapisnici Srpskog Geološkog Društva/1971–1972: 1–171.

Löffler, H. & D. Danielopol (eds), 1977. Aspects of Ecology and Zoogeography of Recent and Fossil Ostracoda. Junk Publishers, The Hague.

Lowndes, A. G., 1935. The sperms of freshwater ostracods. Proceedings of the Zoological Society London 1935: 35–48.

Maddocks, R. (ed.), 1983. Applications of Ostracoda. Department of Geosciences, University Houston.

Martens, K. & D. J. Horne, 2000. Preface: Ostracoda and the four pillars of evolutionary wisdom. In Horne, D. J. & K. Martens (eds), Evolutionary biology and ecology of Ostracoda, Hydrobiologia 419: vii–xi.

McKenzie, G. & P. Jones (eds), 1993. Ostracoda in the Earth and Life Sciences. Balkema, Rotterdam, The Netherlands.

Müller, G. W., 1889. Die Spermatogenese der Ostracoden. Zoologische Jahrbücher, Abteilung für Anatomie und Ontogenie der Tiere 3: 677–726.

Neale, J. W. (ed.), 1969. The Taxonomy, Morphology and Ecology of Recent Ostracoda. Oliver & Boyd, Edinburgh.

Neale J. W., 1988. Ostracoda—A historical perspective. In Hanai, T., N. Ikeya & K. Ishizaki (eds), Evolutionary Ecology of Ostracoda. Its Fundamentals and Applications. Elsevier, Amsterdam, 3–15.

Oertli, H. J. (ed.), 1971. Colloques sur la Paléoécologie des Ostracodes/Colloquium on the Paleoecology of Ostracodes. Bulletin du Centre de Recherches Pau—SNPA, 5, supplement, 1–953.

Puri, H. S. (ed.), 1964. Ostracods as Ecological and Palaeoecological Indicators. Pubblicazioni della Stazione Zoologica di Napoli 33: 1–612.

Reger, J. F., 1970. Some aspects of the fine structure of filiform spermatozoa (Ostracod, *Cypridopsis* sp.) lacking tubule substructure. In Baccetti, B. (ed.), Spermatologia Comparata. Academia Nazionale dei Lincei, Rome, 237–245.

Reger, J. F. & N. T. Florendo, 1969. Studies on motile, nontubulecontaining, filiform spermatozoa of the ostracod *Cypridopsis*. I. Spermiogenesis. Journal of Ultrastructure Research 28: 235–249.

Ríha, J. (ed.), 1995. Ostracoda and Biostratigraphy. A. A. Balkema, Rotterdam, The Netherlands; Brookfield, VT.

Sohn, I. G., 1977. Zoogeography of ostracodologists. In Löffler, H. & D. Danielopol (eds), Aspects of Ecology and Zoogeography of Recent and Fossil Ostracoda. Junk Publishers, The Hague, 3–12.

Swain, F. M. (ed.), 1975. Biology and paleobiology of Ostracoda. Bulletins of American Paleontology, 65: 1–687.

Whatley, R. & C. Maybury (eds), 1990. Ostracoda and Global Events. British Micropalaeontological Society Publication Series, Chapman and Hall, London, NY.

Wingstrand, K. G., 1988. Comparative spermatology of the Crustacea Entomostraca. 2. Subclass Ostracoda. Biologiske Skrifter 32, The Royal Danish Academy of Sciences and Letters, Copenhagen: 1–149.

Zenker, G. F. W., 1854. Monographie der Ostracoden. Archiv für Naturgeschichte 20: 1–87.

Zissler, D., 1966. Zur Feinstruktur der Ostracoden-Spermien. Naturwissenschaften 53: 561.

Zissler, D., 1970. Zur Spermiohistogenese im Vas Deferens von Süßwasser-Ostracoden. Cytobiologie 2: 83–86.

Hydrobiologia (2007) 585:13–27
DOI 10.1007/s10750-007-0625-6

Ostracod recovery in the aftermath of the Permian–Triassic crisis: Palaeozoic–Mesozoic turnover

S. Crasquin-Soleau · T. Galfetti · H. Bucher · S. Kershaw · Q. Feng

Abstract During the earliest Triassic, the neritic environments were completely devastated and the recolonization of biotopes was very slow. Besides some small foraminifera and bivalves, ostracods are among the few neritic organisms that were able to survive and/or to thrive in the inhospitable environments after the disaster events. But the Permian–Triassic boundary marks also a great change in the ostracod assemblages. The Palaeozoic ostracods left room for the "modern" fauna. New data on the Early Triassic neritic fauna in South China (Sichuan and Guangxi Provinces) and bibliographic synthesis on other areas yield a first description of the timing of this turnover. First "typically modern" forms appear already in the Late Permian. The Early Triassic (Griesbachian to Spathian) ostracod faunas display a mixture of Palaeozoic and Mesozoic taxa. Completion of the Palaeozoic–Mesozoic turnover could be located in the Middle Triassic (Anisian).

Keywords Ostracods · Palaeozoic – Mesozoic turnover · mass extinction · recovery · Late Permian · Early Triassic

Guest editors: R. Matzke-Karasz, K. Martens & M. Schudack
Ostracodology Linking Bio- and Geosciences

S. Crasquin-Soleau (✉)
CNRS, UMR 5143, Université Pierre et Marie Curie, Laboratoire de Micropaléontologie, T.46-56, E.5, case 104, 75252 Paris cedex 05, France
e-mail: crasquin@ccr.jussieu.fr

T. Galfetti · H. Bucher
Paläontologisches Institut und Museum, Universität Zürich, Karl Schmid-Strasse 4, 8006 Zürich, Switzerland

S. Kershaw
Department of Geography and Earth Sciences, Brunel University, Uxbridge, Middlesex UB8 3PH, UK

Q. Feng
University of GeoSciences, Wuhan, Hubei Province 430074, P.R. China

Introduction

The end-Permian mass extinction led to drastic change in marine diversity (Fig. 1). The protracted upper Permian biodiversity decline lasted some 10 Ma. The Triassic recovery can be divided into three phases (Erwin, 1993). The mass extinction is followed by a phase of biotic poverty during the Scythian. Most of the Early Triassic deposits are characterized by abundant microbial limestone, generally interpreted as disaster form (Schubert & Bottjer, 1992). Early Triassic communities exhibit low diversity. Based on the ages proposed by Gradstein et al. (2005), the lag phase lasts around 6 million years

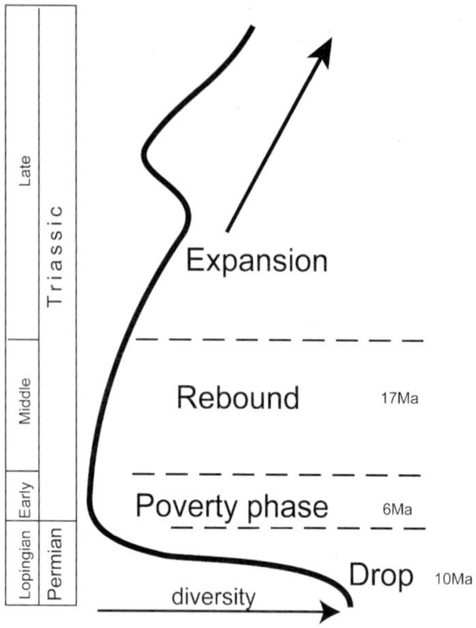

Fig. 1 Divisions of the post-extinction recovery (modified after Erwin, 1993)

(between 251 Ma-PT boundary- and 245 Ma-end of Scythian). We shall see later that recent data of Ovtcharova et al. (2006) revise this estimate slightly. The rebound phase characterizes the Middle Triassic with a return to normal marine fauna. The true expansion of Mesozoic marine faunas took place during the Late Triassic.

Ostracods are known to inhabit all aquatic environments. Despite its remarkable adaptive potential, the sub-class was deeply affected, as are all other neritic fauna, by the end-Permian mass extinction. This period marks the great change in the ostracod group evolution. The Palaeozoic fauna were replaced by the "modern" assemblages, which grew during the Late Triassic. The transition between Palaeozoic and Meso-Cenozoic fauna is poorly documented. In order to understand the changes in ostracod taxa throughout the extinction and recovery, this paper synthesizes the current state of knowledge of all ostracod faunas described from different areas, and aims to identify major features and trends in ostracods throughout the interval. In this paper, we take into account only the neritic forms. Palaeopsychrospheric ostracods, deep benthic inhabitants below the thermocline, are not considered in this work because the basinal environment is very conservative and the evolution and disappearance of this biotope, somewhere in the Anisian, is a problem per se.

The Palaeocopids, ostracods with straight dorsal border, were considered for a long time to have gone extinct during the Late Permian and their occurrence was thought to be restricted to the Palaeozoic. Gramm (1995) figured an undetermined *Kirkbyocopina* of early Anisian age from Primor'ye (Russian Far East). If recorded specimens indubitably belong to Kirkbyocopina, then the dating of the sample seems more questionable. No other fossils were associated with these ostracods. Crasquin-Soleau et al. (2004a, b) and Crasquin-Soleau & Kershaw (2005) showed that Palaeocopids survived into the earliest Triassic.

The recent studies on ostracods around the Permo-Triassic boundary in Western Taurus, Turkey (Crasquin-Soleau et al., 2004a, b), Saudi Arabia (Crasquin-Soleau et al., 2005) and South China in Sichuan Province (Crasquin-Soleau & Kershaw, 2005) and Guangxi Province (Crasquin-Soleau et al., 2006 and this paper) allow further precision to the mapping. Some unpublished data on Early and Middle Triassic of Rumania are taken into account.

Late Permian ostracod fauna (examples illustrated on Fig. 2, 1–6)

We do not discuss here the progressive disappearance of Palaeozoic forms within the Permian. This topic will be dealt with in a separate paper. The first Mesozoic forms appear in the Late Permian (Fig. 2). Some typical strongly shelled and ornamented Bairdiidae, which comprise an important part of the Late Triassic Tethyan fauna, are recognized in the Wuchiapingian and Changhsingian (Fig. 3). For example, *Sinabairdia nodosa* Becker & Wang, 1992 is described in the Wuchiapingian of Sichuan as well as *Ceratobairdia sinensis* Becker & Wang, 1992 in the Changhsingian of Zhejiang Province (Becker & Wang, 1992).

In the latest Permian of Hubei Province, Chen & Shi (1982) present four species of *Ceratobairdia* Sohn, 1954, three of *Petasobairdia* Chen, 1982 and three of *Mirabairdia* Kollmann, 1963.

In the Changhsingian of Meishan section, Shi & Chen (1987) recognized three species of *Mirabairdia* Kollmann, 1963 and two of *Lobobairdia* Kollmann, 1963. These Triassic genera have their maximum development during Ladinian–Carnian time interval. The genera *Petasobairdia* Chen, 1982 (very close to *Ceratobairdia* Sohn, 1954, described also by Kristan-Tollmann (1970) in the Late Triassic of Alps) and *Abrobairdia* Chen, 1982 (very close to the Triassic genus *Mirabairdia* Kollmann, 1963) are represented by four and two species, respectively. The systematics of the ostracods of Meishan section still need an important systematic revision which is currently in progress. As noted by Chen & Shi (1982, p. 146), the Triassic genera *Parurobairdia* and *Mirabairdia* seem to be a transitional type between Late Palaeozoic to Early Mesozoic genera. This group of strongly shelled and ornamented Bairdiidae is called "*Petasobairdia–Ceratobairdia–Mirabairdia–Parurobairdia fauna*" by Chen & Shi (1982). This confirms the phyletic proximity of all these forms through the PT boundary and the necessity of the systematic revision.

Some other genera which are first representatives of Mesozoic inhabitants are recognized by different authors:

- By Kozur (1985) in the Bükk Mountains: *Judahella bogschi bogschi* Kozur, 1985 in the Changhsingian, *Callicythere mazurensis* (Styk, 1972) in the early Wuchiapingian, *Gruendelicythere* (*Trodocythere*) *permica* Kozur, 1985 and *Fueloepicythere pulchra* Kozur, 1981 in the middle Wuchiapingian;
- By Gerry et al. (1987) and Honigstein et al. (2006) in the Late Permian of Israel: five species of *Arqoviella* Gerry & Honigstein, 1987. As it was underlined by these authors this genus shows typical Mesozoic features;
- By Crasquin-Soleau et al. (2004a, b) in the Lopingian (Wuchiapingian–Changhsingian) of Antalya Nappes (Western Taurus): *Petasobairdia nantongensis* Chen 1987, *Petasobairdia* cf *subnantongensis* Chen, 1987 *sensu* Crasquin-Soleau et al. 2004a, b *Arqoviella tahtaliensis* Crasquin-Soleau, 2004a, b *Callicythere lysi* Crasquin-Soleau, 2004, gen. et sp. indet.
- By Crasquin-Soleau et al. (2005) in the Lopingian Khuff Formation of Saudi Arabia with four species of *Arqoviella* genus.

Scythian (Early Triassic) and early Anisian (Middle Triassic) ostracod fauna (examples illustrated on Fig. 2, 7–32)

The Early Triassic neritic ostracods are poorly known. Psychrospheric faunas are known from early Anisian of Rumania (Crasquin-Soleau & Gradinaru, 1996).

Some neritic species were recognized (or just quoted) in the Early Triassic (Induan–Olenekian) and early Anisian of Australia (Jones, 1970), Pakistan (Sohn, 1970), Nepal (Bunza & Kozur, 1971), Greece (Kozur, 1971b; Ardens et al., 1979), Germanic Basin (Kozur, 1973b), Israel (Hirsch & Gerry, 1974), Kashmir (Agarwal, 1979, 1980, 1981; Agarwal et al., 1980) and South China (Wang, 1978; Wei Ming, 1981; Hao, 1992, 1994).

Recent works on Permian–Triassic sections in Western Taurus (Crasquin-Soleau et al., 2004a, b), in Saudi Arabia (Crasquin-Soleau et al., 2005) and South China (Crasquin-Soleau & Kershaw, 2005; Crasquin-Soleau et al., 2006 and this paper) give new data on the latest Palaeozoic and earliest Mesozoic ostracods. It was evidenced that a transitional interval existed for this group. Some forms of Mesozoic affinities were discovered in the latest Permian mixed with typical Palaeozoic forms (Crasquin-Soleau et al., 2004a, b). In the earliest Triassic, some survivors are associated with the newcomers. The presence of survivors was first evidenced by Jones (1970) in Perth Basin (Australia) and Sohn (1970) in Salt Range (Pakistan). It was confirmed in South China by Wang (1978) in Guizhou and North Yunnan Provinces, Wei (1981) in Sichuan Province, Hao (1992, 1994) in Guizhou Province. More recently, Palaeozoic survivors were recognized in Western Taurus (Crasquin-Soleau et al., 2004a, b) and South China in Sichuan Province (Crasquin-Soleau & Kershaw, 2005). The problem is to date the final disappearance of Palaeozoic forms and complete conquest of environments by typical Triassic inhabitants.

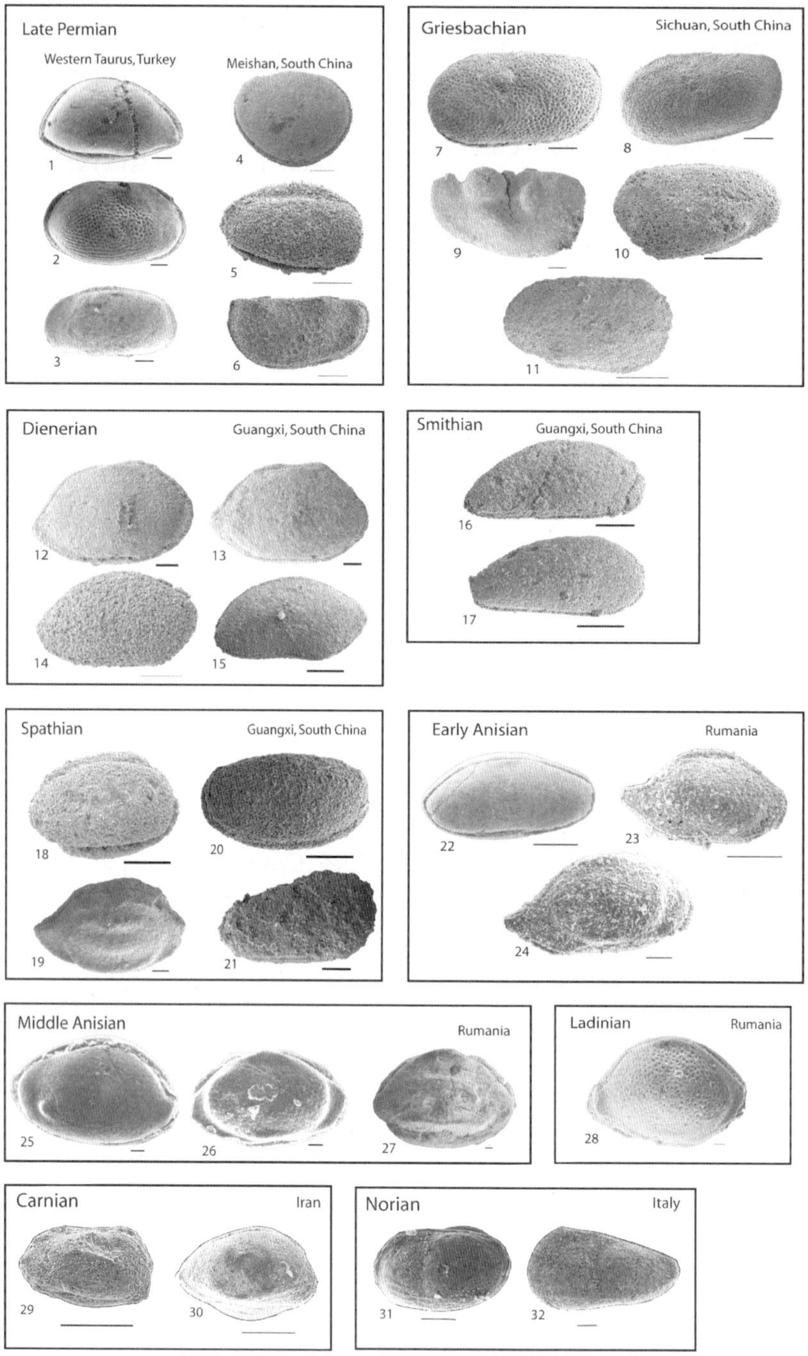

The ostracods analyzed in this paper are dated by ammonoids from Griesbachian up to Spathian. The Table 1 compiles the available data on Early Triassic ostracods, from bibliography and from personal works. Unpublished data on Early–Middle Triassic ostracods from Dobrogea (East Rumania) are added. In this table, only the species which have typical Palaeozoic or Mesozoic affinities are reported. Species which are representative of panchronic genera (as smooth *Bairdia* or *Paracypris*, ...) are not quoted.

◀ **Fig. 2** Examples of ostracod associations during the different phases of extinction—recovery patterns at Permian–Triassic boundary. PF, Palaeozoic form; MCF, Meso-Cainozoic form (**1–6**: latest Permian): (**1**) PF, *Acratia changxingensis* (Shi, 1987) from Western Taurus (Turkey; Crasquin-Soleau et al., 2004; (**2**) MCF: *Arqoviella tahtalensis* Crasquin-Soleau, 2004 from Western Taurus (Turkey; Crasquin-Soleau et al., 2004; (**3**) MCF, *Callicythere lysi* Crasquin-Soleau, 2004 from Western Taurus (Turkey; Crasquin-Soleau et al., 2004; (**4**) PF, *Samarella* sp.1 from Meishan section (level 15) (Crasquin-Soleau et al., unpubl.); (**5**) PF, *Microcheilinella* sp.1 from Meishan section (level 15) (Crasquin-Soleau et al., unpubl.); (**6**) PF, *Kirkbyidae* sp.1 from Meishan section (level 22) (Crasquin-Soleau et al., unpubl.); (**7–11**) Griesbachian of Sichuan (South China; Crasquin-Soleau & Kershaw, 2005); (**7**) PF, *Langdaia laolongdongensis* Crasquin-Soleau & Kershaw, 2005; (**8**) PF, *Langdaia suboblonga* Wang, 1978; (**9**) PF, *Hollinella* sp. 1; (**10–11**) MCF, *Callicythere postiangulata* Wei, 1981; (**12–15**) Dienerian of Guangxi (South China; Crasquin-Soleau et al., 2006); (**12–13**) MCF, *Ptychobairdia luciae* Crasquin-Soleau, 2006; (**14**) *Bairdia fengshaensis* Crasquin-Soleau, 2006; (**15**) *Bairdia wailiensis* Crasquin-Soleau, 2006; (**16–17**) Smithian of Guangxi (South China; Crasquin-Soleau et al., 2006); (**16**) PF, ?*Acratia nostriaca* Monostori, 1994; (**17**) *Paracypris jinyaensis* Crasquin-Soleau, 2006; (**18–21**) Spathian of Guangxi (South China; Crasquin-Soleau et al., 2006); (**18**) PF, *Microcheilinella* cf. venusta Chen, 1958; (**19**) MCF, *Ptychobairida aldae* Crasquin-Soleau, 2006; (**20**) PF, *Carinaknightina*? sp. *sensu* Crasquin-Soleau et al., 2006; (**21**) MCF, *Kerocythere*? sp. *A sensu* Crasquin-Soleau et al., 2006; (**22–24**) Early Anisian of Dobrogea (Rumania; Crasquin-Soleau & Gradinaru, 1996); (**22**) *Bairdiacypris galbruni* Crasquin-Soleau & Gradinaru, 1996; (**23**) MCF, *Urobairdia fauconnierae* Crasquin-Soleau & Gradinaru, 1996; (**24**) MCF, *Urobairdia uzumensis* Crasquin-Soleau & Gradinaru, 1996; (**25–27**) Middle Anisian of Dobrogea (Rumania; Crasquin-Soleau, unpublished data); (**25**) MCF, *Ogmoconchella* sp.2; (**26**) MCF, *Ptychobairdia* sp.6; (**27**) MCF, *Ptychobairdia* sp.5; (**28**) Ladinian of Dobrogea (Rumania; Crasquin-Soleau, unpublished data). MCF, *Lobobairdia* cf. *salinaria* Kollmann, 1963; (**29–30**) Carnian of Zagros (Iran; Crasquin-Soleau & Teherani, 1995); (**29**) MCF, *Moscovitschia* cf. *interrupta* Kristan-Tollmann, 1983 *sensu* Crasquin-Soleau & Teherani 1995; (**30**) MCF, *Metacyteropteron*? *zagrosensis* Crasquin-Soleau & Teherani 1995; (**31–32**) Norian of Northern Italy (Crasquin-Soleau et al., 2000); (**31**) MCF, *Rhombocythere dimorphica* Crasquin-Soleau et al., 2000; (**32**) MCF, *Kerocythere quattervalsi* Crasquin-Soleau et al., 2000

Griesbachian

Griesbachian ostracod fauna is recognized in South China (Guizhou, Sichuan, Yunnan and Guangxi Provinces), Western Taurus, and Tibet (Table 1). Many Palaeozoic forms are still present, with genera like *Hollinella, Carinaknightina* and *Langdaia*. Even if the *Hollinella* specific attributions are wrong in Hao (1992, 1994) and Wang (1978), this genus is recognized without ambiguity (Crasquin-Soleau et al., 2004a). Few genera present Late Triassic features: *Callicythere* in Sichuan, *Kerocythere* in Guangxi, both in South China, *Lutkevichinella, Judahella, Hungarella, Monoceratina* in Tibet (see Table 1).

TRIASSIC	Late	Rhaetian	
		Norian	
		Carnian	
	Middle	Ladinian	
		Anisian	
	Early = Scythian	Olenekian	Spathian
			Smithian
		Induan	Dienerian
			Griesbachian
PERMIAN	Late = Lopingian	Changsingian	
		Wuchiapingian	

Fig. 3 Late Permian—Triassic stratigraphic subdivisions (from Gradstein et al., 2005). No vertical scale

Dienerian

Before this work, we had no data on Dienerian ostracods. In the Jinya/Waili section (Guangxi Province, South China—Crasquin-Soleau et al., 2006), we recognized three species in the Dienerian: *Bairdia fengshanensis* Crasquin-Soleau, 2006, *Bairdia wailiensis*, Crasquin-Soleau, 2006 and *Ptychobairdia luciae* Crasquin-Soleau, 2006. This last species presents typical Mesozoic characters.

Smithian

As for the Dienerian, the only Smithian available data come from the *Flemingites* beds of Jinya/Waili section (Crasquin-Soleau et al., 2006). We found four species (*Bairdia fengshanensis* Crasquin-Soleau, 2006, ?*Acratia nostorica* Monostori, 1994, *Bythocypris*? sp.3 and *Paracypris jinynensis*

Table 1 (1) Hao (1992, 1994)—South China; (2) Crasquin-Soleau & Kershaw (2005)—South China; (3) Crasquin-Soleau et al. (2004a, b) —Turkey; (4) Crasquin-Soleau et al. (2006)—South China; (5) Wang (1978)—South China; (6) Wei Ming (1981)—South China; (7) Jones (1970)—Australia; (8) Sohn (1970)—Pakistan; (9) Kozur (1973a, b) —Germanic Basin; (10) Crasquin-Soleau (unpublished data)—Rumania; (11) Crasquin-Soleau & Gradinaru (1996)—Rumania; (12) Gramm (1995)—Far East Russia

Species	Guizhou Griesbachian parvus zone 1	Sichuan Griesbachian parvus zone 2	Taurus Griesbachian parvus – staeschei zone 3	Guangxi Griesbachian 4	W Guizhou NE Yunnan Griesbachian 5	Sichuan Griesbachian 6	Australia Griesbachian 7	Pakistan Late Griesbachian 8	Guangxi Dienerian 4	Guangxi Smithian 4
Hollinella cf. *plana* Jiang *sensu* Hao, 1992	◆									
Hollinella unispinata Hao, 1992	◆									
Hollinella tingi (Patte) *sensu* Wang and Hao	◆				◆	◆				
Hollinella cf *tingi* (Patte) *sensu* Hao, 1992	◆									
Carinaknightina carinata Sohn, 1970	◆									
Carinaknightina zhenfengensis Hao, 1992	◆									
Langdaia suboblonga Wang, 1978	◆				◆					
Acratia symmetrica Hao, 1992	◆									
Hollinella sp.1 *sensu* CS & Kershaw, 2005		◆								

Table 1 continued

Species	Guizhou Griesbachian parvus zone 1	Sichuan Griesbachian parvus zone 2	Taurus Griesbachian parvus – staeschei zone 3	Guangxi Griesbachian 4	W Guizhou NE Yunnan Griesbachian 5	Sichuan Griesbachian 6	Australia Griesbachian 7	Pakistan Late Griesbachian 8	Guangxi Dienerian 4	Guangxi Smithian 4
Langdaia laolong dongensis CS & Kershaw 2005		♦								
Langdaia suboblonga Wang, 1978		♦				♦				
Acratia sp.1		♦								
Callicythere postiangusta Wei Ming, 1981		•								
Revyia curukensis Crasquin-Soleau 2004			♦							
Paraparchitacea indet			♦							
"*Healdia*" sp.A *sensu* CS et al., 2006				♦						
Kerocythere? sp.A *sensu* CS et al., 2006				•						
Triassinella cuneiformis Wang, 1978					•					
Carinaknightina neutrum Jones, 1970							♦			
Hollinella sp.							♦			

Table 1 continued

Species	Guizhou Griesbachian parvus zone	Sichuan Griesbachian parvus zone	Taurus Griesbachian parvus – staeschei zone	Guangxi Griesbachian	W Guizhou NE Yunnan Griesbachian	Sichuan Griesbachian	Australia Griesbachian	Pakistan Late Griesbachian	Guangxi Dienerian	Guangxi Smithian
	1	2	3	4	5	6	7	8	4	4
Truncobairdia beaglensis Jones, 1970							•			
Carinaknightina discarinata Sohn, 1970								♦		
Carinaknightina aff. *carinata* Sohn, 1970								♦		
Carinaknightina carinata Sohn, 1970								♦		
Lutkevichinella? ornata Sohn, 1970								•		
Judahella? sp.								•		
Hungarella? sp.								•		
Monoceranita? sp.								•		
Ptychobairdia luciae Crasquin-Soleau, 2006									•	
?Acratia nostrica										♦

Table 1 continued

Species	Pakistan Spathian 8	Guangxi Spathian 4	Germanic Basin Spathian 9	Dobrogea Rumania Spathian 10	Germanic Basin Early Anisian 9	Dobrogea Rumania Early Anisian 10 and 11	Far East Russia Early Anisian 12
Kirkbyidae gen. indet. sp.	♦						
Microcheilinella sp. *sensu* Sohn, 1970	♦						
Reubenella? sp. *sensu* CS et al., 2006; Sohn, 1970	•						
Carinaknighiina? sp. *sensu* CS et al., submitted		♦					
Microcheilinella? cf. *venusta sensu* CS et al., 2006		♦					
Microcheilinella? sp.1 *sensu* CS et al., submitted		♦					
Ogmoconcha? sp.A *sensu* CS et al., 2006		•					
Ptychobairdia aldae CS, 2006		•					
Ptychobairdia sp.A *sensu* CS et al., 2006		•					
Ptychobairdia luciae CS, 2006		•					
Lutkevichinella (*Cytherissinella*) *bisulcata* (Kozur)			•				
Lutkevichinella (*Cytherissinella*) *nodosa* (Kozur)			•				
Judahella pulchra pulchra (Knüpfer & Kozur)			•		•		
Judahella tsorfaria Sohn, 1968			•		•		
Lutkevichinella simplex simplex Kozur			•		•		
Acratia sp.2				♦			
Ogmoconchella sp.5				•			

Table 1 continued

Species	Pakistan Spathian	Guangxi Spathian	Germanic Basin Spathian	Dobrogea Rumania Spathian	Germanic Basin Early Anisian	Dobrogea Rumania Early Anisian	Far East Russia Early Anisian
	8	4	9	10	9	10 and 11	12
cf. *Langdaia subobloga* Wang, 1978					♦		
Cultella? *levis* Kozur					•		
Cytherella jenesis Kozur					•		
Lutkevichinella (*Cytherissinella*) *lata lata* Kozur					•		
Triassocypris pusilla (Kozur)					•		
Triassocypris tenuis (Kozur)					•		
Urobairdia sp.1						•	
Urobairdia sp.2						•	
Ptychobairdia sp.1						•	
Ptychobairdia sp.2						•	
Metacytheropteron sp.2						•	
*Urobairdia fauconnierae*CS & Gradinaru, 1996						•	
Hungarella sp.2						•	
Lobabairdia cf *salinaria* Kollmann, 1963						•	
Metacytheropteron sp.						•	
Kirkbyocopina sp.							♦
	♦	Palaeozoic affinities					
	•	Mesozoic affinities					

Crasquin-Soleau, 2006) which are smooth Bairdiacea. These species belong to panchronic genera (smooth Bairdiacea) and are not informative for our purpose.

Spathian

In the Spathian, the ostracod assemblage becomes diversified again. Faunas were published by Sohn (1970—dating reviewed by Pakistani-Japanese Research Group (1985) and Wignall & Hallam (1993). Ostracods known from the Salt Range occur in the lower part of Mittiwali member of Mianwali Formation (late Griesbachian) and in the Narmia Member of Mianwali Formation (Spathian). We add here data on Luolou Formation, Guangxi Province, South China (Crasquin-Soleau et al., 2006).

Kirkbyidae indet, *Microcheilinella* and *Carinaknightina* (see Table 1) are late Palaeozoic forms documented in the Spathian of Tibet and Guangxi Province.In the Germanic Basin (Kozur, 1973a), all the representatives have Mesozoic affinities.

Early Anisian

Available data come from the Germanic Basin (Kozur, 1973a), from Dobrogea in Rumania (Crasquin-Soleau & Gradinaru, 1996 and Crasquin-Soleau, unpublished data) and from Far East Russia (Gramm, 1995). Two imprecisely determined species exhibit Palaeozoic affinities: cf. *Langdaia subobloga* Wang, 1978 (Kozur, 1973a) and Kirkbyocopina sp. (Gramm, 1995). These two species are the latest Palaeozoic representatives.

Middle (late Anisian) and Late Triassic ostracod fauna (Fig. 3):

The Tethyan Upper Triassic ostracods were studied particularly by Méhes (1911—Hungary), Anderson (1964—Great Britain), Kollmann (1963, 1960—Austria), Kristan-Tollmann (1970, 1973, 1983, 1986, 1991—Alps, South China, Papua-New Guinea, Iran), Sohn (1968 —Israel), Bolz (1969, 1970a, b—Alps), Will (1969 —Germany), Kozur & Nicklas (1970—Alps), Monostori (1994—Hungary).

The Upper Triassic neritic fauna is very characteristic and well known in the Tethyan domain. Most of the forms are massive and thick-shelled, as exemplified by representatives of Cytherellidae (as *Reubenella* Sohn, 1968, *Leviella* Sohn, 1968, ...), Bairdiidae (as *Carinobairdia* Kollmann, 1963, *Cornutobairdia* Kristan-Kollmann, 1970, *Lobobairdia* Kollmann, 1963, *Dicerobairdia* Kollmann, 1963, *Nodobairdia* Kollmann, 1963...), Kerocytheridae (as *Kerocythere* Kozur & Nicklas, 1970, ...), Mesozoic Healdiidae (as *Hungarella* Méhes, 1911; *Ogmoconchella* Gründel, 1964; *Hermiella* Kristan-Tollmann, 1977; *Triadohealdia* Kristan-Tollmann; 1971, *Soëlla* Kristan-Tollmann et al., 1987, ...), Cytheruridae (as *Metacytheropteron* Oertli, 1957, ...), Judahellidae (*Judahella* Sohn, 1968), Glorianellidae (*Lutkevichinella* Schneider, 1956, ...).

For the Middle Triassic, all the published neritic ostracods of upper Anisian and Ladinian age are typical Mesozoic forms. Palaeopsychrospheric assemblages (not taken in account in this paper) were published by Kozur (1970) from late Anisian of Hungary.

Data on upper Anisian neritic ostracods are found in Sohn (1968—Israel), Gramm (1975—South Primorye, Far East Russia), Kristan-Tollmann (1983—South China) and Monostori (1995—Hungary).

For the end of Middle Triassic, we can mention the work in Israel by Sohn (1968), in the Alps by Kozur (1971a–c), in Southern Spain by Kozur et al. (1974), in Himalaya by Agarwal & Kumar (1981), in NE Iran by Kristan-Tollmann (1991), in India by Goel et al. (1984), in Alaska by Sohn (1987), in Slovenia by Kolar-Jurkovsek (1991).

Discussion and conclusion

This analysis emphasizes that the first occurrences of ostracods exhibiting modern features are recorded from the Late Permian of South China, Western Taurus, Israel and Saudi Arabia. They include *Callicythere*, *Arqoviella*, and representatives of the "*Petasobairdia–Ceratob-*

airdia–Mirabairdia–Parurobairdia fauna" of Chen & Shi (1982).

Palaeozoic forms survive in the Early Triassic, as indicated by the occurrences of *Hollinella, Carinaknightina, Langdaia, Microcheilinella, Acratia*. The lastest representatives are from the early Anisian.

It is also important to note that the Early Triassic ostracod assemblages from the Luolou Formation, Guangxi Province, South China (Crasquin-Soleau et al., 2006) do not differ significantly from Late Permian ones. Data obtained in Early Triassic strata in Pakistan (Sohn, 1970), in Western Taurus (Crasquin-Soleau et al., 2004a, b) and in Eastern Sichuan (Crasquin-Soleau & Kershaw, 2005) show exactly the same features: composition similar to Late Palaeozoic assemblages and open marine environments. Twittchett et al. (2004) stated that "the hypothesis that the apparent delay in the recovery after end-Permian mass extinction event was due to widespread and prolonged benthic oxygen restriction and in the absence of anoxia, marine recovery is much faster", a statement which may well apply to our data. But, contrary to those authors, the "pre-extinction fauna" of Late Palaeozoic aspect occurs on the borders of Neo-Tethys, at least in South China, Tibet, and Western Taurus. The "mixed fauna interval" extending from latest Permian to Spathian separates Palaeozoic ostracod communities from the modern ones. For the moment, we have only few data on ostracod fauna from the Late Permian–Triassic interval and in a next future we expect to provide further data and quantitative distributions. Nevertheless, we can try to compare our results with the recovery of other well-documented groups as ammonoids, conodonts and brachiopods.

The conodonts have an explosive radiation during the Smithian when the fauna contains four-times as many conodont genera as those known in the

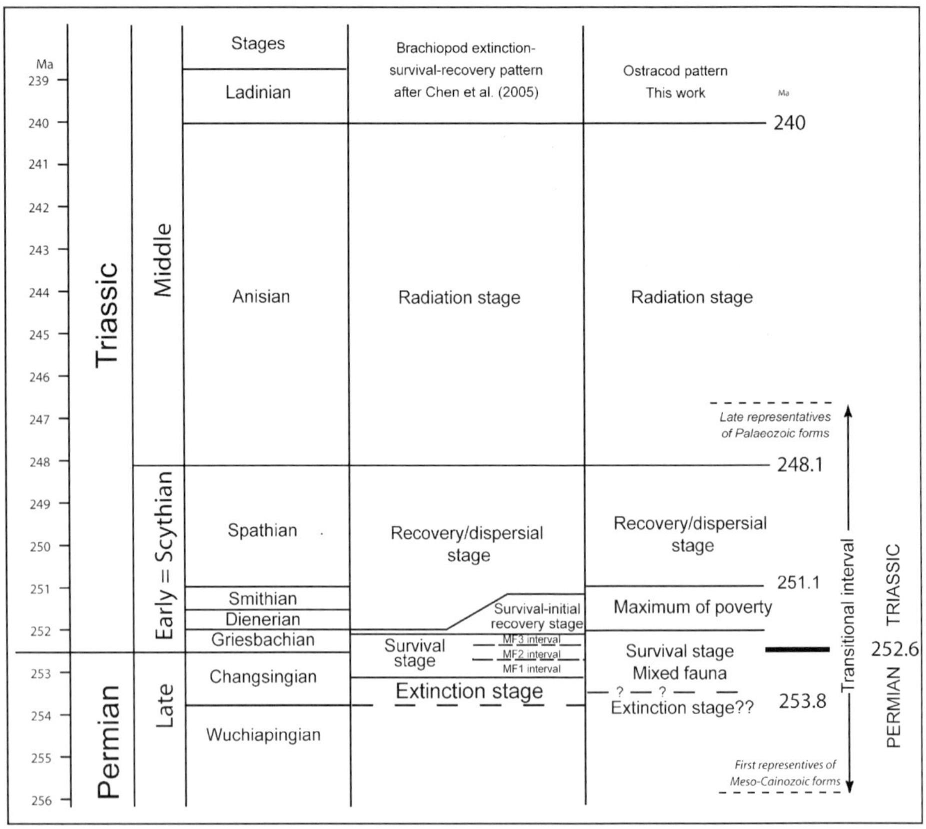

Fig. 4 Comparison between extinction and recovery patterns of brachipods and ostracods through Permian–Triassic boundary events. Brachiopod data after Chen et al. (2005). MF: mixed fauna (datations from Ovtcharova et al., 2006)

Late Dienerian (Orchard, 2005). The ammonoids return to full diversity in the Spathian (Brayard et al., 2006) i.e., 1–3 Ma after the PTB. For the Brachiopods, Chen et al. (2005) described different phases of extinction-survival-recovery pattern across the end-Permian extinction (Fig. 4): an extinction stage during the early Changhsingian, three survival stages (mixed fauna 1, 2 and 3 intervals) from late Changhsingian to middle Griesbachian, a survival-initial recovery stage in late Griesbachian, three phases of recovery/dispersal stage from Dienerian to the end of the Spathian and finally the radiation stage from the Anisian.)

At the present date, we do not have enough data to precisely date the end of the "extinction stage". The survival stage exists also for the ostracods. It seems to be longer than for brachiopods (Palaeozoic survivors still exist in the Spathian). The survival-initial recovery stage may begin earlier for the ostracods than for the Brachiopods (Mesozoic representatives are present from the Late Permian). The radiation stages seem to coincide for the two groups during the Anisian.

The final turnover of ostracods from Palaeozoic to Mesozoic faunas took place later during the Anisian.

If we consider a 252.6 ± 0.2 Ma age for the PTB (Mundil et al., 2004) and a late Spathian *N. haugi* Zone age of 248.1 ± 0.4 Ma (Ovtcharova et al. 2006) a minimal duration of ca. 4.5 ± 0.6 Ma can be inferred for the Early Triassic. The recovery phase is significantly shorter than previous estimates.

This study is a first step in the knowledge of ostracod fauna recovery after the events of the Permian–Triassic boundary. Detailed analysis of reference sections are in progress and may lead to a quantitative approach.

Acknowledgements This work was supported by the Swiss NSF project no. 200020-105090/1 (HB), by the PRA ST03-01 of AFCRST (Association Franco-Chinoise pour la Recherche Scientifique et Technique) (S.C. and F.Q.) and ECLIPSE 2 Programme (CNRS-INSU) (S.C., S.K. and F.Q.).We are most grateful to Dr. Avraham Honigstein (Geological Survey of Israel), Dr. Michael Schudack (University of Berlin, Germany) and the anonymous reviewer for their critical review and their help in the improvement of the manuscript. We are indebted to Renate Matzke-Karasz for her friendly work on this special volume.

References

Agarwal, P. N., 1979. Scythian Ostracodes from the 'Khunamuh Foramation', Khreuh, Kashmir Himalayas. Current Sciences 48: 442–443.

Agarwal, P. N., 1980. A record of Lower Triassic Ostracodes from Guryul Ravine, Kashmir. Geoscience Journal 1: 93–94.

Agarwal, P. N., 1981. An account of marine Triassic Ostracodes and their possible utility in biostratigraphy, with special reference to *Judahella* SOHN, in certain Lower Triassic sections in Kashmir Himalaya. Proceeding of IX Indian Colloquium on Micropalaeontology and Stratigraphy 146–155.

Agarwal, P. N. & S. Kumar, 1981. An ostracod assemblage from the Kalapani limestone, Malla Johar Area, Kumaun Himalaya In Sinha, A. K. (ed.), Comtemporary geoscientific research in Himalaya. Dehra Dun 1: 217–224.

Agarwal, P. N., S. N. Singh & S. Ashok, 1980. Scythian Ostracodes from Kashmir Himalayas. Journal of the Palaeontological Society of India 23/24: 110–114.

Anderson, F. W., 1964. Rhaetic Ostracoda. Bulletin of the Geological Survey of Great Britain 21: 133–174.

Ardens, R., J. P. Colin & H. Kozur, 1979. Sur la présence de Scythien supérieur fossilifère dans la chaîne du Vardoussia (Grèce continentale). Conséquences paléogéographiques. Comptes rendus sommaires de la Société Géologique de France 3: 132–135.

Becker, G. & S. Q. Wang, 1992. Kirkbyacea and Bairdiacea (Ostracoda) from the Palaeozoic of China. Palaeontographica A 224: 1–54.

Bolz, H., 1969. Der "bairdoppilate" Verschluss und Skulptur–Unterschiede bei Bairdien (Ostrac.) der alpinen Obertrias. Senckenbergiana lethaea 50: 411–431.

Bolz, H., 1970a. Einige Cytherelloidea - Arten (Ostracoden) aus der alpinen Obertrias. Senckenbergiana lethaea 51: 239–263.

Bolz, H., 1970b. Late Triassic Bairdiidae and Healdiidae. Bulletin du Centre de Recherche Pau - SNPA 5 supplément, 717–745.

Brayard, A., H. Bucher, G. Escarguel, F. Fluteau, S. Bourquin & T. Galfetti, 2006. The Early Triassic ammonoid recovery: Paleoclimatic significance of diversity gradients. Palaeogeography, Palaeoclimatology, Palaeoecology 239: 374–395.

Bunza, G. & H. Kozur, 1971. Beiträge zur Ostracodenfauna der tethyalen Trias. Geologisch-paläontologische Mitteilungen Innsbruck 1: 1–76.

Chen, Z. Q., K. Kaiho & A. D. George, 2005. Early Triassic recovery of the brachiopod faunas from the end-Permian mass extinction: A global review. Palaeogeography, Palaeoclimatology, Palaeoecology 224: 270–290.

Chen, D. Q. & C. G. Shi, 1982. Latest Permian ostracoda from Nantong, Jiangsu and from Mianyang, Hubei. Bulletin of Nanjing Institut of Geology and Palaeontology, Academia Sinica 4: 105–152 (in Chinese with English abstract).

Crasquin-Soleau, S., F. Berra & R. Rettori, 2000. A Late Triassic Ostracod assemblage from the Quattervals Nappe (Austroalpine, Northern Italy). Rivista Italiana di Paleontologia e Stratigrafia 106: 181–190.

Crasquin-Soleau, S. & E. Gradinaru, 1996. Early Anisian ostracode fauna from the Tulcea unit (Cimmerian North Dobrogean Orogen, Romania). Annales de Paléontologie 82: 59–116.

Crasquin-Soleau, S. & S. Kershaw, 2005. Ostracod fauna from the Permian - Triassic boundary interval of South China (Huaying Mountains, eastern Sichuan Province): palaeoenvironmental significance. Palaeogeography, Palaeoclimatology, Palaeoecology 217: 131–141.

Crasquin-Soleau, S., J. Marcoux, L. Angiolini & A. Nicora, 2004a. Palaeocopida (Ostracoda) across the Permian - Triassic events: new data from South-Western Taurus (Turkey). Journal of Micropalaeontology 23: 67–76.

Crasquin-Soleau, S., J. Marcoux, L. Angiolini, A. Nicora & Y. Bertho, 2004b. New ostracode fauna from Permian - Triassic boundary in Turkey (Taurus, Antalya Nappes). Micropaleontology 23: 67–76.

Crasquin-Soleau, S. & K. Teherani, 1995. Première découverte d'ostracodes triasiques dans la Formation de Khanet Kat, Montagne de Michparvar (Ouest Iran). Revue de Micropaléontologie, 38: 27–36.

Crasquin-Soleau, S., D. Vaslet & Y. M. Le Nindre, 2005. Ostracods from Permian -Triassic boundary in Saudi Arabia (Khuff Formation). Palaeontology 48: 853–868.

Crasquin-Soleau, S., T. Galfetti, H. Bucher & A. Brayard, 2006. Early Triassic ostracods from Guangxi Province, South China. Rivista Italiana di Paleontologia e Stratigrafia 112: 55–75.

Erwin, D. H., 1993. The Great Palaeozoic Crisis: Life and Death in the Permian. Columbia University Press, New York.

Gerry, E., A. Honigstein, B. Derin & A. Flexer, 1987. Late Permian Ostracodes of Israel. Taxonomy, distribution and paleogeographical implications. Senckbergiana lethaea 68: 197–223.

Goel, R., H. Kozur & S. S. Srirvastava, 1984. Middle Anisian (Pelsonian) Ostracoda from Spiti (Himachal Pradesch), India. Geoscience Journal 5: 53–62.

Gradstein, F. M., J. G. Ogg & A. G. Smith, 2005. A Geological Time Scale 2004. Cambridge University Press, Cambridge.

Gramm, M. N., 1975. Marine Triassic Cytheracea (Ostracoda) from South Primorye (Soviet Far East). Palaeontographica A 151: 102–110.

Gramm, M. N., 1995. Ostracodes of the suborder Kirkbyocopina from the Middle Triassic of Southern Primor'ye (Far East). Paleontological Journal 29: 154–156.

Hao, W. C., 1992. Early Triassic Ostracods from Guizhou. Acta Micropalaeontologica Sinica 9: 37–44 (in Chinese with English abstract).

Hao, W. C., 1994. The development of the Late Permian - Early Triassic ostracod fauna in Guizhou Province. Geological Review 40: 87–93.

Hirsch, F. & E. Gerry, 1974. Conodont and ostracode biostratigraphy of the Triassic in Israel. Schriftenreihe der Erdwissenschaftlichen Kommissionen, Österreichische Akademie der Wissenschaften 2: 107–114.

Honigstein, A., A. Rosenfeld & B. Derin, 2006. Late Permian ostracodes from Israel: New subsurface data. Micropaleontology 51: 405–422.

Jones, P. J., 1970. Marine Ostracoda (Palaecopida, Podocopida) from the Lower Triassic of the Perth Basin, Western Australia. Bulletin of Bureau of Mineral Ressources, Geology and Geophysics 108: 115–142.

Kolar-Jurkovsek, T., 1991. Microfauna from the Upper Triassic (Norian) of Karavanke Mts. (Slovenia), Symposium on Triassic Stratigraphy, Lausanne, Switzerland, 15 pp.

Kollmann, K., 1960. Ostracoden aus der alpinenTrias Österreichs. I.- *Parabairdia* n.g. und *Ptychobairdia* n.g. (Bairdiidae). Jahrbuch der Geologischen Bundesanstalt Wien 5: 79–105.

Kollmann, K., 1963. Ostracoden aus der Trias. II.- Weitere Bairdiidae. Jahrbuch der Geologischen Bundesanstalt Wien 106: 121–203.

Kozur, H., 1970. Neue Ostracoden-Arten aus dem Obersten Anis des Bakonyhochlandes (Ungarn). Berichte des Naturwissenschaftlich-Medizinischen Vereins in Innsbruck 58: 384–428.

Kozur, H., 1971a. Die Bairdiacea der Trias. Teil I: Skulpturierte Bairdiidae aus mitteltriassischen Flachwasserablagerungen. Geologisch-paläontologische Mitteilungen Innsbruck 1: 1–27.

Kozur, H., 1971b. Die Bairdiacea der Trias. Teil 2: Skulpturierte Bairdiidae aus mitteltriassischen Tiefschelfablagerungen. Geologisch-paläontologische Mitteilungen Innsbruck 1: 1–21.

Kozur, H., 1971c. Die Bairdiacea der Trias. Teil 3: Einige neue Arten triassischer Bairdiacea und Bemerkungen zur Herkunft der Macrocyprididae (Cypridacea). Geologisch-paläontologische Mitteilungen Innsbruck 1: 1–18.

Kozur, H., 1973a. Die Bedeutung triassischer Ostracoden für stratigraphische und paläoölogische Untersuchungen. Mitteilungen der Gesellschaft der Geologie- und Bergbaustudenten in Österreich 21: 632–660.

Kozur, H., 1973b. Einige Bemerkungen zur Systematik der Ostracoden und Beschreibung neuer Platycopida aus der Trias Ungarns und der Slowakei. Geologisch-paläontologische Mitteilungen Innsbruck 2: 1–27.

Kozur, H., 1985. Neue Ostracoden-Arten aus dem Oberen Mittelkarbon (Höheres Moskovian), Mittel- und Oberperm des Bükk-Gebirges (N-Ungarn). Geologisch-paläontologische Mitteilungen Innsbruck 2: 1–145.

Kozur, H., W. Kampschuur, C. W. H. Mulder-Banken & O. J. Simon, 1974. Contribution to the Triassic ostracode faunas of the Betic zone (Southern Spain). Scripta Geologica 23: 1–56.

Kozur, H. & L. Nicklas, 1970. Ostrakoden aus dem Plattenkalk-Niveau des Hauptdolomites (Rhätikon). In Mostler, H. (ed.), Beiträge zur Mikrofazies und Stratigraphie von Tirol und Vorarlberg 1971 (Festband des Geologischen Institutes, 300-Jahr-Feier Universität Innsbruck 1970), 309–320.

Kristan-Tollmann, E., 1970. Einige neue Bairdien (Ostracoda) aus der alpinen Trias. Neues Jahrbuch fur Geologie und Paläontologie. Monatshefte 135: 268–310.

Kristan-Tollmann, E., 1973. Zur Ausbildung des Schliessmuskelfeldes bei triadischen Cytherellidae (Ostracoda). Neues Jahrbuch fur Geologie und Paläontologie. Monatshefte 6: 351–373.

Kristan-Tollmann, E., 1983. Ostracoden aus dem Oberanis von Leidapo bei Guiyang in Südchina. Schriftenreihe der Erdwissenschaftlichen Kommissionen, Österreichische Akademie der Wissenschaften 5: 121–176.

Kristan-Tollmann, E., 1986. Beobachtungen zur Trias am Südostende der Tethys – Papua/Neuguinea, Australien, Neuseeland. Neues Jahrbuch fur Geologie und Paläontologie. Monatshefte 4: 201–222.

Kristan-Tollmann, E., 1991. Ostracods from the Middle Triassic Sina Formation (Aghdarband Group) in NE Iran. Abhandlungen der Geologischen Bundesanstalt 38: 195–200.

Méhes, G., 1911. Über Trias-Ostrakoden aus dem Bakony. Resultate der wissenschaftlichen Erforschung des Balatonsees 1: 1–38.

Monostori, M., 1994. Ostracod evidence of the Carnian salinity crisis in the Balaton Highland, Hungary. Neues Jahrbuch fur Geologie und Paläontologie. Monatshefte 193: 311–331.

Monostori, M., 1995. Environmental significance of the Anisian Ostracoda fauna from the Forras Hill near Felsoors (Balaton Highland, Transdanubia, Hungary). Acta Geologica Hungarica 39: 37–56.

Mundil R., K. R. Ludwig, I. Metcalfe & P. R. Renne, 2004. Age and Timing of the Permian Mass Extinctions: U/Pb Dating of Closed-System Zircons. Science 305: 1760–1763.

Orchard, M. J., 2005. On the explosive radiation of Lower Triassic Conodonts: A new multielement perspective. Albertiana 33: 65.

Ovtcharova, M., H. Bucher, U. Schaltegger, T. Galfetti, A. Brayard & J. Guex, 2006. New Early to Middle Triassic U–Pb ages from South China: Calibration with ammonoid biochronozones and implications for the timing of the Triassic biotic recovery. Earth and Planetary Science Letters 243: 463–475.

Pakistani-Japanese Research Group, 1985. Permian and Triassic Systems in the Salt Range and Sughar Range, Pakistan. In Nakazawa, K. & J. M. Dickins (eds), The Tethys. Her paleogeography and paleobiogeography from Palaeozoic to Mesozoic. Tokai University Press, Tokyo, 221–312.

Shi, C.-G. & D.-Q. Chen, 1987. The Changhsingian ostracodes from Meishan Changxing, Zhejiang. Stratigraphy and Palaeontology of Systemic Boundaries in China; Permian and Triassic Boundary 5: 23–80 (in Chinese with English abstract).

Schubert, J. K. & D. J. Bottjer, 1992. Early Triassic stromatolites as post-extinction disaster forms. Geology 20: 883–886.

Sohn, I. G., 1968. Triassic Ostracodes from Makhtesh Ramon, Israel. Geological Survey of Israel Bulletin 44: 1–71.

Sohn, I. G., 1970. Early Triassic Marine Ostracodes from the Salt Range and Surghar Range, West Pakistan. In Kummel, B. & C. Teichert (eds), Stratigraphic boundary problems: Permian and Triassic of West Pakistan. Department of Geology, University of Kansas, Special publication, Lawrence, 149–206.

Sohn, I. G., 1987. Middle and Upper Triassic Marine Ostracoda from the Shublik Formation, Northeastern Alaska. U.S. Geological Survey Bull., Shorter contribution to Paleontology and Stratigraphy (1664), C1–C47.

Twitchett, R. J., L. Krystyn, A. Baud, J. R. Wheeley & S. Richoz, 2004. Rapid marine recovery after the end-Permian mass-extinction event in absence of marine anoxia. Geology 32: 805–808.

Wang, S. -Q., 1978. Late Permian and Early Triassic ostracods of Western Guizhou and Northeastern Yunnan. Acta Palaeontologica Sinica 17: 307–308 (in Chinese with English abstract).

Wei, M., 1981. Early and Middle Triassic Ostracods from Sichuan. Acta Palaeontologica Sinica 20: 501–510 (in Chinese with English abstract).

Wignall, P. B. & A. Hallam, 1993. Griesbachian (Earliest Triassic) palaeoenvironmental changes in the Salt Range, Pakistan and southeast China and their bearing on the Permo-Triassic mass extinction. Palaeogeography, Palaeoclimatology, Palaeoecology 102: 215–237.

Will, H. J., 1969. Untersuchungen zur Stratigraphie und Genese des Oberkeupers in Nordwestdeutschland. Beihefte zum Geologischen Jahrbuch 54: 1–240.

Hydrobiologia (2007) 585:29–41
DOI 10.1007/s10750-007-0626-5

OSTRACODA (ISO15)

The influence of El Niño 1997–98 on pelagic ostracods in the Humboldt Current Ecosystem off Peru

Ruth Castillo · Tarsicio Antezana · Patricia Ayón

Abstract Planktonic ostracods are deep dwelling mesozooplankton that are diel migrators and detritus feeders. In the Humboldt Current region off Peru they are shown to be strongly influenced by El Niño conditions. Species composition and distribution were compared in day and night (vertically hauled 50–0 m Hensen Net) samples that were collected from three latitudinal zones (3°40' S–6°S; 9°S–12°S and 15°S–18°S) during 1998 (El Niño year) and in 2001 (a year with usual upwelling regime). Unexpectedly, ostracods diversity and abundance were much higher during El Niño than during the upwelling regime. Tropical species that normally occurred north of 6°S, spread southwards and increased in abundance; *Euconchoecia aculeata* was often dominant, particularly in coastal waters, and the frequency of occurrence of others increased throughout the entire region (e.g. *Porroecia porrecta, Archiconchoecia striata*). Conversely, *Conchoecetta giesbrechti*, a species associated with subsurface oxygen minimum waters, was more frequently found and in higher abundance during the upwelling regime in 2001. Thus, El Niño conditions apparently favor the occupation of the epipelagic zone by many halocyprid species advected in from tropical waters, possibly because the higher oxygen content in the subsurface waters enables these diel migrating species to survive better. Additional factors may be a consequence of the switch to a picoplankton production cycle, which results in more detritus being retained in the upper waters, and the sharp reduction in planktivorous fishes such as the anchovy which may reduce the predation pressure they experience.

Guest editors: R. Matzke-Karasz, K. Martens & M. Schudack
Ostracodology – Linking Bio- and Geosciences

R. Castillo (✉)
ISATEC Program, Universität Bremen, Fahrenheitstrasse 6, Bremen 28359, Germany
e-mail: rutycastillo@hotmail.com

T. Antezana
Departamento de Oceanografía, Facultad de Ciencias Naturales y Oceanográficas, Universidad de Concepción, Casilla 160-C, Concepción, Chile

P. Ayón
Instituto del Mar del Perú (IMARPE), Esq. Gamarra y Valle s/n, Apartado 22, Callao, Peru

Keywords Pelagic ostracods · Halocyprididae · Humboldt Current Ecosystem · El Niño Southern Oscillation · *Euconchoecia*

Introduction

The Humboldt Current System (HCS) is one of the most productive ecosystems of the world's ocean. Strong and persistent upwelling of cold subsurface waters, enriched with nutrients, boosts primary production particularly by diatoms (Zuta & Guillén, 1970; Alheit & Bernal, 1993). The HCS is a complex system of poleward and equatorward surface and subsurface currents, and is characterized by coastal upwellings. Latitudinal changes in faunal composition result from the switches between equatorial and subpolar water masses (Balech, 1954). It is also strongly influenced by seasonal, annual and decadal variability (Chávez et al., 2003; Alheit & Ñiquen, 2004) with "El Niño Southern Oscillation" as the dominant source of this climatic variability (Chávez et al., 2003). El Niño occurs every 2–7 years and is characterized by the deepening of the thermocline to levels below the zone from which the persistent upwellings are sourced, so cool nutrient-rich waters no longer reach the wind-mixed layer. During this event, the tropical waters extend further south and also further inshore into coastal waters, and they drastically modify the pelagic fauna of the region (see Arntz & Fahrbach, 1996 and Fiedler, 2002 for a review).

The El Niño of 1997–1998 was one of the most intense ever recorded (McPhaden, 1999). Its influence in Peru was first detected in March 1997, reached its peak intensity in November and December 1997, and persisted until May 1998 (Jaimes, 1999). It deepened the thermocline off Peru to >100 m, reduced the intensity of upwelling, diminished productivity and affected the distribution and abundance of anchovy, causing a strong impact on the fisheries. After the El Niño, there was a 2 month transition period between August and September 1998 that ended in a very cold La Niña event, when upwelling-favorable winds strengthened.

The influence of El Niño on zooplankton has generally been reported to be a sharp reduction in macrozooplankton biomass (large copepods and euphausiids) but an increase in the biomass of small copepods and gelatinous zooplankton (Carrasco & Santander, 1987; Lavaniegos et al., 2002; Lavaniegos et al., 2003; Escribano et al., 2004; Zamon & Welch, 2005). However, the adaptive responses of individual taxa are not well known. Based on the general warming of sea surface temperatures throughout the region during the El Niño, it was predicted that equatorial species would extend southwards to higher latitudes. Conversely, species normally associated with cooler waters were expected to retreat either southwards or into deeper waters, where temperatures remain cool or dissolved oxygen concentrations increase.

Planktonic ostracods have received scant attention in oceanic ecological studies, in spite of their relatively high abundances, their ubiquitous occurrence at all oceanic depths, and their importance in recycling material in pelagic ecosystems (Angel, 1999). There have been very few previous studies of them in the region (Müller, 1895; Poulsen, 1973; Martens 1979, 1981) and none of their responses to El Niño. The aim of this work was to compare the normal distributional patterns of ostracods in the epipelagic waters of Humboldt Current off Peru during the upwelling regime, with the patterns observed during the 1997–1998 El Niño. We examine their distributions in three latitudinal zones, which not only have different continental shelf characteristics, but also where the distribution of water masses changes drastically during El Niño events. This study extends the results of Martens (1979, 1981) from Chilean waters.

Materials and methods

Samples were collected by IMARPE (Peruvian Institute of the Sea) research vessels RV *Humboldt* (Cruise 9803–05) in March–May 1998 towards the end of 1997–98 El Niño event, and RV *Olaya-SNP2* (Cruise 0102–04) in March–April 2001 at the end of a cold event.

The region was divided according to hydrographic, ecological and fisheries criteria into three latitudinal zones: 1. north of 6°S, 2. between 6°–14°S, and 3. south of 14°S (Zuta & Guillén, 1970; Carrasco & Lozano, 1989). Samples for each year were selected randomly in order to illustrate changes latitudinally and longitudinally (Fig. 1).

Zooplankton samples were collected using Hensen nets (0.333 mm mesh, mouth area 0.332 m^2) hauled vertically from 50 to 0 m at 1 ms^{-1}, and preserved in 2% formaldehyde. Halocyprid ostracods were picked from the total sample, identified and counted. Adults and juveniles (A-1 to A-4) were enumerated separately. For the very abundant species, appropriate size

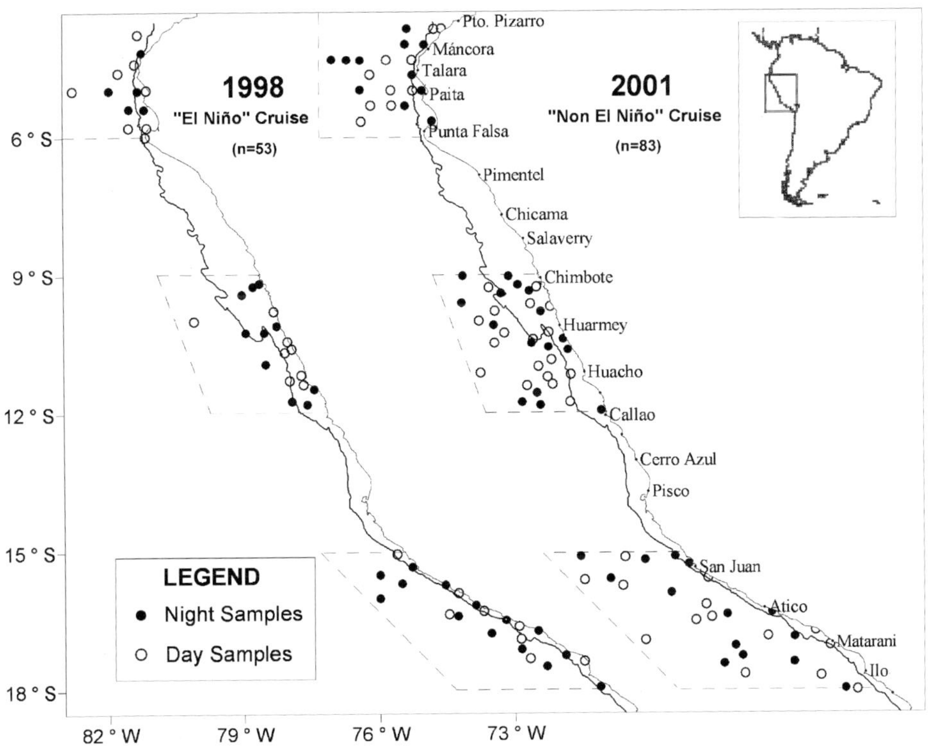

Fig. 1 Day and night samples from IMARPE cruises (1998 and 2001) examined in this study

aliquots were taken before enumeration, following the Huntsman Marine Laboratory Method (Van Guelpen et al., 1982). Species identification was according to Angel (1969a, b, 1971, 1999), Gooday (1981), Martens (1979), Poulsen (1969, 1973) and Müller (1895). Because of the poor state of knowledge of the taxonomy of the Halocyprididae, particularly in the Eastern Pacific, there were uncertainties in the identifications of some species (Table 1), which will only be removed when a thorough revision of the group supported by molecular studies is carried out. Specimens of the identified species will be deposited at the IMARPE Collection. Details may be obtained directly from the first author.

Abundances were standardized to numbers of individuals per 100 m^3 of water filtered, assuming there had been no significant clogging of the nets. Samples collected from over the continental shelf (i.e. inside the 200 m isobath) were considered to be coastal and those collected outside were considered to be oceanic. Day and night differences in species abundances were tested by Mann–Whitney's U Test (abbreviated as UMW), to estimate the influence of vertical migration and any daytime net avoidance. Diversity was estimated by calculating several indices (Shannon and Wiener based on log 2, Margalef's richness and Pielou's evenness).

Temperature, salinity and dissolved oxygen data provided by IMARPE was used to calculate Spearman rank correlation coefficients for the most frequent species.

Results

Changes in the physical environment between 1998 and 2001

During both years SST (Sea Surface Temperature) showed a gradient decreasing from north to south; in 1998 the range was: 30.1–16.4°C (Vásquez & Tello, 1998), while in 2001: 28.0–13.8°C. In the northern sector the temperature disparities between the 2 years were almost 4°C.

Vásquez & Tello (1998) found the thermocline deeper during 1998, tilting up from 125 m in the

Table 1 A list of the planktonic ostracods of the Humboldt Current in the upper 0–50 m layer off Peru during 1998 and 2001

N	Species[1]	Size (mm)[2]		Usual Depth Range (m)[2]		Frequency (%)		Mean abundance (N° ind/100 m^3)	
		Females	Males			1998	2001	1998	2001
1	*Euconchoecia aculeata* (Scott 1894)	1.10–1.30	1.00–1.20	0–300	a	75	14	5388[3]	20
2	*Archiconchoecia striata* Müller 1894*	0.54–0.60	0.62	0–200		77	29	445	24
3	*Proceroecia decipiens* (Müller 1906)	1.40–1.70	1.20–1.30	50–100	a	77	4	150	6
4	*Porroecia porrecta pacifica* Martens 1979	1.43–1.59	1.27–1.37	0–300	b	89	46	140	41
5	*Conchoecia aff magna* Claus 1874	1.83–1.99	1.74–1.94	<200	b	55	6	48	3
6	*Orthoconchoecia agassizi* (Müller 1906)	4.30–4.80	4.20–4.60	30–400		34	8	12	1
7	*Discoconchoecia tamensis* (Poulsen 1973)*		1.10		a	26	4	6	2
8	*Conchoecetta giesbrechti giesbrechti* (Müller 1906)	2.53–2.96	2.29–2.54	0–1250	b	23	31	6	13
9	*Orthoconchoecia striola striola* (Müller 1906)	2.94–3.22	2.58–2.78	0–600	b	11	6	5	2
10	*Metaconchoecia* aff *acuta* (Gooday 1981)*	0.84–0.93	0.84–0.93	50–500	c	25	1	4	0
11	*Metaconchoecia teretivalvata* (Iles 1953)	0.74–0.92	0.76–0.86	0–400		13	0	3	0.1
12	*Metaconchoecia inflata* (Gooday 1981)*	0.97–1.16	0.99–1.16	200–500	c	6	5	0.8	4
13	*Proceroecia microprocera* (Angel 1971)	0.86–1.04	0.76–0.90	100–500		2	0	0.3	0
14	*Halocypris inflata* Dana 1849	1.40–1.80	1.32–1.60	25–50		4	8	0.2	1
15	*Porroecia pseudoparthenoda* (Angel 1972)	1.72–1.92	1.56–1.72	0–300		2	5	0.1	0.3
16	*Metaconchoecia kyrtophora* (Müller 1906)	0.76–0.86	0.72–0.82	200–400		2	1	0.1	0.1

Frequency represents percentage of samples that contained the species. (n = 83 in 2001, and n = 53 in 1998)

[1] Asterisks following the species name denote uncertainties on the identifications. *A. striata* might be the recently described species *A. instriata* (Chavtur & Stovbun, 2003). *M. inflata lata* (n ssp) described also lately by Chavtur, 2003 might be the subspecies occurring in the Pacific. *D. tamensis* has been rarely reported in the literature after its description. Finally, our *M. aff acuta*, looks similar to the one reported by Gooday, 1981. However, present available data are not sufficient to make a clear distinction for these species

[2] Size and depth ranges taken from Angel (19999 or from [a]Poulsen (1969, 1973), [b]Martens (1979), [c]Gooday (1981)

[3] Species abundances are highlighted when differences in abundance between years was significant (U Mann–Whitney Test, $P < 0.001$, values adjusted for ties)

north to ca 80 m in the south. In 2001, as is typical of "non-El Niño" years, it was strong and shallow (i.e. <50 m) throughout all three regions.

During 1998, Superficial Equatorial Water (SEW, 19–26°C, S 34.0–34.8) had extended to 7°S and Superficial Subtropical Water (SSW, >17°C, S > 35.1) reached the coastal area south of this latitude, no mixing with cold upwelled waters was observed, except in a few, small upwelling centers south of 14° (Vásquez & Tello, 1998). In 2001, Superficial Tropical Water (STW, >26°C, S < 34.0) was restricted to a small coastal area north of 4°30' S being replaced by SEW, which covered the coastal region to the north of 6°S, and further south (to 8°S) 40 nm offshore. Superficial Subtropical Water was found 90 nm offshore south of 6°S. In the coastal waters along the entire latitudinal range SSW was mixed with upwelled cold coastal waters (CCW, 15–19°C, S 34.8–35.05).

Halocyprid species composition and abundance

Sixteen species of halocyprid ostracods belonging to ten genera were identified from the samples (Table 1). All but two of the species were found during both cruises—the exceptions were *Metaconchoecia* aff *acuta* and *Proceroecia microprocera*, which were not sampled in 2001. Table 1 shows the adult size ranges and the preferred depth ranges of the species based on published data. It can be seen that the adult sizes varied by nearly an order of magnitude from the smallest *Archiconchoecia striata* to the largest *Orthoconchoecia agassizi*, and that most species have broad depth ranges being predominantly mesopelagic.

Overall the total abundance of ostracods in individual samples was significantly higher in 1998 than in 2001 (UMW Test, $P < 0.0001$, Table 1). In

2001, only 31% of the samples contained >100 individuals/100 m^3, whereas in 1998, 92% contained >100 individuals/100 m^3. The percentage of samples containing no ostracods was lower in 1998 (3%) compared to 2001 (32%).

The relative abundance of individual species varied considerably between years. In 1998, the halocyprid populations were dominated by a single species *Euconchoecia aculeata*, which contributed 87% of all the specimens sampled. In 2001, the relative abundances of species were more even with six species contributing 91% of the specimens; *Porroecia porrecta* as the most abundant. Ostracods were more widespread in 1998 than in 2001, as indicated by their higher frequencies of occurrence. *Porroecia porrecta* was the most consistently species caught in both years, while *A. striata, Proceroecia decipiens, E. aculeata* and *Conchoecia* aff *magna* were collected at >70% of the stations in 1998 (Table 1).

Day and night changes

In both years, more species were collected per sample at night. The maximum species richness in a nighttime sample was identical between years (1998 = 12; 2001 = 12), but much lower by day (1998 = 8; 2001 = 4). All species, except for *E. aculeata*, occurred at higher abundances at night (Fig. 2). Even so, day–night differences were only highly significant (UMW test at $P < 0.0007$) for adults of five species: *C.* aff *magna*, *Discoconchoecia tamensis*, *P. decipiens*, *O. agassizi* and *O. striola* (for the latter two species, adults were present only at night). There were significant differences (UMW test, $P < 0.007$) in the abundances of the juveniles of the same five species (but juveniles of *O. striola* occurred at night). When UMW test was applied separately for each year, results were significant only for 1998 abundances and diversity (Fig. 2, Table 2), although the patterns were maintained during 2001 with the species mentioned before occurring exclusively during night. Therefore, only nighttime data of these species were used for the subsequent results.

Onshore–offshore distributions

In 1998, there was little difference between the onshore–offshore abundances and distribution of the species. Only two species: *P. porrecta* and *O. agassizi* were significant in the UMW test ($P = 0.018$ for both). However in 2001, the oceanic samples were richer in both species and individuals (Table 2). Of the 14 species sampled that year only four occurred in coastal waters, and of these, *E. aculeata*, was restricted to the northern zone and was the only frequent species that

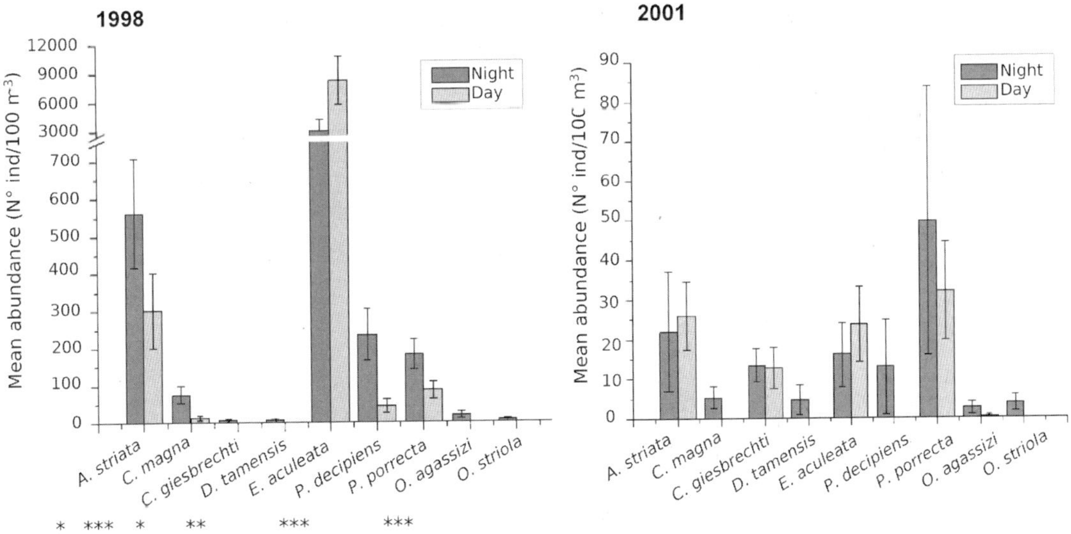

Fig. 2 Mean abundance and standard error of ostracods during night and day for each year. The asterisks indicate species whose abundances differed significantly when compared through U Mann–Whitney Test (* = $P < 0.05$; ** = $P < 0.01$; *** = $P < 0.0001$). N° of samples in 1998: Night = 29, Day = 24; in 2001: Night = 41, Day = 42

Table 2 Mean and standard deviation of the total abundance and diversity indexes for 1998 and 2001

	1998			2001		
	Day (n = 24)	Night (n = 29)	UMW Test	Day (n = 42)	Night (n = 41)	UMW Test
Total abundance	8735 ± 12505	4116 ± 6235	ns	98 ± 125	137 ± 301	ns
Diversity index (H' log2)	0.69 ± 0.58	1.29 ± 0.77	**	0.48 ± 0.55	0.50 ± 0.79	ns
Richness index (D)	0.36 ± 0.24	0.75 ± 0.28		0.26 ± 0.20	0.35 ± 0.44	
Evenness index (J)	0.43 ± 0.31	0.51 ± 0.25		0.73 ± 0.22	0.66 ± 0.29	
	Coast (n = 31)	Ocean (n = 22)	UMW Test	Coast (n = 26)	Ocean (n = 57)	UMW Test
Total abundance	7568 ± 10784	4291 ± 7983	ns	38 ± 75	153 ± 264	***
Diversity index (H' log2)	0.98 ± 0.78	1.06 ± 0.70	ns	0.09 ± 0.23	0.67 ± 0.73	***
Richness index (D)	0.58 ± 0.34	0.55 ± 0.31		0.10 ± 0.14	0.34 ± 0.36	
Evenness index (J)	0.47 ± 0.29	0.47 ± 0.27		0.55 ± 0.28	0.71 ± 0.25	
	North (n = 13)	Center (n = 18)	South (n = 22)	North (n = 21)	Center (n = 34)	South (n = 28)
Total abundance	13675 ± 11921	7408 ± 10447	813 ± 799	175 ± 272	29 ± 56	181 ± 289
Diversity Index (H' log2)	0.37 ± 0.49	1.07 ± 0.66	1.36 ± 0.70	0.74 ± 0.88	0.18 ± 0.40	0.67 ± 0.64
Richness Index (D)	0.36 ± 0.33	0.57 ± 0.22	0.70 ± 0.34	0.39 ± 0.48	0.16 ± 0.17	0.30 ± 0.25
Evenness Index (J)	0.21 ± 0.26	0.44 ± 0.22	0.65 ± 0.21	0.70 ± 0.24	0.73 ± 0.29	0.68 ± 0.26

U Mann–Whitney Test significance is also shown

did not differ significantly in the UMW test (P = 0.12).

Latitudinal distributions

The latitudinal distributions changed between years. In 1998, the overall species richness was similar between the three latitudinal zones (north-10, central-12, south-13). *Euconchoecia aculeata* (Fig. 3) occurred in all the three zones, but was the most frequent and dominant species in the northern and central zones. *Archiconchoecia striata* and *P. decipiens* were frequent and abundant in the central zone. In the south, *A. striata* became the most numerous species with *P. porrecta* and *P. decipiens* as subdominants (Fig. 3).

In 2001, the northern zone was the richest in species (13 species). *Euconchoecia aculeata* and *P. decipiens* were the most abundant. In the central zone, which encompassed the main

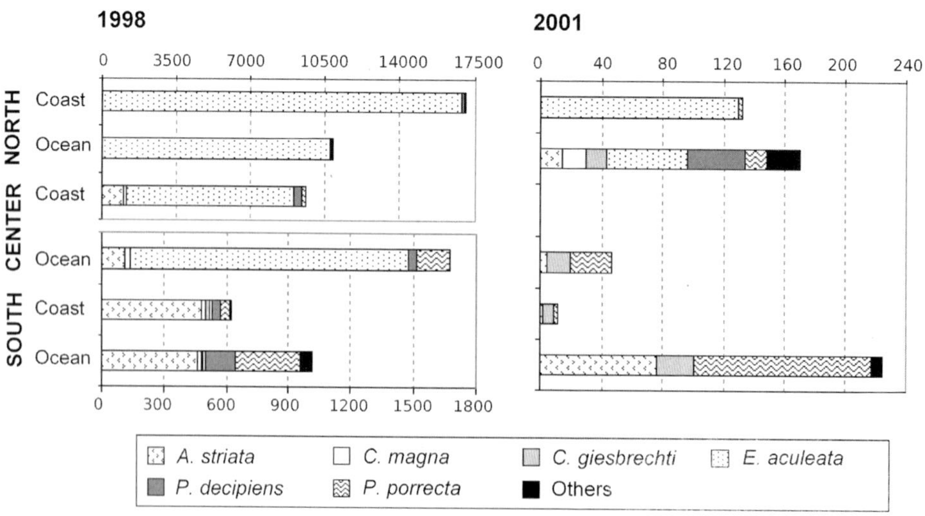

Fig. 3 Offshore–onshore and latitudinal mean abundances (N° ind/100 m^3) of the most abundant ostracods in 1998 and 2001. Note that two different scales were used for the 1998 plot

upwelling centers, species richness and abundance was sharply reduced, especially in the wide coastal area (no ostracods found), and the four species collected in the oceanic region were never abundant (Fig. 3, Table 2). In the south, eight species were found; *P. porrecta*, *A. striata* and *Conchoecetta giesbrechti* were the most abundant (Fig. 3).

Diversity indexes

The mean for Shannon diversity indices (H' $\log_2$) for 2001 night samples was 0.50 (range 0–2.82), considerably lower than the mean of 1.29 (range 0–2.60) for the 1998 night samples. This trend for diversity to be higher in 1998 than in 2001 was maintained (Table 2), except in the northern zone, where diversity was higher in 2001, due to the lowest evenness in this zone during 1998.

Distribution of juvenile instars

Juveniles (instars A-1 to A-4) contributed up to 38% of all individuals examined. Relative proportion of juveniles to adults varied between species and between years. However, these proportions are subject to sampling bias because the earlier instars of the smaller species could have been extruded through the meshes of the net, unless there was significant clogging. In 1998, abundances of juveniles exceeded those of the adults in five species (*C.* aff *magna*, *C. giesbrechti*, *P. decipiens*, *P. porrecta*, *O. striola*) (Fig. 4). Abundances of the juveniles of *A. striata* never exceeded 20% of the adult abundance, reflecting the poor retention by the net of these tiny animals. For the larger species, retention of the ultimate juvenile instar (A-1) should have been as efficient as for the adults, and so when they were markedly less abundant than the adults, either the juveniles were segregated bathymetrically, occupying depths deeper than the adults or the dynamics of the life cycle were generating the difference—for example the duration of the instar might be short compared with the longevity of the adults.

In 2001, juvenile instars were generally less abundant than adults in most species—*P. porrecta* and *C. giesbrechti* were exceptions. Juveniles contributed <20% of individuals in four species (*A. striata, C.* aff *magna, O. agassizi, O. striola*) (Fig. 4); the latter three are large species and there would have been no retention artifacts as in *A. striata*

Correlation analyses

Spearman rank correlations coefficients between the abiotic data and the abundances of adults and juveniles of the four most common species for both years are shown in Table 3. Oxygen content at higher depths was available only for some

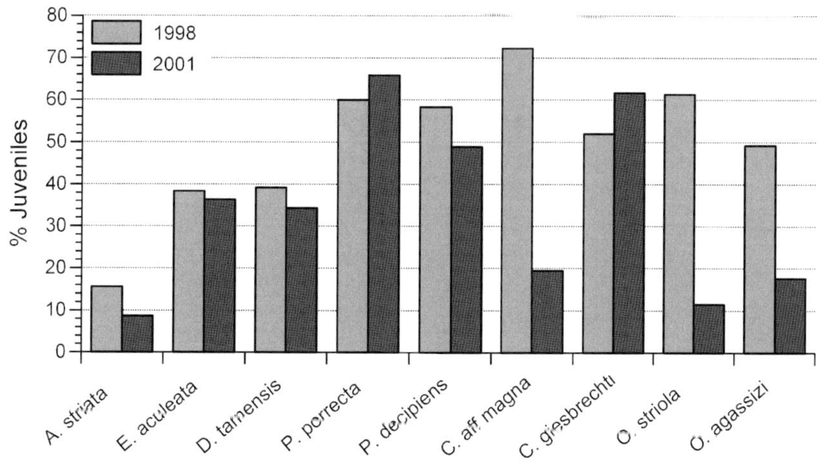

Fig. 4 Relative proportion of juveniles of the most frequent and abundant ostracods in 1998 and 2001 off Peru. Species are arranged in a size scale: smallest species to the left, and biggest species to the right

Table 3 Significant ($P < 0.05$) Spearman correlation coefficients between abiotic data and abundance of adults and juveniles of the most frequent species

		1998			2001					
		SST (n = 53)	SSS (n = 51)	Oxygen content at surface (n = 47)	SST (n = 83)	SSS (n = 68)	Oxygen content			
							At surface (n = 54)	At 25 m (n = 16)	At 50 m (n = 17)	At 100 m (n = 11)
A. striata	Total	**–0.37**	**0.39**	ns	ns	ns	ns	ns	ns	ns
	Juv	**–0.36**	0.31	ns	ns	ns	ns	ns	0.50	ns
C. giesbrechti	Total	**–0.42**	ns	–0.29	ns	ns	ns	ns	ns	ns
	Juv	**–0.46**	ns	ns	ns	ns	ns	ns	ns	ns
E. aculeata	Total	0.72	ns	ns	0.31	**–0.60**	ns	ns	ns	ns
	Juv	0.71	0.36	ns	0.31	**–0.59**	ns	ns	0.50	ns
P. porrecta	Total	ns	**0.66**	ns	0.23	ns	ns	ns	0.53	ns
	Juv	ns	**0.61**	ns	**0.32**	ns	ns	ns	0.56	ns
Total Abund.	Total	**0.53**	ns	ns	0.26	ns	–0.29	0.61	**0.81**	0.69
	Juv	0.58	ns	ns	**0.32**	ns	–0.31	**0.70**	**0.78**	0.65
Diversity	Total	**–0.40**	**0.45**	ns	ns	ns	ns	ns	0.49	ns

Bold numbers show highly significant correlations ($P < 0.01$)

stations in 2001, but still shows very strong correlations on the abundances of adults and juveniles of ostracods.

Discussion

Halocyprid ostracods are generally more abundant and diverse in waters deeper than 100 m; the maximum abundance of the group is often between 200–400 m and their maximum species richness at 1000 m (Angel, 1999). However, in this region the intense Oxygen Minimum Zone (OMZ, <1 ml O_2 l^{-1}) may disrupt this 'normal' pattern. In this study the sampling was restricted to the upper 50 m, and so 16 species identified are probably a fraction of the species inhabiting the region. Nevertheless, we observed substantial changes in the abundances, distributions and diversity of ostracods within the study area and between an El Niño year and a normal year with extensive upwelling.

A few previous studies (without identifying species) have reported changes in ostracod abundances during El Niño. Santander & Carrasco (1985) reported that ostracods were among the important groups contributing to the zooplankton populations during El Niño 1982–83. Lavaniegos et al. (2002) also referred a slight (though not significant) increase in ostracod abundances in the Gulf of California during El Niño 1997–1998. Jiménez-Pérez & Lara-Lara (1988) reported them as the third most abundant group of zooplankton during El Niño 1982–1983, (after cladocerans and copepods) in the same area of the Gulf of California.

During El Niño, there were major changes in the hydrographic regime in the region particularly in the northern and central zones. At the surface, cool upwelling waters became restricted to localized upwelling centers and were much reduced close into the coast. Tropical ostracod species, spread throughout the region consistent with the spread of warm water masses. The populations became dominated by the 'blooms' of two species. *E. aculeata*, a species strongly correlated with the temperature (Table 3), occurred in enormous numbers in the northern and central zones (Figs. 3, 5a), and *A. striata*, (associated with salinity and lower temperatures, Table 3), became extremely abundant in the central and southern zones (Figs. 3, 5b). The overwhelming numerical dominance of these two species was responsible for the sharp decline in evenness of the ostracod populations during 1998 relative to 2001 (Table 2).

During a typical upwelling year like 2001, highest diversity of ostracods was found in the

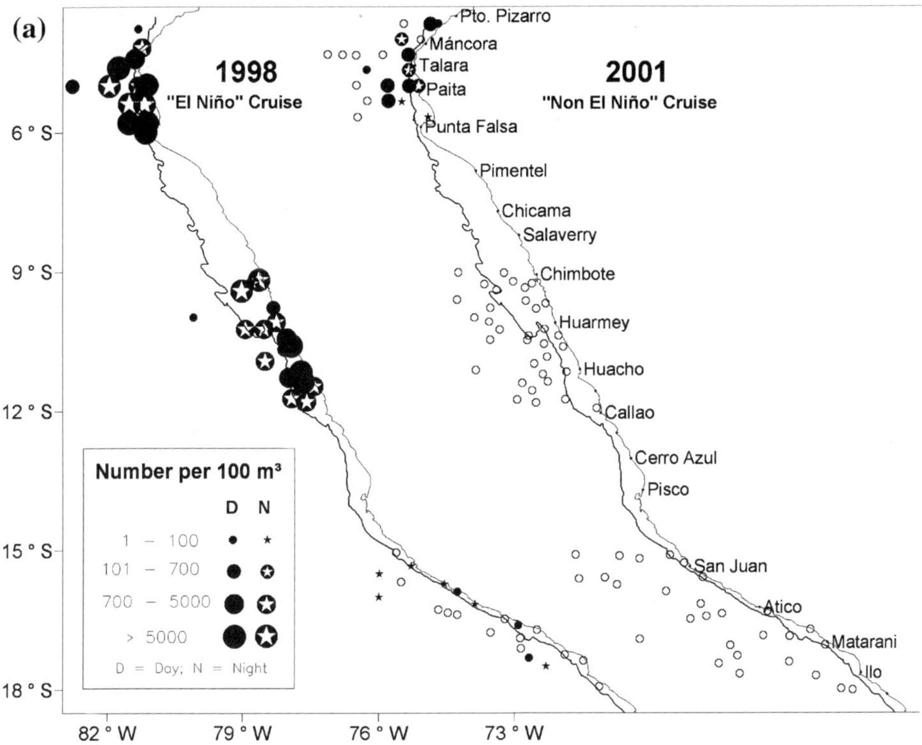

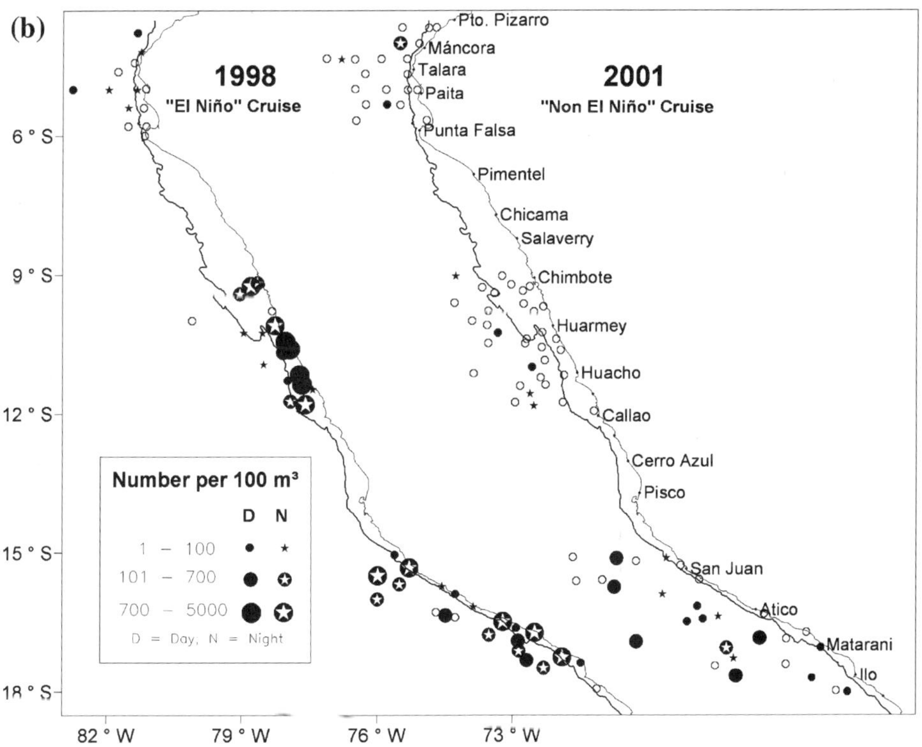

Fig. 5 Distribution of *Euconchoecia aculeata* and *Archiconchoecia striata* during 1998 and 2001 (**a**) *E. aculeata* (**b**) *A. striata*

northern zone (3.6°S–6°S), which is influenced by both equatorial (SEW) and subtropical waters (SSW), and north of 4°S by tropical water masses. In the central and southern zones, and particularly in the coastal areas, ostracod species richness and abundance was very low. *P. porrecta* was the most common and abundant species, and was found throughout the region, but specially related to high salinity (SSW, more oceanic) waters (Table 3). *Conchoecetta giesbrechti* although never abundant, occurred more frequently in the southern and coastal zones during this year, and was the only species that showed a negative correlation with the surface oxygen content (Table 3). Martens (1981) considered this species to be an indicator of Equatorial Subsurface Water (ESSW), which is the water mass that underlies the wind-mixed layer and makes the predominant contribution to the water that is upwelled in this region (Zuta & Guillén, 1970). Thus, this species may be an indicator of upwelling and so would be suppressed by El Niño conditions.

The absence of ostracods in central coastal waters and the general reduction in species numbers and abundances in coastal waters during 2001 (Fig. 3, Table 3) might be related to the development of the Oxygen Minimum Zone. The OMZ occurs throughout the Eastern Tropical Pacific, and during intense coastal upwelling periods, it can reach the surface in the Humboldt Current region (Judkins, 1980; Giesecke & González, 2004). Over the wide continental shelf off northern central Peru, biological and chemical processes combine with the influence of the circulation to create the intense oxygen deficiency observed in coastal waters (Richman & Smith, 1981). A few species are adapted to, and favored by oxygen deficiency (Longhurst, 1967; Antezana, 2002), so usually the OMZ has a strong impact on the distributions of the pelagic fauna. During 2001, not only were ostracods absent from the surface waters, but the biovolumes of zooplankton were exceptionally low (Aronés, 2004). Poulsen (1977) concluded that ostracods are poorly adapted to living in waters with oxygen contents less than 2–3 ml/l, so the observed depletion of ostracods during upwelling is not unexpected. Moreover, the extent to which the very shallow OMZ was impinging over the shelf in 2001 (Fig. 6) may have been enhanced by the unusual environmental conditions that prevailed during the summer. Subtropical Surface Water (SSW; >17°C, S > 35.1) which was over 70 nm offshore, was usually much closer inshore.

Conversely, the invasion by ostracods of coastal waters during the El Niño may have been the result of advection from the north and the presence of highly-oxygenated SSW associated with the deepening of the thermocline.

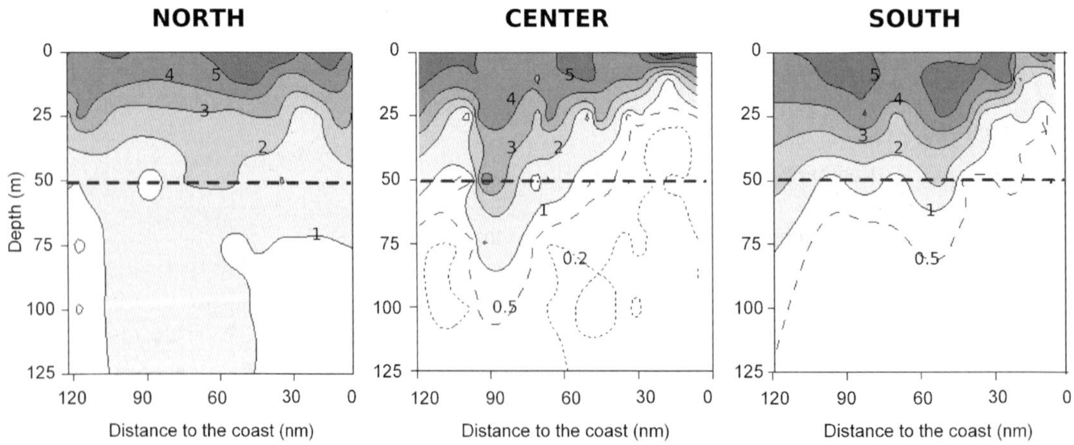

Fig. 6 Vertical and offshore distribution of oxygen content in the three regions during 2001. Available oceanographic data from the main legs was integrated and combined into one single graph for each region. The dashed line at 50 m shows the depth from which vertical hauls were towed

Was the observed increase in diversity and abundance of the ostracods during the El Niño solely the result of advection or could it have been involved a population response to more favorable conditions? In Table 1, species whose bathymetric range normally encompasses the upper 50 m (12), were supposed to occur in comparable abundances. Insignificant differences (UMW test, $P > 0.05$) were found in the abundances of five species between the two interannual regimes: *C. giesbrechti*, *H. inflata*, *M. teretivalvata*, *O. striola* and *P. pseudoparthenoda* (Table 1). However, apart from *C. giesbrechti* none of these species was ever abundant. So, the other seven species were the ones contributing to the increase in ostracod abundances and to zooplankton biomass during El Niño.

A population response may result in increases in the proportions of juvenile instars, although this would be dependent on aspects of population dynamics that are generally unavailable for halocyprids, such as the duration of instars and rates of developments. However, as discussed above populations of some of the species did comprise higher proportions of juveniles as would be expected from a population response, but these data can hardly be regarded as convincing. There are aspects of the biology of the two species whose populations demonstrated the largest responses, *Euconchoecia aculeata* and *Archiconchoecia striata* that need to be considered.

Both these species are likely to have high intrinsic rates of population growth, but for different reasons. The genus *Euconchoecia* is more neritic than other halocyprid genera, and its species have an unusual mode of reproduction (Tseng, 1975, 1976). Like cladocerans, the females brood eggs within the carapace, keeping a consistent number of developing embryos within the brood pouch, each one at a different stage of development. This is because the female lays the eggs one at a time and releases each embryo separately as it reaches maturity. This mode of reproduction probably enables *Euconchoecia* populations to respond rapidly to environmental improvements. *Archiconchoecia striata* has a different strategy whereby it may achieve a high rate of population growth. Adults only have five pairs of claw setae on the caudal furca, compared to the eight pairs that are normal in halocyprids. Newly hatched embryos of halocyprids have just two pairs of caudal furca claws and add an additional pair at each ecdysis. Thus, after the sixth juvenile instar the normal adult has attained its compliment of eight pairs of caudal furcal claws. The implication is that *A. striata* has abbreviated development, attaining maturity neotenously after maybe just three instars (Angel, personal communication, Dec. 2005). So, this very small species probably has a much shorter life cycle than the other species, enabling it to respond much faster to improving conditions. Both species are small and so are probably favored by the switch in primary production from being dominated by diatoms under the 'normal' upwelling conditions to being dominated by picoplankton as in El Niño conditions (Iriarte & González, 2004). Although this switch leads to a reduction in overall productivity during El Niño, most of the organic material produced is retained within the wind-mixed layer, and there is an increase in gelatinous zooplankton that generates the detrital material on which the ostracods feed. An additional factor may be a consequence of the well reported collapse of the regional fisheries for anchovy during El Niño, which may also signal a relaxation in the predation pressure experienced by these small epipelagic zooplankton species.

During upwelling, sedimentation of large quantities of organic matter resulting from the enhanced high primary production, results in substantial reductions in subthermocline concentrations in dissolved oxygen. Poulsen (1973) stated that the majority of halocyprid species are intolerant of low oxygen environments, so during upwelling sharp reductions in ostracod populations would be expected as the OMZ intensifies. Conversely, when the upwelling relaxes and the OMZ weakens, the ostracods can re-invade the subthermocline waters and increasing abundances of diel migrating halocyprids will be encountered in the epipelagic zone at night, which will tend to be advected onshelf by the circulation. This, combined with cooler SSTs which may be unsuitable for the warm water species, may explain the rather unexpected reductions in ostracod abundances and species richness we observed during upwelling.

Conclusions

During typical upwelling conditions in 2001, when sea surface temperatures were cool, the thermocline was intense and shallow, and the subthermocline water was depleted in dissolved oxygen, planktonic ostracods were neither abundant nor frequent in the upper surface layer (0–50 m). However, during the 1997–98 El Niño, intrusions of tropical warm water masses moved south from equatorial latitudes, the thermocline deepened and the deep waters became more oxygenated, allowing plankton ostracods to be abundant in the surface layer, especially in the coastal environments, where they are not usually present. Almost all the species of ostracods that were observed increased in abundance and frequency of occurrence during the El Niño event. But the response of the tropical species *E. aculeata* was overwhelming and it represented up to 87% of total ostracods captured during 1998. This population outburst was probably partially the result of advection but also the ability of this species to generate swarms rapidly, making it very effective in colonizing improving environments by reproducing very quickly and being well adapted to feed on the products of primary productivity dominated by pico- and nanoplankton.

Acknowledgments We are thankful to Instituto del Mar del Perú for facilitating the use of the data for this work, which is a summary of the thesis of R. C. at Univ. Nac. Agraria (Lima). Special acknowledgments go to Martin Angel for the basis on identification, for his valuable comments and improvements on the manuscript but specially for his encouragement and promptness to help the first author since the very beginning of this work. We also thank Luis Vásquez for his help with the analysis of environmental data, Jaime Mendo for the discussions and advices, and Kirsten Pohlmann for her valuable suggestions on the statistical analysis. Finally, we appreciate the comments provided by Marc Taylor, Gonzalo Olivares and one anonymous reviewer.

References

Alheit, J. & P. Bernal, 1993. Effects of physical and biological changes on the biomass yield of the Humboldt Current Ecosystem. In Sherman, K. & L. Alexander, B. Gold (eds), Large marine ecosystems: stress, mitigation and sustainability. American Association for the Advancement of Science, Washington DC, 53–68.

Alheit, J. & M. Ñiquen, 2004. Regime shifts in the Humboldt Current ecosystem. Progress in Oceanography 60: 201–222.

Angel, M. V., 1969a. The ostracod *Conchoecia porrecta* Claus redescribed and compared with *C. spinirostris* Claus. Crustaceana 17: 35–44.

Angel, M. V., 1969b. The redescription of three halocyprid ostracods, *Conchoecia hyalophyllum* Claus, *C. magna* Claus and *C. parthenoda* Müller from the North Atlantic. Crustaceana 17: 45–63.

Angel, M. V., 1971. *Conchoecia* from the North Atlantic. The 'procera' group. Bulletin of the British Museum of Natural History (Zoology) 21: 259–283.

Angel, M. V., 1999. Ostracoda. In Boltovskoy, D. (ed.), South Atlantic Zooplankton. Backhuys Publishers, Leiden, The Netherlands; 815–868.

Antezana, T., 2002. Adaptive behaviour of *Euphausia mucronata* in relation to the oxygen minimum layer of the Humboldt Current. Oceanography of the Eastern Pacific II: 29–40.

Arntz, W. E. & E. Fahrbach, 1996. El Niño, Experimento climático de la naturaleza. Fondo de Cultura Económica, México, 312 pp.

Aronés, K. J., 2004. Zooplancton e ictioplancton del mar peruano durante el verano 2001. Informe Instituto del Mar de Perú 32: 47–56.

Balech, E., 1954. División zoogeográfica del litoral Sudamericano. Revista de Biología Marina (Chile) 4: 184–195.

Carrasco, S. & O. Lozano, 1989. Seasonal and long term variations of zooplankton volumes in the Peruvian Sea, 1964-1987. In Pauly, D., P. Muck, J. Mendo & I. Tsukayama (eds), The Peruvian upwelling ecosystem: dynamics and interactions, ICLARM Conference Proceedings 18, IMARPE, GTZ and ICLARM: 82–85.

Carrasco, S. & H. Santander, 1987. The El Niño Event and its influence on the zooplankton off Peru. Journal of Geophysical Research, 92(C13): 14405–14410.

Chávez, F. P., J. Ryan, S.E. Lluch-Cota & M. Ñiquen, 2003. From anchovies to sardines and back: multidecadal change in the Pacific Ocean. Science 299: 217–221.

Chavtur, V. G., 2003. Morphology and distribution of some new pelagic ostracods of genus Metaconchoecia (Halocyprida: Halocyprididae) from the North Pacific. Zootaxa 229: 1–102.

Chavtur, V. G. & G. G. Stovbun, 2003. Revision of pelagic ostracods of the subfamily Archiconchoecinae. Publications of the Seto Marine Biological Laboratory 39: 136–219.

Escribano, R., G. Daneri, L. Farías, V. A. Gallardo, H. E. Gonzales, D. Gutiérrez, C. B. Lange, C. E. Morales, O. Pizarro, O. Ulloa & M. Braun, 2004. Biological and chemical consequences of the 1997–1998 El Niño in the Chilean coastal upwelling system: a synthesis. Deep Sea Research II 51: 2389–2411.

Fiedler, P. C., 2002. Environmental change in the eastern tropical Pacific Ocean: review of ENSO and decadal variability. Marine Ecology Progress Series 244: 265–283.

Giesecke, R. & H. E. González, 2004. Feeding of *Sagitta enflata* and vertical distribution of chaetognaths in relation to oxygen concentrations. Journal of Plankton Research 25: 475–486.

Gooday, A. J., 1981. The *Conchoecia skogsbergi* species complex (Ostracoda, Halocyprididae) in the Atlantic Ocean. Bulletin of the British Museum (Natural History) Zoology 40: 137–209.

Iriarte, J. I. & H. E. González, 2004. Phytoplankton size structure during and after the 1997/98 El Niño in a coastal upwelling area of the northern Humboldt Current System. Marine Ecology Progress Series 269: 83–90.

Jaimes, E., 1999. Condiciones meteorológicas a nivel global y local, cambio climático y "El Niño 1997–98". In Tarazona, J. & E. Castillo (eds), El Niño 1997–98 y su impacto sobre los ecosistemas marinos y terrestres, Revista Peruana de Biologia, 1–8.

Jiménez-Pérez, L. C. & R. Lara-Lara, 1988. Zooplankton biomass and copepod community structure in the Gulf of California during the 1982–1983 El Niño Event. CalCOFI Report XXIX: 122–128.

Judkins, D. C., 1980. Vertical distribution of zooplankton in relation to the oxygen minimum off Peru. Deep Sea Research 27A: 475–487.

Lavaniegos, B. E., L. C. Jiménez Pérez & G. Gaxiola Castro, 2002. Plankton response to El Niño 1997–1998 and La Niña 1999 in the southern region of the California Current. Progress in Oceanography 54: 33–58.

Lavaniegos B. E., G. Gaxiola-Castro L. C. Jiménez-Pérez M. R. González-Esparza T. Baumgartner & J. García-Córdova, 2003. 1997–98 El Niño effects on the pelagic ecosystem of the California Current off Baja California, Mexico. Geofísica Internacional 42: 483–494.

Longhurst, A. R., 1967. Vertical distribution of zooplankton in relation to the eastern Pacific oxygen minimum. Deep Sea Research 14: 51–63.

McPhaden, M. J., 1999. Genesis and evolution of the 1997–98 El Niño. Science 283: 950–954.

Martens, J. M., 1979. Die pelagischen Ostracoden der Expedition Marchile I (Südost-Pazifik) II: Systematik und Vorkommen (Crustacea: Ostracoda: Myodocopa). Mitteilungen aus dem Hamburgischen Zoologischen Museum und Institut 76: 303–366.

Martens, J. M., 1981. Die pelagischen Ostracoden der Marchile I—Expedition (Südost Pazifik), 1: Verbreitung, Zoogeographie und Bedeutung als Indikatoren für Wasserkörper. (Crustacea, Ostracoda: Myodocopida). Studies on Neotropical Fauna and Environment 16: 57–97.

Müller von, G. W., 1895. Reports on the dredging operations off the west coast of Central America to the Galapagos to the west coast of Mexico and in the gulf of California, in charge of Alexander Agassiz, carried on by the US Fish Commission steamer ALBATROS, during 1891, Liet. Commander Tanner, Z. L. XIX. DIE OSTRACODEN. Bulletin of the Museum of Comparative Zoology XXVII: 155–169.

Poulsen, E. M., 1969. Ostracoda-Myodocopa. 3a. Halocypriformes-Thaumatocypridae and Halocypridae. Dana Report 75: 1–100.

Poulsen, E. M., 1973. Ostracoda-Myodocopa. 3b. Halocypriformes-Halocypridae. Conchoecinae. Dana Report 84: 1–223.

Poulsen, E. M., 1977. Zoogeographical remarks on marine pelagic Ostracoda. Dana Report 87: 1–34.

Richman, J. G. &. S. L. Smith, 1981. On the possible enhancement of oxygen depletion in the coastal waters of Peru between 6°S and 11°S. Boletin Instituto Mar del Perú, 22–33.

Santander, H. & S. Carrasco, 1985. Cambios en el zooplancton durante El Niño 1982-1983 en el área de Chimbote. In Tresierra, A. (ed.), Anales del I Congreso Nacional de Biología Pesquera (28 Jun–1 de Julio de 1984).

Tseng, W. Y., 1975. Biology of the pelagic ostracod *Euconchoecia elongata* Müller. Reports of the Taiwan Laboratory of Fish Biology 27: 1–183.

Tseng W. Y., 1976. Development of the pelagic ostracod, Euconchoecia elongata Müller, 1906. In Proceedings of the Fifth International Symposium on evolution of Post-Paleozoic Ostracoda. Hartmann, G. (ed.), Abhandlungen des naturwissenschaftlichen Vereins in Hamburg 18/19: 201–213.

Van Guelpen, L., D. F. Markle & D. J. Duggan, 1982. An evaluation of accuracy, precision and speed of several zooplankton subsampling techniques. Journal du Conseil International pour l' Exploration de la Mer 40: 226–236.

Vásquez, L. & E. Tello, 1998. Condiciones oceanográficas durante el Crucero BIC Humboldt 9803–05 de Evaluación Hidroacústica de recursos pelágicos desde Tumbes a Tacna. Informe Instituto del Mar de Perú 135: 20–33.

Zamon, J. E. & D. W. Welch, 2005. Rapid shift in zooplankton community composition on the northeast Pacific shelf during the 1998–1999 El Niño—La Niña event. Canadian Journal of Fisheries and Aquatic Sciences 62: 133–144.

Zuta, S. & O. Guillén, 1970. Oceanografía de las Aguas Costeras del Perú. Boletín Instituto del Mar del Perú 2: 154–324.

Hydrobiologia (2007) 585:43–56
DOI 10.1007/s10750-007-0627-4

OSTRACODA (ISO15)

A new, interstitial species of *Terrestricythere* (Crustacea: Ostracoda) and its microdistribution at Orito Beach, northeastern Sea of Japan

Shimpei F. Hiruta · Shin-ichi Hiruta · Shunsuke F. Mawatari

Abstract We describe *Terrestricythere proboscidis* sp. nov. from a littoral interstitial habitat at Orito Beach, Matsumae, southern Hokkaido, Japan, a site characterized by coarse sand with many pebbles. This is the fifth known species of the ostracod superfamily Terrestricytheroidea, the first interstitial species of the superfamily, and the first record of the superfamily from Japan. The new species is clearly distinguished from its four known congeners by its small carapace with a transversely flat ventral margin, and the seventh limb consisting of four podomeres; its hemipenis has a long, S-curved process. We also report here the habitat, microdistribution, and locomotory behavior of the new species. Quantitative sampling revealed that the species lives in an interstitial environment. *T. proboscidis* is euryhaline. Behavioral observations showed that *T. proboscidis* cannot swim, but uses its appendages to push itself among sediment grains. An analysis of variance showed temperature and grain size, as well as temperature-depth and temperature-grain size interactions, to be significant influences on variation in population density. Sediment moisture content is correlated with these variables and likely is a primary factor in this species' microdistribution.

Keywords *Terrestricythere* · Morphology · Interstitial habitat · Microdistribution · Japan · Hokkaido

Guest editors: R. Matzke-Karasz, K. Martens & M. Schudack
Ostracodology – Linking Bio- and Geosciences

S. F. Hiruta (✉) · S. F. Mawatari
Laboratory of Systematics and Evolution Division of Biological Sciences, Graduate School of Science, Hokkaido University, Sapporo 060-0810, Japan
e-mail: s-hiruta@sci.hokudai.ac.jp

S.-i. Hiruta
Department of Biology, Hokkaido University of Education, Kushiro Campus, Kushiro 085-8580, Japan

Introduction

Four species of the ostracod genus *Terrestricythere* have previously been reported, two of them from the northwestern Pacific (Fig. 1). One of the NW Pacific species is *Terrestricythere ivanovae* Schornikov, 1969, upon which Schornikov (1969) based a new genus, new family, and new superfamily. This species was described from the supralittoral zone at Kurilsk, Iturup Island, Kurile archipelago, among small pebbles kept moist by mist, rain, and sea spray. The other NW Pacific species is *T. pratensis*, described by Schornikov (1980) from a monsoonal salt marsh with brackish groundwater on an island near Vladivostok. In the same paper, Schornikov (1980) also reported finding *T. ivanovae* near where he collected

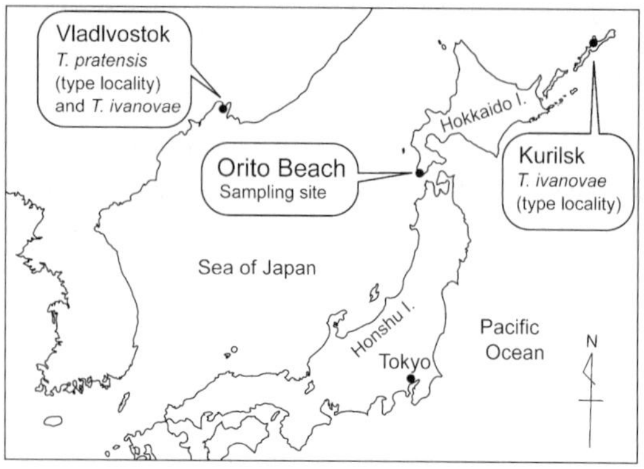

Fig. 1 Map showing Japan and the margin of continental Asia, indicating known localities of species of *Terrestricythere*, including the study area at Orito beach

T. pratensis. Recently, Horne et al. (2004) described an additional species, *T. elisabethae*, from a littoral estuarine habitat in southern England, and illustrated a single specimen of a fourth, undescribed species of *Terrestricythere* from an intertidal rock pool in southern England.

During a survey (Hiruta, 2003) of marine interstitial animals of Hokkaido Island, Japan, we collected an undescribed species of *Terrestricythere* from a littoral interstitial habitat near Matsumae on the Japan Sea coast, southwestern Hokkaido, Japan (Figs. 1, 2). This is only the fifth species known for the genus, and indeed, for the entire superfamily Terrestricytheroidea. The other four known species are surface dwellers; hence, the new species, uniquely living in an interstitial habitat, provides an excellent opportunity for comparative studies of its adaptations to interstitial life. To establish that the new species is unambiguously an interstitial animal, we conducted a quantitative survey of its horizontal and vertical distribution in the sediment. Finally, we performed an analysis of variance to investigate which of several physical parameters might influence this distribution.

Material and methods

Fieldwork was conducted for three days from 30 April to 2 May 2005 at Orito Beach, an open-coast site (41°26′09.5″ N, 140°03′48.2″ E) facing the Sea of Japan and characterized by very coarse sand with many pebbles (Fig. 2). The Sea of Japan is an inland sea, so that tidal change is relative small. In Matsumae, which is located about 1.5 km southeast from the beach, the maximum annual change of tidal level is 46 cm. A stream called Daijinnai River flows through the sampling area to the sea (Fig. 3). Qualitative samples of sand were collected from the intertidal zone near the water's edge at a depth of more than 20 cm. Specimens were extracted from the sand by decanting and sieving with freshwater through a 0.04-mm-mesh sieve net. Appendages were dissected and mounted in Hoyer's solution on glass slides and drawn with the aid of a camera lucida. Carapaces (with soft parts) were treated with hexamethyldisilazane (HMDS) (Nation, 1983) for examination by SEM at 15 kV accelerating voltage. Type specimens are deposited in the Zoological Institute, Faculty of Science, Hokkaido University (ZIHU).

To study the microdistribution of the species, quantitative samples were taken from eight stations along two transects, A and B (Fig. 3; for transect profiles, see Fig. 10). Transect A began with sampling station A/1 at the water's

Fig. 2 Photograph of study area at Orito beach, with arrows indicating sampling stations along transect A

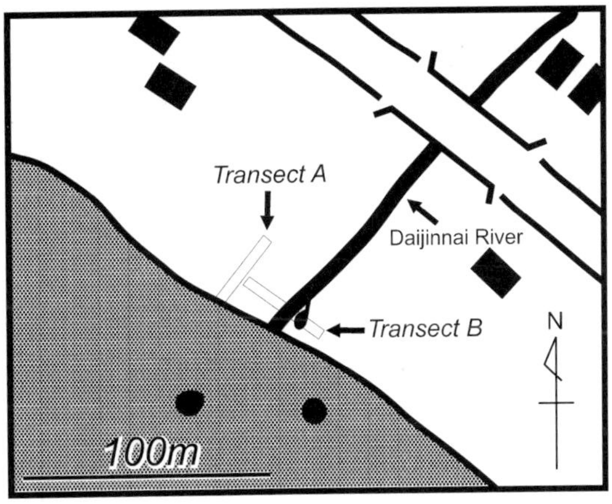

Fig. 3 Diagram of the study area, showing the location of the transects A and B. Transect B crosses the Daijinnai river and a small connected pond

edge and ran northeastward to A/4 about 20 m inland. Transect B ran along the shore in a northwest to southeast orientation perpendicular to transect A and crossed the Daijinnai River and a small pond, with sampling station B/1 closest to Transect A and B/4 farthest. The beach profile (see Fig. 10) along each transect was determined by measuring the beach height above a reference point at irregular intervals, using a surveyor's rod and transit. Sampling station A/1, at water's edge at high tide on 1 May 2005, was used as the reference point. According to tidal data of Matsumae, this day's high tide is 28 cm at 9:58 am, which is 16 cm lower than annual maximum high tide.

At each station, a 2-l sand sample was collected at each 10-cm depth interval from the surface to a maximum depth of about 80 cm. Animals were removed by stirring the sediment sample in water in a bucket and then filtering the water through a fine-mesh net. This was done until no more animals were found. For each 10-cm interval at each station, sand temperature was measured on 1 May 2005. Distance from the surface to groundwater was measured for every sampling station except A/4, where the groundwater level was too deep to reach by digging; groundwater depth is indicated in Fig. 10 by the deepest temperature measurement taken for each station, except A/4.

A particle-size analysis was performed for sand samples from which animals were extracted. A well mixed, 200- to 300-ml subsample was taken from each sediment sample, air-dried in the laboratory, and passed through a series of sieves of mesh sizes 4, 2, 1, 0.5, 0.25, and 0.125 mm, and the mass of the fraction of the sample retained by each mesh size recorded. The particle-size distribution of the sediment was expressed as the mass percentage of each size fraction, for each depth interval at each sampling station.

For statistical analysis, a grain-size index was calculated for each sediment sample in the following manner. Median particle diameters in the six fractions were considered to be 6, 3, 1.5, 0.75, 0.375 and 0.1875 mm. From these values, the median particle volume in each fraction was calculated in mm^3. The mass percentage value of each fraction in a sample was divided by the median particle volume of the fraction, and the resulting values were summed across the fractions. The resulting sum constitutes a grain-size index whose value decreases with increasing contributions by the large-particle-size fractions, and increases with increasing contributions by the small-particle-size fractions.

A three-way limited analysis of variance was performed using JMP IN 5.1.2 (SAS Institute) software. Depth (DEPTH), temperature (TEMP), and grain-size index (GRAIN) were utilized as explanatory variables and number of individuals in a sample as the response variable.

Results

Taxonomic description

Superfamily Terrestricytheroidea Schornikov, 1969
[nom. trans. Hartmann & Puri, 1974]

Family Terrestricytheridae Schornikov, 1969

Genus *Terrestricythere* Schornikov, 1969

Type species. *Terrestricythere ivanovae* Schornikov, 1969: 495–497, Figs. 1–16.

Diagnosis. Carapace with curved dorsal margin, LV larger than RV. LV with small lobe-like extension overlapping RV towards anterior margin. Hinge visordont with two teeth on RV situated towards anterior and posterior margins and joined by ridge running along dorsal margin. Lobes interlock and rotate in two pits on hinge of

LV so that LV dorsally overrides RV. Basic adductor muscle scar pattern of five in approximately vertical row, second and third from bottom diagonally offset from each other (second displaces anteriorly, third posteriorly, from imaginary vertical center line). Antennula with five or six articulated podomeres. Antennal exopodite a plate with seven or eight long setae. Mandibular palp with rake setae and branchial plate with one ray. Female fifth limb endopodite with three articulated podomeres; modified into clasping palps in males. Sixth limb with four robust podomeres, with a series of short claws and setules along anterior edge. Seventh limb with four or five podomeres, terminating in two slender, hooked claws, one longer than the other. Furcal rami each reduced to three small setae. Eye absent.

Remarks. We amend the diagnosis of the genus *Terrestricythere* by Horne et al. (2004) to include species with four or five podomeres on the seventh limb, because the species described herein has four podomeres.

Terrestricythere proboscidis sp. nov.
(Figs. 4, 5, 6, 7, 8, 9).

Type specimens. Holotype, ZIHU03189, male, carapace 328 μm long; parts mounted on 17 slides in Hoyer's solution; collected 1 May 2005 at type locality (below) by S. F. Hiruta. Allotype, ZIHU03190, female, parts mounted on 17 slides in Hoyer's solution; collection data as for holotype. Paratypes, three males; three females (ZIHU03191-03195) dissected and mounted on 16–18 slides each; collection data as for holotype.

Type locality. Intertidal zone of Orito Beach (41°26′09.5″ N, 140°03′48.2″ E) on the Sea of Japan coast, southern Hokkaido, Japan; interstitial habitat near water's edge, influenced by the nearby Daijinnai River.

Etymology. The specific name is derived from Latin *proboscidis* (f.) meaning trunk of an elephant, in reference to the long, curved process on the hemipenis.

Diagnosis. Carapace small, about 0.3 mm long. Ventral surface of carapace flat. Seventh limb with four podomeres. Hemipenis with long, S-curved process.

Description of male. Holotype, ZIHU-03189. Carapace (Fig. 8a, b) 328 μm long, 170 μm high; dorsal margin well arched in lateral view, with distinct anterior and posterior angles; anterior margin more sharply rounded than posterior one; ventral margin almost straight. Left valve larger than right, overlapping the latter dorsally. Dorsal view elongate-oval, anterior more sharply rounded than posterior. Hinge of visordont type (Horne et al., 2004); right valve with two rounded teeth (Fig. 4i), one on anterodorsal margin and other on posterior margin, just dorsal to posterior angle, the teeth joined by ridge running along dorsal margin between them. Two sockets located in corresponding positions on left valve (Fig. 4j). Inner lamella broader anteriorly than posteriorly, reaching close to teeth and sockets; ventral part of inner lamella narrow. Five elongate adductor muscle scars present just anterior to middle of valves; around six or seven small muscle scars anterodorsally (Fig. 8c). Surface of valves smooth in general appearance, but covered with very shallow, polygonal reticulation (Fig. 4f), many (more than 60) pits without a sensillum located in mid-lateral region (Fig. 4f), and about 100 normal pores on each valve; pores on ventral surface with bushy sensillum, but those of lateral and dorsal surface with simple sensillum; about 50 marginal pore canals with a bushy sensillum (Fig. 4e). Each valve with three unusually long sensilla originating from mid-lateral region.

Antennula (Fig. 5a) five-segmented. First podomere subquadrate. Second podomere with one long and one shorter ventral-apical setae, one mid-ventral setulose seta, and one long dorsal-subapical seta. Third podomere subquadrate, with two apical setae having setules about 10-μm long at base. Fourth podomere elongate, with four setae mid-dorsally and six medial-apical setae, with many setules at bases of setae. Fifth podomere with four apical setae.

Antenna (Fig. 5b) biramous, robust. First podomere (coxa) broad, with one long anterior-apical setulose and one shorter posterior-apical setae. Second podomere (basis) with one anterior-apical setulose and one short posterior-apical setae. Exopodite protrudes from anterior-apical corner of basis and overlaps 3rd podomere; with one seta and one larger setulous seta; also from

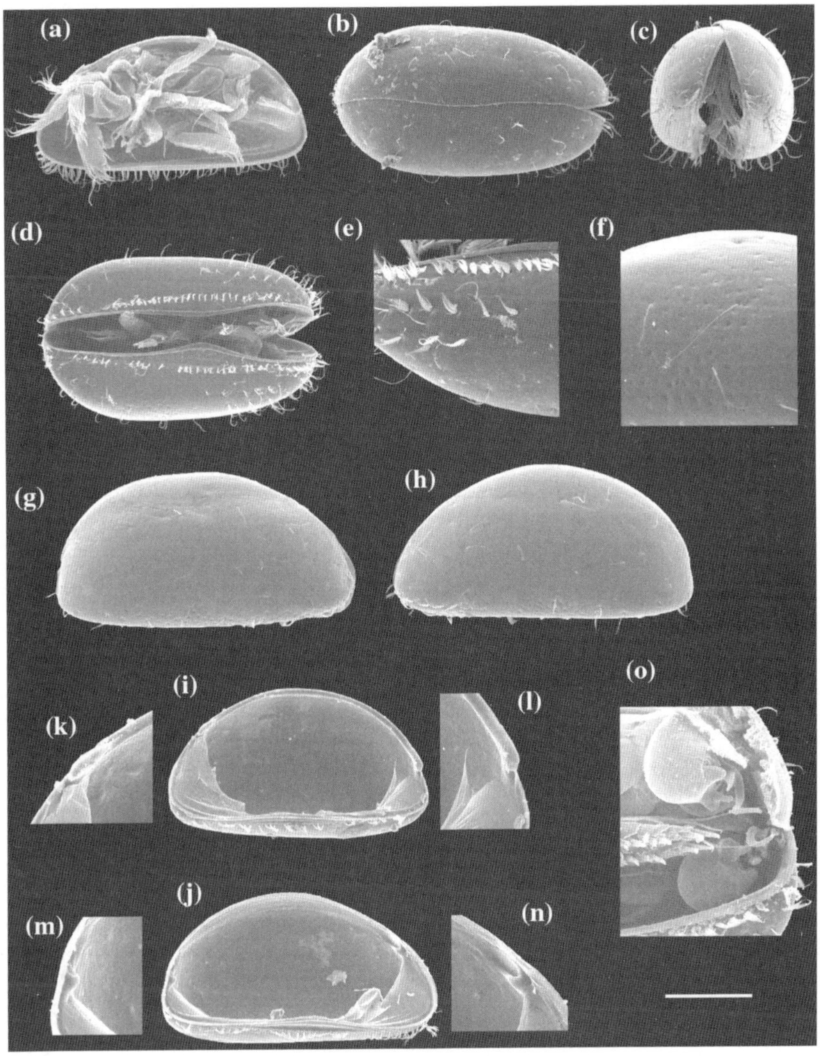

Fig. 4 *Terrestricythere proboscidis* sp. nov. SEM images. **a–n**, from females. **a**, Lateral view of female with left valve removed; **b**, dorsal view of open carapace; **c**, anterior view of open carapace; **d**, ventral view of open carapace; **e**, anterior–ventral sensilla of carapace; **f**, mid-lateral part of left valve showing many pits without sensilla and shallow reticulation; one of three long, simple setae is also shown; **g**, external view of right valve; **h**, external view of left valve; **i**, internal view of right valve; **j**, internal view of left valve; **k**, anterior hinge lobe of right valve; **l**, posterior hinge lobe of right valve; **m**, posterior hinge socket of left valve; **n**, anterior hinge socket of left valve; **o**, male, ventral view of hemipenis. Scale bar: **e**, **f**, and **k–o**, 0.05 mm; others, 0.1 mm

exopodite emerge eight long apical setae, with small loosely curved terminal hooks, that extend past terminal claws. Endopodite consists of three podomeres. Endopodite 1st podomere with one larger and one smaller claw on posterior-apical corner and three posterior-proximal aesthetascs. Endopodite 2nd podomere with one smaller and three larger anterior-apical claws, two claws and two aesthetascs on the posterior edge, and one large comb-like claw posterior-apically. Apex of endopodite 3rd podomere with two large claws, one of them comb-like as on 2nd podomere, and two smaller claws and one aesthetasc.

Mandible (Fig. 6a–c). Small coxa with one seta on outer edge and five teeth; also apically with two setulous setae. Branchial plate (exopodite) on outer edge of the basis consisting of one long seta with flat end. Endopodite on outer edge of basis consisting of three podomeres. First podomere with two outer-apical setae. Second podo-

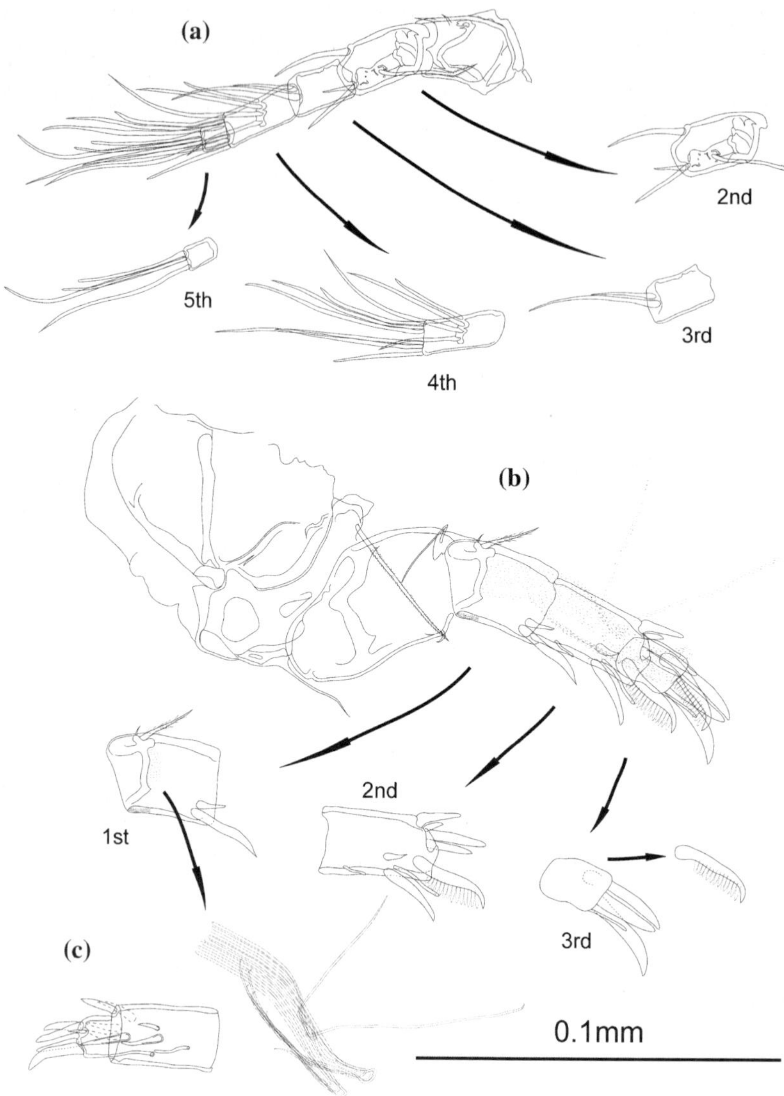

Fig. 5 *Terrestricythere proboscidis* sp. nov. **a**, **b,** holotype, ZIHU03189; **c**, paratype, female, ZIHU03190. **a**, antennula, with endopodite 2nd–5th podomeres enlarged; **b**, antenna, with endopodite podomeres enlarged; **c**, two terminal endopodite podomeres of female antenna

mere with one long and one short seta on outer-apical edge. Inner-apical edge with one long and one short setae and one stout setulous claw. Third podomere with three setulous claws, each of them different in size. Eight simple, rake setae on basis or 1st endopodite podomere.

Maxillula (Fig. 6d, e). Endopodite two-segmented. First podomere with one stout, blunt setulous seta on outer edge and two setae at outer apical corner. Terminal podomere with two claws and two setae, one of which setulous, stout, and blunt. Three endites with several setae. Branchial plate with two large, broad, setulous setae and many finer setulous setae arranged around posterior edge.

Fifth limb (Fig. 6f, g) four-segmented. Protopodite with exopodite inserted on posterior-distal edge and represented by one long setulous seta, and with four setae on anterior-apical edge, two of them setulous. First podomere with three spines, two long setae, and many setules on anterior edge. Second podomere small

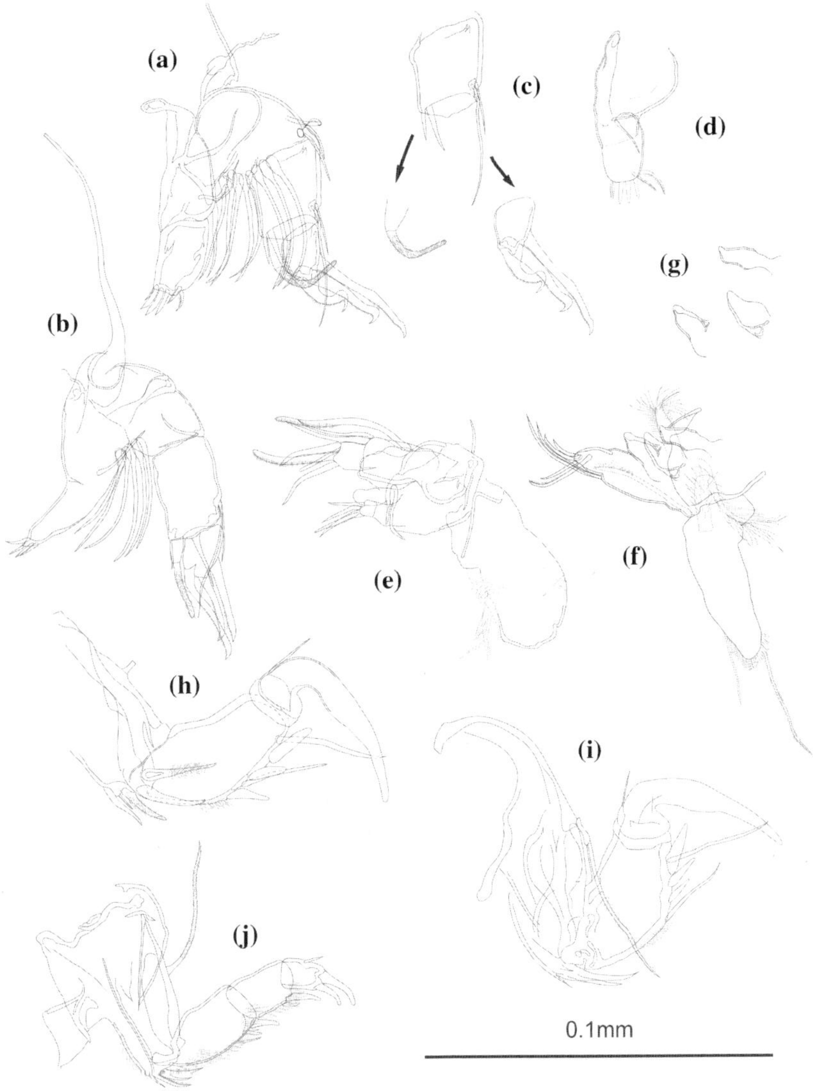

Fig. 6 *Terrestricythere proboscidis* sp. nov. **a–c**, **e**, **h**, **i**, holotype, male, ZIHU03189; **j**, paratype (allotype), female, ZIHU03190; **d**, **f**, **g**, paratype, female, ZIHU03191; **a**, **b**, mandibles; **c**, endopodite 2nd and 3rd podomeres; **d**, coax; **e**, **f**, maxillula; **g**, three endopodal endites; the numerous setae have been omitted from the endites; **h**, **i**, fifth limbs; long setulous seta on exopodite broken in **h**; anterior-apical setae difficult to recognize in **i** because of its rotation; **j**, fifth limb of adult female

and rectanglar, almost fused with 3rd podomere. Third podomere with one small posterior-apical seta and apically fused with large claw with thickened middle.

Sixth limb (Fig. 7a) five-segmented and robust. Protopodite with exopodite consisting of one long setulous seta inserted on posterior-distal edge, and with two setae and many setules on anterior edge and one seta apically at anterior corner. First endopodal podomere with two long and two short claws at anterior-apical corner. Second and third podomeres both with one long and two short claws at anterior-apical corner and numerous setules on anterior edge. Fourth podomere with two terminal claws and one short seta on posterior edge.

Seventh limb (Fig. 7b) four-segmented and slender. Protopodite with one seta on anterior edge. Exopodite seta apparently absent. Endopodite 1st podomere proximally curved. Second

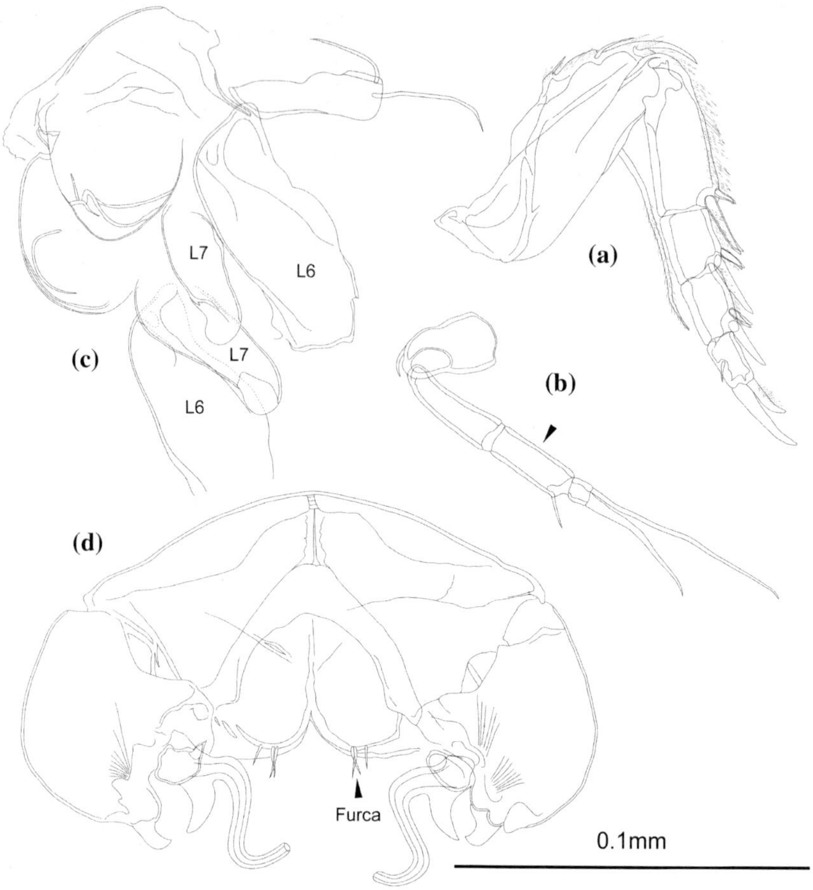

Fig. 7 *Terrestricythere proboscidis* sp. nov. **a**, **b**, **d**, holotype, male, ZIHU03189; **c**, paratype (allotype), female, ZIHU03190. **a**, sixth limb; **b**, seventh limb; **c**, genital lobe; **d**, hemipenis, with furca indicated

podomere with one anterior-apical seta. Third podomere with one long, curved claw and one longer and thinner, curved claw.

Furca (Fig. 7d) reduced to a pair of protuberances each bearing three small setae, situated between bases of hemipenis.

Hemipenis (Fig. 7d) consists of three clasping processes and rounded outer lobe. Largest clasping process biramous. Shorter ramus curved outward. Long, branched, hollow ramus "S"-shaped. Another short process curving toward large biramous process lies on outer edge of outer lobe. The third process round with slender protuberance and situated on inner edge.

Description of female. Allotype, ZIHU03190. Carapace, antennula, mandible, maxillula, sixth limb, and seventh limb similar to those of adult male.

Antenna (Fig. 5c). Distal two podomeres without comb-like claws. However, the number of setae and claws same as adult male. Other parts of antenna similar to those of adult male.

Fifth limb (Fig. 6h) four-segmented. Protopodite with exopodite inserted on posterior-distal edge and consisting of one long, setulous seta. Protopodite with two setae and two setulous setae apically on inner edge. First endopodal podomere with two anterior-apical setae. Second podomere with one large claw and two small setae on anterior-apical corner. Third podomere with one small seta on posterior edge, and one large and one smaller claw apically. Numerous setules situated on anterior edge of 1st–3rd podomeres.

Female genitalia (Fig. 7c) Pair of rounded lobes is situated on posterior end of the body. No furca observed.

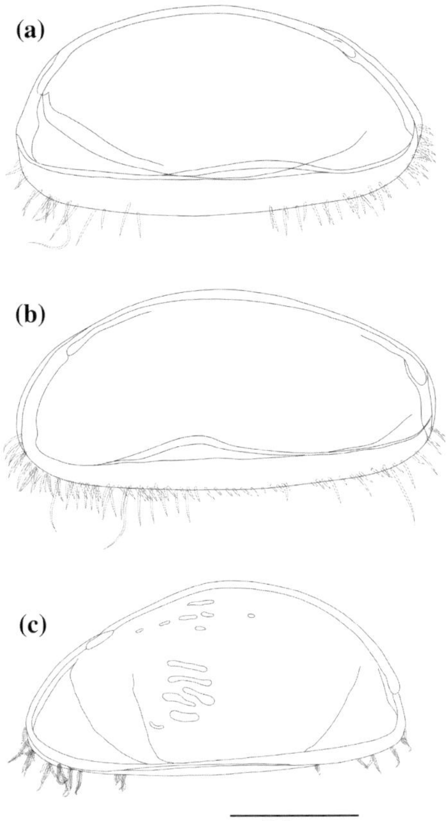

Fig. 8 *Terrestricythere proboscidis* sp. nov. **a** and **b**, holotype, male, ZIHU03189; **c**, paratype (allotype), female, ZIHU03190. **a**, left valve; **b**, right valve; **c**, right valve with muscle scars. Scale bar, 0.1 mm

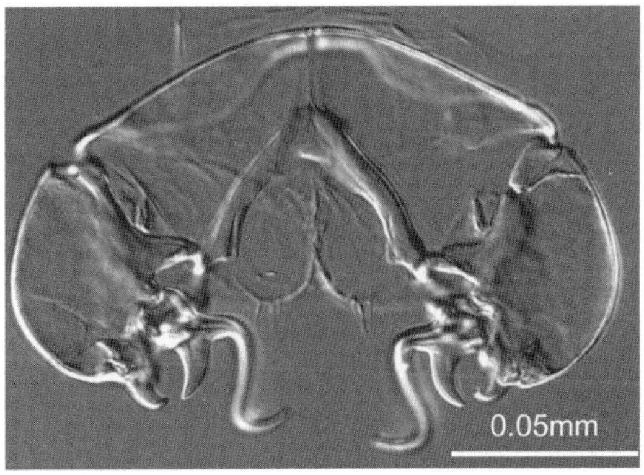

Fig. 9 *Terrestricythere proboscidis* sp. nov. Nomarsky-optical image of holotype, ZIHU03189. Hemipenis and furca

Ecology

The vertical and horizontal distributions of *T. proboscidis* are indicated in Fig. 10. The vertical distribution clearly indicates that the ostracod is an interstitial animal. Only at sampling station B/4 was it found in the 10-cm interval close to the surface; peak abundance at stations where it occurred ranged from the 20-cm (B/4) to the 60-cm (B/1) intervals. At A/2 and A/3, *T. proboscidis* occurred as deep as 60–70 cm, which was the deepest sand sample we could take. We found species of the interstitial ostrocod genera *Microloxoconcha* and *Microcythere* in the same sand samples as *T. proboscidis*.

In terms of horizontal distribution (Fig. 10), peak abundances were found along transect A at stations A/2 and A/3 at intermediate distances from the tide line. Along transect B, ostracods occurred at the two stations of moderate elevation, but not at the two stations where groundwater was close to the surface.

The particle-size distribution of the sediment, expressed as the mass percentage of each size fraction, varied considerably among the samples; however, a general conclusion can be drawn. The particle-size distribution was skewed heavily toward the three largest size fractions (4, 2, 1 mm) (Table 1); the average combined mass percentage of these fractions was 89.1% (range = 65.0–99.6%). No sand sample had the heaviest fraction in any of the three smallest size fractions (0.5, 0.25, 0.125 mm) and the average combined mass percentage of these fractions was only 10.9% (range = 0.4–35.0%). What these data mean is that fine-grained sediments such as mud and silt are scarcely present in the habitat of *T. proboscidis.*

The average grain-size index (Table 1) at a station did not correlate well with the occurrence of *T. proboscidis*, as indicated by both high and low average indices at stations A/2 and A/3, where the greatest densities of the ostracod occurred. However, along transect B, ostracods were absent from stations B/2 and B/3, which also had the two lowest average grain-size indices among all the sampling stations, indicative of particles skewed heavily toward the largest size fractions.

The temperature profile was generally similar among the sampling stations. Sand temperature (Fig. 10) at the surface ranged from 19°C to 22°C

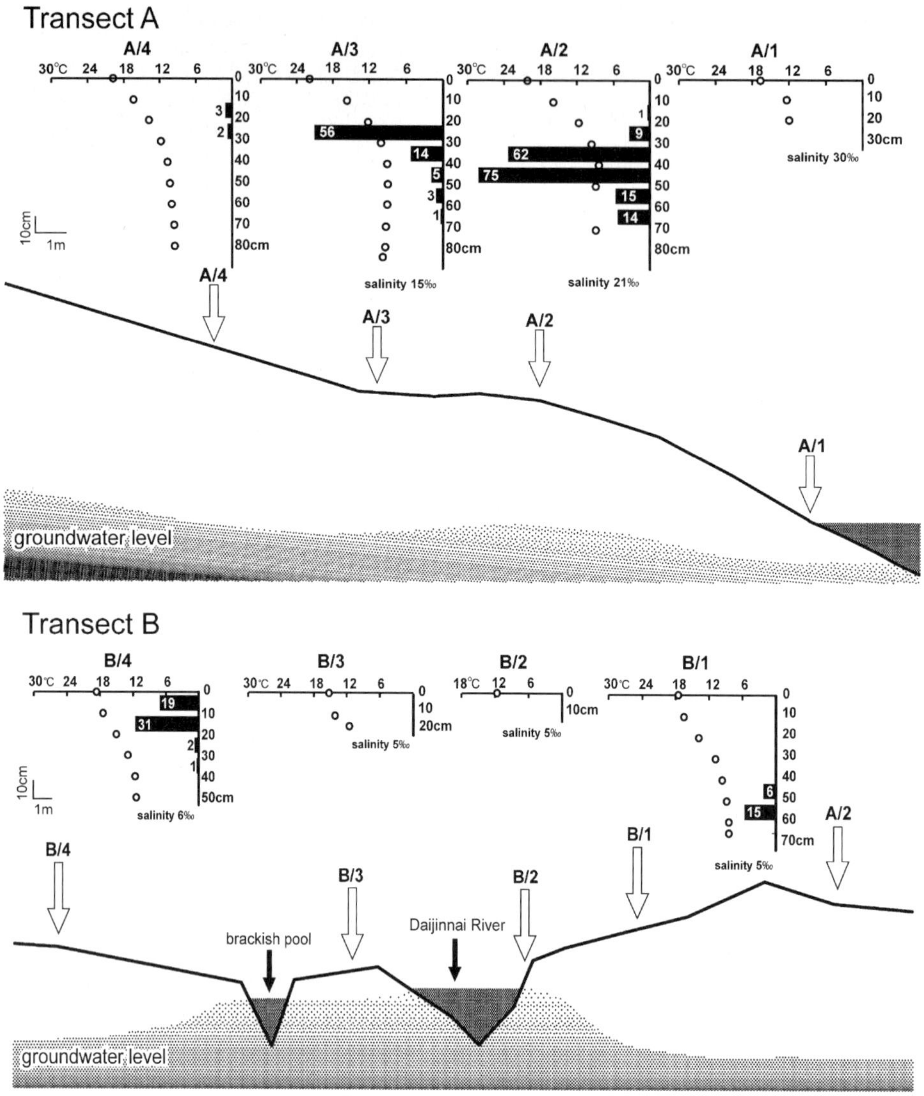

Fig. 10 Microdistribution of *Terrestricythere proboscidis* sp. nov. White arrows show the eight sampling stations. Black bars show the number of individuals counted in a 2-l sand sample from each 10-cm depth interval (bars are with no axis scale, but proportional to values indicated in bars). Open circles are temperature values (horizontal axis) with depth (vertical axis) at each station. The salinity is shown below each station panel, except for station A/4, where the groundwater was too deep to sample. The groundwater level is indicated by stippling in the profile sections. Horizontal and vertical scale bars for the profiles are shown at left

at the more elevated sampling stations and from 12°C to 17°C at stations (A/1, B/2, B/3) where the surface was close to ground- or shore-water level. Sand temperature steadily decreased to roughly 9–12°C at a depth of about 40 cm and remained within this range below that. Sea surface temperature was about 9°C. Temperature in itself did not seem a significant factor in the distriubion of *T. proboscidis*, which was found at 19°C in the surface layer at station B/4 and at temperatures as low as 9°C at most of the other stations where it occurred.

Terrestricythere proboscidis can tolerate a wide range of salinity. Groundwater salinity ranged from 5‰ to 21‰ at stations where the ostracods were collected (see Fig. 10). We also found the

Table 1 Average mass percentage values by station and particle fraction, and overall average mass percentage of the particle fractions. Right-hand column shows the average grain index among samples for each station

Station	Mesh size						Average grain index
	4 mm	2 mm	1 mm	0.5 mm	0.25 mm	0.125 mm	
A/1	21.8	34.9	37.7	5.3	0.3	0.0	9.17
A/2	40.4	20.4	22.6	13.4	3.0	0.1	22.71
A/3	32.7	33.7	25.7	7.2	0.6	0.0	9.64
A/4	20.1	22.2	40.9	14.4	2.3	0.1	26.79
B/1	47.3	27.1	17.5	7.1	1.0	0.1	11.85
B/2	72.7	19.4	7.1	0.8	0.0	0.0	1.21
B/3	70.1	14.9	10.3	4.2	0.6	0.0	6.03
B/4	13.6	25.6	53.4	7.0	0.4	0.0	10.62
Overall average	33.70	25.42	30.00	9.37	1.43	0.06	16.65

Table 2 Analysis of variance values. $R^2 = 0.491571$; adjusted $R^2 = 0.393165$

Source	DF	SS	MS	F	P
Model	6	5900.969	983.495	4.9953	0.0011
Error	31	6103.347	196.882		
Total (adjusted)	37	12004.316			

Table 3 Coefficients from the analysis of variance

Term	Coef	SECoef	T	P
Constant	64.619991	27.51622	2.35	0.0254
TEMP	−4.261986	1.846531	−2.31	0.0278
DEPTH	−0.472703	0.254719	−1.86	0.0730
GRAIN	0.391103	0.183584	2.13	0.0412
TEMP*DEPTH	−0.090627	0.063945	−1.42	0.1664
TEMP*GRAIN	−0.176297	0.080839	−2.18	0.0369
DEPTH*GRAIN	−0.026103	0.010907	−2.39	0.0229

species in the bottom sand of a tidal pool near the transects on 11 October 2004, where the salinity was 7‰. Salinity was low along transect B, influenced by the Daijinnai River and brackish pool. Along transect A, groundwater salinity decreased with increasing distance from the tide line. The salinity of seawater along the shore flanking the study area was about 30‰, and we observed that *T. proboscidis* could survive at this salinity for at least 2 weeks.

The analysis of variance showed that the model significantly explained ($P < 0.05$) (Table 2) the abundance of ostracods in samples where they occurred. Temperature (TEMP) and grain-size index (GRAIN) significantly explained variation in abundance, as well as interactions of TEMP*GRAIN and DEPTH*GRAIN (Table 3). Depth (DEPTH) and the TEMP*-DEPTH interaction did not significantly explain variation in abundance. The low adjusted R^2 values (Table 2) might suggest the involvment of other factors.

T. proboscidis appears to reproduce over an extended period, from at least early spring through fall. Judging from shell size, our samples of *T. proboscidis* included the adult stage of both sexes and A-1, A-2, and A-3 instar juveniles in samples collected in spring (April and May) and fall (September and October). The younger instars (A-8, A-7, A-6) were not detected, suggesting these stages are not of long duration.

We observed that *T. proboscidis* moves in sediment in water mainly by pushing against sediment particles with the antennae and the sixth limbs. Horne et al. (2004) reported that *T. elisabethae* used not only antennae and sixth limbs but also seventh limb. *T. proboscidis* also occasionally move their seventh limbs. However, it seems that seventh limbs don't contribute to their locomotion. In water without sediment, the animals often turned upside down and could do nothing but move their limbs, which indicates they are incapable of swimming. We also placed an individual collected in October 2005 onto wet filter paper to observe its behavior. The animal trapped water between its open valves and the paper and, held upright by the water used its legs to creep along the surface, pulling the trapped water with it.

Discussion

Taxonomy

The new species reported here has the visordont-type hinge characteristic of the genus *Terrestricythere*, the only the genus in the superfamily Terrestricytheroidea. *Terrestricythere proboscidis* sp. nov. represents the fifth known species of the superfamily, the first interstitial species, and the first record of the superfamily from Japan. *Terrestricythere proboscidis* differs from its congeners (*T. ivanovae* Schornikov, 1969; *T. pratensis* Schornikov, 1980; *T. elisabethae* Horne et al., 2004; *T.* sp. (Horne et al., 2004)) in the following characters (characters of the other species in parentheses): (1) adult carapace short, about 0.32 mm long (longer, 0.46–0.53 mm long); (2) carapace with a transversely flat ventral surface and acute anterior and posterior margins (carapace with transversely convex ventral surface and broadly rounded margins); (3) four podomeres on seventh limb (five podomeres); (4) hemipenis with long, S-curved process (process not long, S-curved). *Terrestricythere proboscidis* differs from two of its congeners in the length and shape of the three lateral sensilla on each valve: *T. pratensis* has shorter sensilla, each with a single tip, whereas *T. elisabethae* has longer ones, each with a bifurcate tip.

The interstitial habitat is a stressful environment, the inhabitation of which is correlated with changes in morphology, such as reductions in body size and the number of appendage setae, and the loss of eyes (Maddocks, 1976; Danielopol & Hartmann, 1986). In general, interstitial species tend to be smaller than related species occupying other habitats, and most interstitial ostracods have been reported to have a carapace <0.35 mm in length. Moreover, most interstitial ostracods reported so far have the carapace with a flat ventral surface. Thus, two of the diagnostic characters of *T. proboscidis* seem related to adaptation to an interstitial habitat.

Fewer podomeres on the seventh limb of *T. proboscidis* compared to its congerners may similarly be an adaptation to interstitial life. Horne et al. (2004) observed that the A-2 instar of *T. elisabethae* has four podomeres on the seventh limb, but that the next (A-1) instar has five. They suggested that the third podomere divides into the third and fourth podomeres during the molt. Since there is no sign of fusion in the third podomere of *T. proboscidis*, we speculate that this species has fewer podomeres because either (1) the third podomere fails to divide during molting from A-2 to A-1, or (2) *T. proboscidis* molts two fewer times than the other species, so that A-2 becomes the terminal instar; in this case, the maturity is promoted so that this is an example of progenesis.

Three *Terrestricythere* species have now been reported from the northwestern Pacific region, where the genus was first discovered. Horne et al. (2004) suggested that the genus might have dispersed to Europe relatively recently from a center of origin of the genus in the northwestern Pacific. However, they also pointed out that additional *Terrestricythere* species might have gone undetected in Europe because of their unusual habitat. Clearly, with only two species known from Europe and three from the Pacific, conclusions concerning biogeography are premature and will require additional sampling effort.

Ecology

Terrestricythere proboscidis is an interstitial animal, whereas the other four known *Terrestricythere* species occupy semiterrestrial habitats, either on the surface of the substrate or in an environment intermediate between the surface and the interstitial realm. Nonetheless there are ecological similarities between *T. proboscidis* and its congeners. *Terrestricythere elisabethae* was reported to live under a similarly wide range of salinity, 0–15‰, and to reproduce throughout the year (Horne et al., 2004). The ability of *T. proboscidis* to move out of water is similar to that of *T. elisabethae*, which Horne et al. (2004) observed to move out of water, with water trapped inside and around its carapace.

T. proboscidis appears to be able to function in drier conditions than some other interstitial ostracods. In our study, we also tested the ability of species of *Microloxoconcha* and *Microcythere* to move on wet filter paper. In contrast to *T. proboscidis*, these species did not move, but simply closed their valves. In general, species of *Terrestricythere* may be able to live in the less moist conditions than other ostracods in the same habitats, perhaps due to special structures that allow them to retain water. In this context, Horne et al. (2004) pointed out the role of the densely spaced "shaving brush" sensilla of *T. elisabethae* in the retention of water between the ostracod and the substrate.

Among the factors explaining variation in densities of *T. proboscidis*, particle size was perhaps the most important; both the grain-size index and its interactions with depth and temperature had significant explanatory value. Among all stations, B/2 and B/3 had the lowest average grain-size indices, indicative of well-washed sediments of large particle size; *T. proboscidis* was absent from these stations.

An important variable likely affecting the horizontal and vertical distribution of *T. proboscidis* is moisture content of the sediment, which was not directly measured in this study. Several of the variables examined (temperature, depth, and particle size) are likely correlated with moisture content. Moisture content can be expected to decrease near the surface due to elevated temperature and increased evaporation there, and with increasing average grain size, since coarser sediment would drain faster. On the other hand, moisture content likely increases with proximity to groundwater.

The horizontal distribution of *T. proboscidis* suggests that moisture content of the sand is a significant factor in its distribution. Along transect A, the ostracod was absent or low in abundance at station A/1 closest to the shore (presumably the dampest station along transect A) and A/4 farthest from the shore (presumably the driest). Transect B ran parallel to the shore, with all stations roughly the same distance from the tide line as station A/2, where *T. proboscidis* was most abundant. Stations B/1 and B/4 had moderate abundance, but *T. proboscidis* was absent from stations B/2 and B/3 close to the Daijinnai River and the brackish pool, where moisture content of the sediment was presumably very high.

We speculate that *T. proboscidis* has adapted to live within a specific range of intermediate moisture conditions, not too wet and not too dry. Our behavioral observations indicate that this species cannot swim, but moves by pushing itself among sand grains with its appendages. The filter paper experiment demonstrated it can overcome the surface tension of water and crawl with water trapped beneath its valves. If particle size is large and moisture content high, the animals may have difficulty moving because of their lack of swimming ability. On the other hand, if the sediment is too dry, the animals might have trouble moving because of lack of sufficient moisture and be subject to dessication.

This was a preliminary investigation, the results of which will enable us to design a more rigorous approach to study the ecology of *T. proboscidis.* Future work will focus on elucidation of the life cycle including their ontogeny; further analysis of the factors involved in the species' microdistribution, including moisture content of the sediment; and comparative examination of adaptations to an interstitial existence in the context of the phylogeny of the *Terrestricythere* clade.

Acknowledgements We are grateful to Dr. D. J. Horne (Queen Mary University of London) and Dr. R. J. Smith (Lake Biwa Museum) for valuable advice, and to Dr. M. H. Dick (Hokkaido University) for critical reading of the manuscript and valuable comments.

References

Danielopol, D. L. & G. Hartmann, 1986. Ostracoda. Part I: Stygobiont Ostracoda from Inland Subterrancan Waters. Part II: Die marinen Ostracoden der interstitiellen Lebensräume. In Botosaneanu, L. (ed.), Stygofauna Mundi. A Faunistic, Distributional, and Ecological Synthesis of the World Fauna Inhabiting Subterranean Waters (Including the Marine Interstitial). E. J. Brill, Leiden, 265–294.

Hartmann, G. & H. S. Puri, 1974. Summary of neontological and paleontological classification of Ostracoda. Mitteilungen aus dem Hamburgischen Zoologischen Museum und Institut 70: 7–73.

Hiruta, S., 2003. Notes on marine interstitial animals from Esashi, southern Hokkaido. Kushiro Ronsyu 34: 63–69 [In Japanese, with English abstract].

Horne, D. J., R. J. Smith, J. E. Whittaker & J. W. Murray, 2004. The first British record and a new species of the superfamily Terrestricytheroidea (Crustacea, Ostracoda): morphology, ontogeny, lifestyle and phylogeny. Zoological Journal of the Linnean Society 142: 253–288.

Nation, J. L., 1983. A new method using hexamethyldisilazane for prepararion of soft insect tissues for scanning electron microscopy. Stain Technology 58: 347–351.
Maddocks, R. F., 1976. Pussellinae are interstitial Bairdiidae (Ostracoda). Micropaleontology 22: 194–214.
Schornikov, E. I., 1969. A new family of Ostracoda from the supralittoral zone of the Kuril Islands. Zoologicheskiy Zhurnal 48: 494–498 [In Russian with English abstract].
Schornikov, E. I., 1980. Ostracods in terrestrial biotopes. Zoologicheskiy Zhurnal 59: 1306–1319 [In Russian with English abstract].

Hydrobiologia (2007) 585:57–66
DOI 10.1007/s10750-007-0628-3

OSTRACODA (ISO15)

Non-marine Ostracoda (Crustacea) of Banat district in Serbia

Tamara Karan-Žnidaršič · Brigita Petrov

Abstract In order to enrich the knowledge on ostracods in Serbia, field work was carried out in the central and southern part of the Banat district during 2002 and 2003. Samples from the collection of the Institute of Zoology, Faculty of Biology in Belgrade were also examined. Altogether, 133 samples were investigated. In this research, 23 ostracod species were found. Most of them have a Palaearctic distribution. *Ilyocypris decipiens*, *Bradleycypris obliqua* and *Tanycypris pellucida* are new to the faunas of Serbia and Montenegro, as is the subfamily Cypricercinae. Three species were found again after several decades, including a new and stabile population of *Candona natronophila* Petkovski, 1969. The most frequent species were *Cypridopsis vidua* and *Physocypria kraepelini*. Statistical analysis of ostracod communities and their relations to measured ecological variables is also given.

Keywords Non-marine ostracods · Taxonomy · Banat · Habitats · Ostracod communities

Guest editors: R. Matzke-Karasz, K. Martens & M. Schudack
Ostracodology – Linking Bio- and Geosciences

T. Karan-Žnidaršič (✉) · B. Petrov
Institute of Zoology, Faculty of Biology, University of Belgrade, Studentski trg 16, 11 000 Belgrade, Serbia
e-mail: ktamara@bf.bio.bg.ac.yu

Introduction

It has already been indicated that the diversity and endemicity of crustaceans in the Balkan region and Serbia and Montenegro are exceptionally high (Karan & Petrov, 2000). The areas of greatest richness are Macedonia, Montenegro, and Vojvodina.

Seventy-two species of non-marine ostracods have thus far been found in Serbia and Montenegro. Montenegro is the richest and the most thoroughly investigated area, with 52 ostracod species (Karanović, 1999). On the other hand, the lowland part of the northern province of Vojvodina stands out with the lowest number of records. Excluding Mt. Fruška Gora, where Karanović (1996) noted 21 species, the only ostracod findings in this area were of *Hungarocypris madarasci* (Petkovski, 1964), *Candona natronophila* (Petkovski, 1969), *Potamocypris arcuata* (Meisch & Wouters, 1985), and *Heterocypris vitrea* (Petkovski et al., 2000). The territory of Serbia south of the Sava and Danube rivers has just a few more, but older records. Those are of *Pseudocandona eremita* (Klie, 1940), *Notodromas monacha* (Petkovski, 1959), *Cypria ophtalmica* (Petkovski, 1961), *Trajancypris clavata* (Petkovski, 1964), and two species of *Dolerocypris* (Petkovski, 1960; Mikulić, 1970).

Studies of ostracod communities and their relations to some ecological factors were performed in wetlands and lowland springs in Northern Italy (Rossetti et al., 2004, 2005) and in northeast Germany (Viehberg, 2006). Here we present the results of our surveys and preliminary analysis of ostracods in the Banat district of the northern province of Vojvodina.

Materials and methods

Banat is the southeastern part of the Pannonian Plain, bordered by the Tisza in the west and the Danube in the south.

The investigated area covers the western part of the district, along the left banks of the Tisza and the Danube rivers. It is situated between 44°51′ and 45°36′ of N latitude and between 20°17′ and 20°45′ of E longitude.

The ostracod material analyzed was obtained from two sources. One part was extracted from the Crustacean Collection of the Institute of Zoology, Faculty of Biology in Belgrade. These samples were taken by the second author during several different research projects in the period 1994–2001.

The second, more abundant part of the ostracod material derived from field investigations carried out in the spring, summer and autumn of 2002/2003. Collecting was carried out at 34 localities that included: ① ephemeral pools in open fields, on carriageways, etc. ② permanent ponds and shallow lakes, ③ slowly flowing, man-made canals and small, lowland rivers, and ④ fish-ponds and water bodies connected with them (Table 1). These are common aquatic habitats in this area (Fig. 1).

Samples were taken from littoral zones by sweeping through aquatic macrophytes with a planktonic net (mesh size 250 μm) and from the sediment, using a bottom-sampler. They were preserved by adding 35% formaldehyde in the field, washed (after Martens, 2001) through the same nets and preserved in 70% ethanol.

Altogether, 133 samples were examined, and representative ostracod specimens were drawn, dissected, and identified according to the methods and key of Meisch (2000). Microscope slides were made using Faure's medium (Langeron, 1942). A projection-screen attached to a Carl Zeiss microscope was used for drawing the carapace and appendage morphology. SEM images were taken at the Royal Belgian Institute of Natural Sciences and Faculty of Sciences, University of Novi Sad.

During sampling, water temperature and pH values were measured using a digital thermometer (0.1°C sensitivity) and litmus-paper (0.2 sensitivity), respectively. The presence of vegetation was estimated as a percentage of emersed and submersed macrophytes covering water surface. There were six categories: 0%, <20%, 20–40%,

Table 1 List of localities with marked habitat type, locality codes and site names

Code	Site	Habitat type	Code	Site	Habitat type
1	Melenci–Bašaid 1	①	16	Uzdin 1	③
2	Melenci–Bašaid 2	①	17	Uzdin 2	③
3	Melenci–Bašaid 3	①	21	Sakule 1	③
5	Melenci 2	①	25	Karaš	③
13	Perlez 3	①	26	Japanski cvet	③
15	Perlez–Titel	①	27	Besni Fok	③
6	Velika Rusanda	②	28	Vizelj 1	③
7	Mala Rusanda	②	29	Vizelj 2	③
14	Perlez 4	②	30	Padinska Skela	③
22	Sakule 2	②	33	Sebeš	③
23	Jer	②	34	Pančevo–Dolovo	③
24	Čenta	②	8	Ečka	④
4	Melenci 1	③	18	Uzdin 3	④
9	Carska bara 1	③	19	Uzdin 4	④
10	Carska bara 2	③	20	Uzdin 5	④
11	Perlez 1	③	31	Mika Alas 1	④
12	Perlez 2	③	32	Mika Alas 2	④

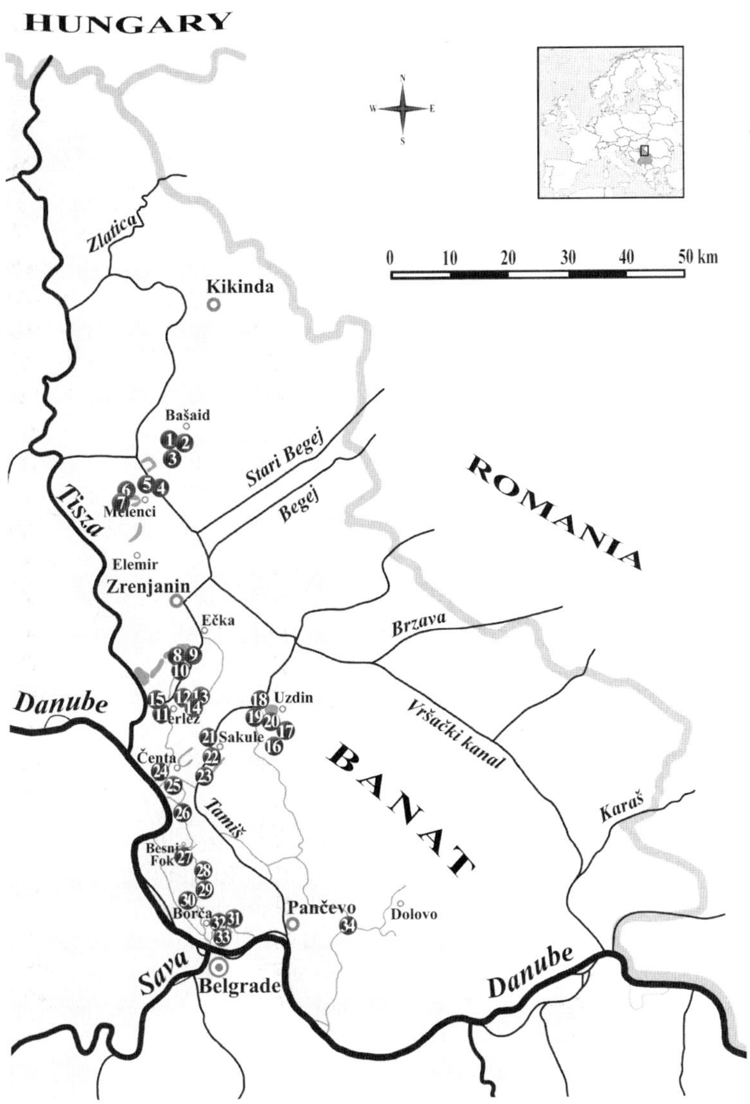

Fig. 1 Map of the investigated area with marked localities where ostracods were found

40–60%, 60–80% and 80–100%. The dominant substratum was assessed according to bottom particle size and the presence of organic matter. It was determined on-site or using the stereomicroscope, and classified as clay, mud, gravel, clay-mud, or clay-sand.

Statistical analyses were performed using MVSP 3.1 (Kovach, 1999) and PAST—Paleontological Statistics, version 1.35 (Hammer et al., 2001). The relations of identified taxa and their presence in the samples were analyzed by Canonial Correspondence Analysis according to pH, temperature, substratum type, and vegetation coverage. The foregoing ecological variables were not available for the ostracod material obtained from the collection, and the samples and species from these localities were therefore excluded from this analysis. Cluster Analysis was applied to ostracod communities using the Jaccard similarity coefficient.

The following abbreviations are used in the text and figures: Temp = water temperature, Substr = type of substratum, Veg = vegetation coverage; abbreviations of species are as in Table 2.

Table 2 List of ostracod species with localities presented as in Fig. 1 and the total number of localities where each species was found (Σ)

Species recorded	Abbreviations	Locality codes	Σ
Limnocythere inopinata	*Lino*	6, 7	2
Candona natronophila	*Cnat*	13, (14 ?)	2
Fabaeformiscandona fabaeformis	*Ffab*	10, 14	2
Pseudocandona pratensis	*Ppra*	1	1
Candonopsis kingsleii	*Ckin*	14	1
Cypria ophtalmica	*Coph*	27, 34	2
Physocypria kraepelini	*Pkra*	4, 9, 26, 27, 30, 31, 32, 33, 34	9
Cyclocypris laevis	*Clev*	13, 14, 34	3
Cyclocypris ovum	*Covu*	14	1
Ilyocypris gibba	*Igib*	12, 13, 14, 23	4
Ilyocypris monstrifica	*Imon*	8, 16, 17, 21, 22, 23, 32	7
Ilyocypris decipiens	*Idec*	12, 20, 21, 22, 23,	5
Notodromas monacha	*Nmon*	14	1
Cypris pubera	*Cpub*	1, 14, 15	3
Eucypris virens	*Evir*	1, 2, 3, 13	4
Tonnacypris lutaria	*Tlut*	1, 2	2
Bradleycypris obliqua	*Bobl*	4, 27	2
Tanycypris sp.	*Tany*	26, 28, 30*	3
Heterocypris incongruens	*Hinc*	1, 5	2
Heterocypris rotundata	*Hrot*	11, 13, 18	3
Heterocypris vitrea	*Hvit*	7	1
Dolerocypris sinensis	*Dsin*	27	1
Cypridopsis vidua	*Cvid*	4, 12, 18, 19, 24, 25, 26, 27, 28, 29	10

Results

During this research, 23 ostracod species were identified. They belong to five podocopid families, 11 subfamilies and 18 genera (Table 2).

Notes on selected species

Candona natronophila (Petkovski, 1969)

Both males and females were found in ephemeral pool 13 on saline land, but only females and juveniles were collected at locality 14. So far, bisexual populations of this species were recorded only near Elemir in Vojvodina, and in several places in Macedonia (Petkovski et al., 2002). Males of *C. natronophila* possess a distinguishing character of the M-process, which is distally curved in the form of a small hook, as described by Petkovski (1969). Petkovski et al. (2002) pointed out that females are not very different from the females of *Candona altoides* or other similar species of the *neglecta*-group. For this reason, it can only be assumed that specimens from the latter site also belong to the given taxon, and the locality is therefore labelled with a question mark in Table 2. In our samples, adults were most abundant in the autumn and larvae in the spring, with the highest number recorded in late May, just before the pond dried up.

Pseudocandona pratensis (Hartwig, 1901)

It was detected in a sample taken from temporal pools at locality 1 in the early spring of 1997. The male/female ratio was approximately 1:1. As an early form (Meisch, 2000), it was absent in new samples from the same locality taken in 2002, since pools were formed only in the autumn. This is the first record of this species in Serbia, while Karanović (1999) registered its presence in Montenegro.

Ilyocypris gibba (Ramdohr, 1808)

Although it is widely distributed, previously recorded in Skadar Lake basin (Petkovski, 1961; Karanović, 1999), *I. gibba* was found for the first time in Serbia at four localities: 12, 13, 14, and 23. The largest number of specimens was collected from the ephemeral pool marked as locality 13, where both forms of carapaces (forma *gibba* and forma *biplicata* according to Meisch, 2000) were observed. In the samples from this habitat, five adults had a peculiarity in the form of an inchoate subdivision of the penultimate segment of the walking leg (T2) or a more or less emphasized notch on the posterior side of the same segment. Martens (1991) has already called attention to the need for an improvement of *Ilyocypris* taxonomy, and the additional consideration of the taxonomical status of populations found in the study area could contribute to this end.

Ilyocypris decipiens (Masi, 1905)

This is the first report of the given species in Serbia. In the sediment samples from five localities (12, 20, 21, 22, 23) only females of *Ilyocypris decipiens* were found. Viewed dorsally, all of them had a somewhat more obese posterior part of the carapace, with grossly developed, rounded posteroventral tubercles, strongly resembling those on the carapace of *I. carinata* (Kovalenko, 1970). All specimens had large and well developed pits on the carapace surface. Some had rounded dorsal tubercles and others had very slightly pointed ones.

Bradleycypris obliqua (Brady, 1868)

Just a single specimen of this species was collected at each of two localities (4 and 27) 60 km apart. Both females were detected among many specimens of *Cypridopsis vidua*, from which they were hard to distinguish at first glance because of the unusual color pattern. Also, being only 0.8 mm in length, they were smaller than the size indicated in the literature (Meisch, 2000). This is the first record of *Bradleycypris obliqua* and the subfamily Cypricercinae as well in Serbia.

Tanycypris pellucida (Klie, 1932)

This is the second genus from the subfamily Cypricercinae that is new for Serbia. Only a few specimens were collected at localities 26 and 28, together with one exuvia (marked with asterisk in Table 1) at locality 30. All of the foregoing sites are canals rich in macrophytes in the littoral zone. The specimens were found in summer and all of them were females or juveniles with a purple to brownish carapace. Apart from carapace color, which is usually reported as yellowish (Bronshtein, 1947; Petkovski, 1964; Victor & Fernando, 1981), other morphological features are as described for *T. pellucida*. Fig. 2

Habitat characteristics and ostracod communities

The water temperature range was 5–35°C, the pH range 6.4–8.2 (Table 3).

Clay was the most common substratum type (65% of the total number of localities). It was sometimes mixed with sand or mud, while strictly muddy or gravelly bottoms were less represented. The aquatic plant coverage was variable in different fishponds and canals, but absent in ephemeral pools. In permanent ponds and shallow lakes, aquatic plants covered the water surface up to 10%, except for locality 14, which was totally overgrown with *Phragmites* stems and rich in submersed plants as well. This pond appeared to be the richest in ostracods, harboring a total of eight species.

The ordination of species and environmental factors on the space defined by CCA axes (Fig. 3) defines four groups of species. The cumulative percentage of two axes included 9.6% and 17.7% of the total variance. The species-environment correlations were 0.87 and 0.83 for axis 1 and axis 2, respectively. For the first axis, the most important factors of species arrangement were temperature (with a canonical coefficient of 0.81) and substratum type (with a canonical coefficient of 0.51). The second axis is strongly correlated with the percentage of vegetation coverage, with a canonical coefficient of 0.98. In general, to judge from the diagram, temperature, substratum type, and vegetation coverage are the most significant factors for the distribution of species in this analysis, while pH values are not of great significance.

Fig. 2 Scanning electron images of selected species: *Limnocythere inopinata* (**a**, **b**), *Pseudocandona pratensis* (**c**, **d**), *Candona natronophila* (**e**–**h**), *Fabaeformiscandona fabaeformis* (**i**), *Candonopsis kingsleii* (**j**). *Cyclocypris laevis* (**k**, **l**), *Cyclocypris ovum* (**m**, **n**), *Ilyocypris monstrifica* (**o**, **p**), *Ilyocypris decipiens* (**q**, **r**), *Heterocypris incongruens* (**s**), *Tonnacypris lutaria* (**t**). All adult specimens. Scale bar 400 μm for **a**–**d** and **i**–**r**; 534 μm for **e**–**h**; 533 μm for **s**; 636 μm for **t**; **a**, female, lateral view; **b**, female, dorsal view; **c**, female, dorsal view; **d**, male, lateral view, **e**, female, lateral view; **f**, male, lateral view; **g**, female, dorsal view; **h**, male, dorsal view; **i**, female, dorsal view; **j**, female, dorsal view; **k**, dorsal view; **l**, lateral view; **m**, dorsal view; **n**, lateral view; **o**, female, dorsal view; **p**, male, lateral view; **q**, female, dorsal view; **r**, female, lateral view; **s**, female, dorsal view; **t**, female, dorsal view

The first group, situated on the upper part of the diagram, encompasses the following six species: *Fabaeformiscandona fabaeformis*, *Candonopsis kingsleii*, *Cyclocypris laevis*, *Cyclocypris ovum*, *Notodromas monacha* and *Cypris pubera*. They were mostly found in ponds and slowly flowing canals, and all were present at locality 14. The opposite group (on the lower part) includes *Cypria ophtalmica*, *Ilyocypris monstrifica*, *Ilyocypris decipiens* and *Dolerocypris sinensis*, which

Table 3 Temperature and pH ranges measured in different habitat types

Habitat type	Temperature (°C)	pH
①	5–25	6.4–7.8
②	21.5–35	7.2–8.2
③	10.5–31	6.8–8.2
④	27–31	7.0–7.8

were usually found on clayey substrates and in warmer waters. The species on the left side of the ordination diagram (*Candona natronophila*, *Ilyocypris gibba*, *Eucypris virens*, *Tonnacypris lutaria*, *Heterocypris incongruens*, and *H. rotundata*) were typical of ephemeral pools. The group on the right side includes the two most common species, namely *Cypridopsis vidua* and *Physocypria kraepelini*, and two species from the subfamily Cypricercinae, all collected in the canals and fishponds with a high percentage of macrophyte coverage.

Cluster Analysis of ostracod communities distinguishes four groups of localities (Fig. 4), but similarity percentages inside these clusters are low. The first group is diverse in regard to ostracod species and contains all ephemeral pools and a few other localities, including locality 14, which is the one with the greatest diversity. The second group comprises localities that are relatively close together and mostly dominated by *Ilyocypris* species. The third group mostly consists of canals and fishponds and is represented by *P. kraepelini* and *C. vidua*. Including two localities, the last group is represented by *L. inopinata* and *H. vitrea*, which do not occur in other samples. These neighboring sites also differ from all other localities in their water chemistry characteristics. According to Petrović (1981), previous assessment of some environmental conditions showed high alkalinity (with pH values higher than 8.2) and salinity (in the range of 3.5–13.5‰). This is also in accordance with the results of Mezquita et al. (2001), who linked *L. inopinata* with high nitrate content.

Discussion

The general distribution of 21 species recorded in the Banat district is known. Most of them have a Palaearctic or Holarctic distribution, while *Ilyocypris gibba*, *Cypria ophtalmica*, *Heterocypris incongruens* and *Cypridopsis vidua* are cosmopolitan. Although *Candona natronophila* has been found only here and in Macedonia thus far, there are no endemic species.

The lowland part of Serbia harbors 25 ostracod species detected to date, while the ostracod fauna of neighboring Hungary comprises a total of 86 species (Meisch, 2000). Comparing these numbers, we can safely assume that many more ostracod species are present in this region, especially since Banat is on the border of the Balkan Peninsula, with its great biodiversity.

Three ostracod species – *Ilyocypris decipiens*, *Bradleycypris obliqua*, *Tanycypris pellucida* – were found for the first time in Serbia and Montenegro. Also, the subfamily Cypricercinae

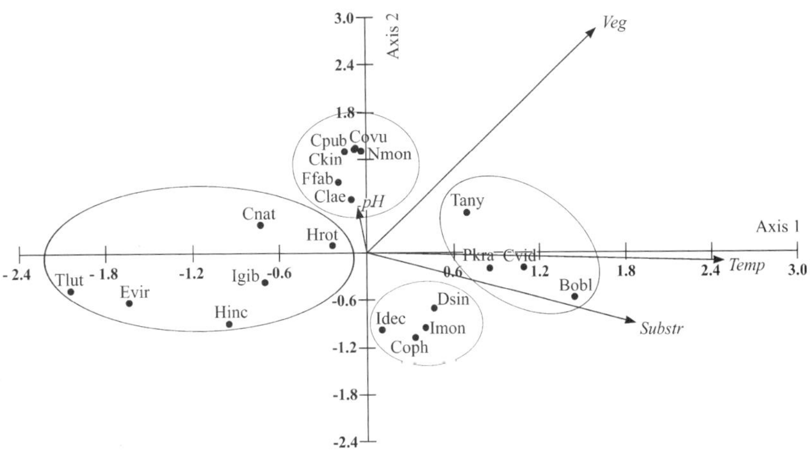

Fig. 3 Canonial Correspondence Analysis representing ordination of ostracod assemblages and environmental factors (vector scaling = 3.40) on the space defined by the first two canonical axes. Abbreviations of species are as in Table 2. Temp: water temperature. Substr: type of substratum. Veg: vegetation coverage

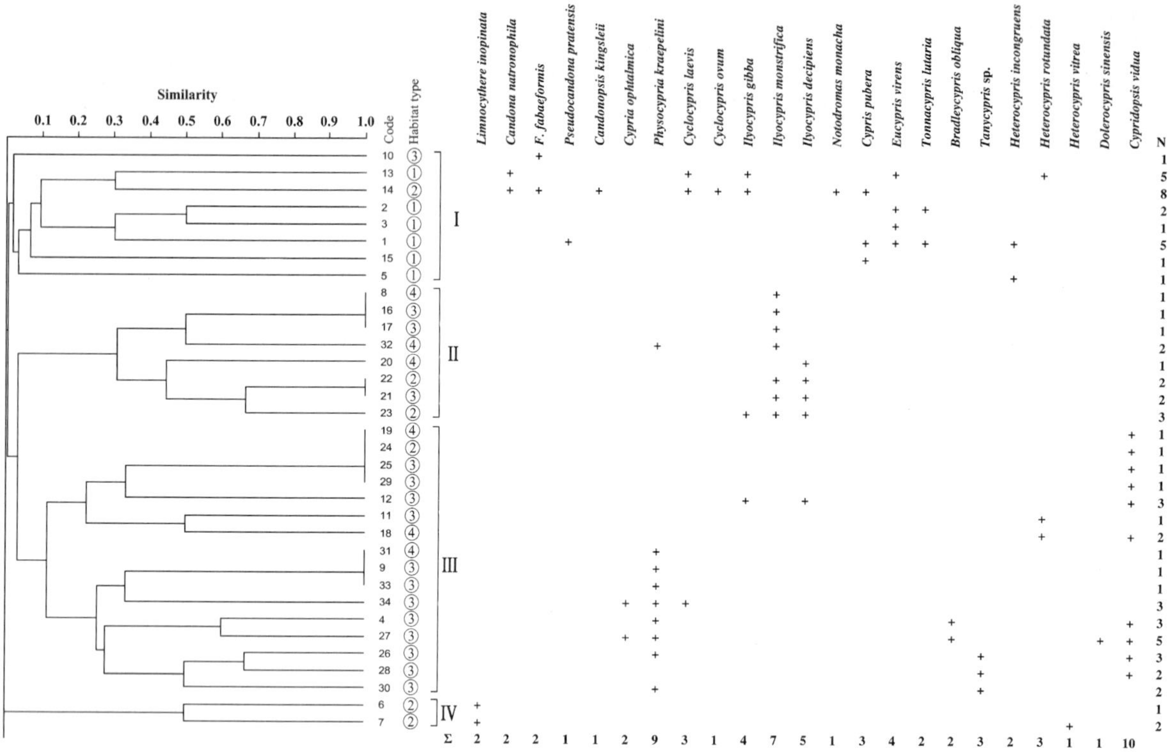

Fig. 4 Dendrogram of the Cluster Analysis presents the similarity level between ostracod communities. The occurrence of ostracod species in each locality is also indicated. N: number of species per site. Σ: total species record

is a new taxon for this region. The genus *Tanycypris* comprises nine species (Broodbakker, 1984), mainly distributed in South America, South Africa and Eastern Asia, while the presence of *Tanycypris pellucida* was previously recorded in Central and Southeast Asia (Bronshtein, 1947), Macedonia (Petkovski, 1964), and Italy (Rossi et al., 2003) mostly in the ricefields. *Notodromas monacha* was found again in Serbia after almost 50 years, while *Dolerocypris sinensis* and *Candona natronophila* were rediscovered after more than three decades.

In all the species with mixed reproduction, only parthenogenetic populations were found. Although hundreds of specimens of *Limnocythere inopinata* were collected, no males were found; since males can be rare (Meisch, 2000), more abundant samples should be taken in order to register their presence.

The highest number of ostracod species is detected in permanent ponds and canals, while fish-ponds had only few species. This is probably due to the nature of the latter habitat, as it is designed for cultivation of (predatory) fish and is also under constant influence of man. Ostracod communities overlap in some degree due to relatively similar conditions, such as slow water flow in canals, which brings them closer to the properties of large ponds. The reason for almost complete congruency of fishpond habitats and canals could be the fact that the former are usually connected and fed by canal waters.

Ephemeral pools ranked third in species diversity, but (as expected) these habitats contained the largest numbers of ostracods. As shown by the CCA diagram, this type of aquatic habitats was clearly separated from the others in the community structure, with dominance of *E. virens*, *T. lutaria* and *H. incongruens*.

Comparing the two analyses performed, we note that all species from the first cluster, embracing permanent and ephemeral ponds, are situated on the right side of the CCA ordination diagram. This is probably a consequence of relative proximity of the sites and differences in temperature gradients, which resulted in grouping of early and summer forms with permanent ones. Among others, this cluster contains locality 14, with the largest number of species occasionally occurring in other samples. The great richness of ostracods (and other aquatic invertebrates) in this relatively small pond can be attributed to its permanence, dense macrophyte coverage and also to the lack of fish as predators, although some amphibian larvae (Discoglossidae and Salamandridae) were present. In comparison with other ponds in the area, which are either shallow and located on extremely saline land or used in different ways, this kind of pristine lowland aquatic habitat seems to be the most important for conservation of aquatic invertebrates.

As this was a preliminary study on Ostracoda in a region with a very small number of previous records, new data provided information mostly on common taxa with high tolerance for various ecological conditions. That is probably why the linkage with measured ecological variables analyzed by CCA was low. Considering differences in specific hydrochemical conditions of surface waters and the high level of groundwater in the Banat district (Petrović, 1981), we feel that future studies should be supplemented with additional data on both physical and chemical conditions.

Acknowledgements We are highly grateful to Dr. Koen Martens of the Royal Belgian Institute of Natural Sciences for invaluable help in ostracod taxonomy and instruction in illustration techniques. We also wish to acknowledge the kindness of Dr. Trajan Petkovski of the Natural History Museum in Skopje, who provided advice on ostracod identification. For the scanning electron micrographs, we are obliged to the R.B.I.N.S. and Ricardo Pinto for instruction and assistance in material preparation, as well as to Miša Bokorov of the University of Novi Sad. In addition to them, we are indebted to Dr. Renate Matzke-Karasz of the Department of Earth and Environmental Sciences, Section Palaeontology of the Ludwig-Maximilians-Universität München for helpful suggestions in processing and presenting SEM images; and to Dr. Sonia Fontana of the University of Liverpool's Department of Geography for improvement of the manuscript. Special thanks go to our colleagues Ljiljana Tomović, Imre Krizmanić, Biljana Stojković, and Dragana Miličić, who kindly helped with statistical analyses, as well as to Pavelka Ćirić who helped with English language. The final correction of English grammar was performed by Mr Raymond W. Dooley. The presented results were obtained within the framework of Project 1533 ("Investigations of invertebrate fauna diversity in small, isolated natural habitats on the Pannonian Plain") approved and financed by the Ministry of Science and Technology of Serbia and Montenegro.

References

Broodbakker, N. W., 1984. The distribution and zoogeography of freshwater Ostracoda (Crustacea) in the West Indies, with emphasis on species inhabiting wells. Bijdragen tot de Dierkunde 52: 327–368.

Bronshtein, Z. S., 1947. Fresh-water Ostracoda. Fauna of the USSR Crustaceans, Vol. 2, No.1. Zoologicheskiy Institut Akademii Nauk SSSR, n.s. 31. Russian Translation Series 64, Oxonian Press, New Delhi.

Hammer, Ø., D. A. T. Harper & P. D. Ryan, 2001. PAST: Paleontological Statistics Software Package for education and data analysis. Palaeontolia Electronica 4. http://palaeo-electronica.org/2001_1/past/issue1_01.htm.

Karan, T. & B. Petrov, 2000. A review of the investigations, preliminary-list of species and diversity estimation of freshwater and terrestrial crustaceans of the Balkans and FR Yugoslavia I part. Ekologija 35: 1–44.

Karanović, I., 1996. Fauna ostrakoda (Ostracoda, Crustacea) Fruške Gore (Vojvodina, Jugoslavija). Magistarska teza, Novi Sad.

Karanović, I., 1999. Fauna ostrakoda (Ostracoda, Crustacea) sliva Skadarskog jezera. Doktorska disertacija, Novi Sad.

Klie, W., 1940. Zur Kenntnis der Ostracodenarten *Candona eremita* (Vejd.) und *Candona reducta* (Alm). Mitteilungen über Höhlen- und Karstforschung 1: 24–29.

Kovach, W. L., 1999. MVSP - A Multivariate Statistical Package for Windows, ver. 3.1. Kovach Computing Services, Pentraeth, Wales, UK.

Kovalenko, A. L., 1970. Novyi vid rakushkovyh rakoobraznyh Dnestr *Ilyocypris carinata*. Fauna kainozoya Moldavii, Akademia Nauk Moldavskoi SSSR: 112–116.

Langeron, M., 1942. Précis De Microscopie: technique – epérimentation – diagnostic. Masson et Cie, Paris.

Martens, K., 1991. On a small collection of non-marine ostracods from Mongolia, with the description of a new species (Crustacea, Ostracoda). Miscellanea Zoologica Hungarica 6: 53–60.

Martens, K., 2001. Ostracoda. In Day, J. A. et al. (eds.), Guides to the Freshwater Invertebrates of South Africa. Vol. 3, Crustacea II. Ostracoda, Copepoda & Branchiura, 9–77.

Meisch, C., 2000. Freshwater Ostracoda of Western and Central Europe. In Schwoerbel, J. & P. Zwick (eds) Süßwasserfauna von Mitteleuropa 8/3. Spektrum Akademischer Verlag, Heidelberg, Berlin.

Meisch, C. & K. Wouters, 1985. *Potamocypris arcuata* (Crustacea, Ostracoda) new to Hungary. Miscellanea Zoologica Hungarica 3: 73–76.

Mezquita, F., H. I. Griffiths, M. I. Domínguez & M. A. Lozano-Quilis, 2001. Ostracoda (Crustacea) as ecological indicators: a case study from Iberian Mediterranean brooks. Archiv für Hydrobiologie 150: 545–560.

Mikulić, F., 1970. Neki aspekti distribucije vrsta roda *Dolerocypris* Kaufm. u okolini Beograda. Glasnik Prirodnjačkog Muzeja u Beogradu. Serija B 25: 229–331.

Petkovski, T. K., 1959. Beitrag zur Kenntnis der Ostracodenfauna Jugoslawiens (V). Hidrobiologi. Publications of the Hydrobiological Research Institute, Faculty of Science, University of Istanbul, Seri B, 4: 158–165.

Petkovski, T. K., 1960. Süsswasserostracoden aus Jugoslawien VII. Fragmenta Balcanica, III, 12: 99–106.

Petkovski, T. K., 1961. Zur Kenntnis der Crustaceen des Skadar (Scutari-) Sees. Acta Musei Macedonici Scientiarum Naturalium, VIII, 2: 29–52.

Petkovski, T. K., 1964. Bemerkenswerte Entomostraken aus Jugoslawien. Acta Musei Macedonici Scientiarum Naturalium, IX, 7: 147–182.

Petkovski, T. K., 1969. Einige neue und Bemerkenswerte Candoninae aus dem Ohridsee und einigen anderen Fundorten in Europa (Crustacea – Ostracoda). Acta Musei Macedonici Scientiarum Naturalium, XI, 5: 82–111.

Petkovski, T. K., B. Scharf & D. Keyser, 2000. New and little known ostracods of the genus *Heterocypris* (Crustacea, Ostracoda) from the Balkan Peninsula. Limnologica 30: 45–57.

Petkovski, T. K., B. Scharf & D. Keyser, 2002. New and little known species of the genus *Candona* (Crustacea, Ostracoda) from Macedonia the other Balkan areas. Limnologica 32: 114–130.

Petrović, G., 1981. On the chemistry of some salt lakes and ponds in Yugoslavia. Hydrobiologia 81: 195–200.

Rossetti, G., M. Bartoli & K. Martens, 2004. Limnological characteristics and recent ostracods (Crustacea, Ostracoda) of freshwater wetlands in the Parco Oglio Sud (Northern Italy). Annales de Limnologie – International Journal of Limnology 40: 329–341.

Rossetti, G., V. Pieri & K. Martens, 2005. Recent ostracods (Crustacea, Ostracoda) found in lowland springs of the provinces of Piacenza and Parma (Northern Italy). Hydrobiologia 542: 287–296.

Rossi, V., G. Benassi, M. Veneri, C. Bellavere, P. Menozzi, A. Moroni & K. G. McKenzie, 2003. Ostracoda of the Italian ricefields thirty years on:new sythesis and hypothesis. Journal of Limnology 62: 1–8.

Victor, R., & C. H. Fernando, 1981. Freshwater ostracods (Crustacea: Ostracoda) of the subfamily Dolerocypridinae Triebel, 1961 from Southeast Asia. Zoological Journal of the Linnean Society 72: 107–116.

Viehberg, F. A., 2006. Freshwater ostracod assemblages and their relationship to environmental variables in waters from northeast Germany. Hydrobiologia 571: 213–224.

Hydrobiologia (2007) 585:67–77
DOI 10.1007/s10750-007-0629-2

OSTRACODA (ISO15)

A year round comparative study on the population structures of pelagic Ostracoda in Admiralty Bay (Southern Ocean)

Katarzyna Blachowiak-Samolyk · Martin V. Angel

Abstract The population structures of the three dominant planktonic halocyprid Ostracoda species in Admiralty Bay (King George Island, Antarctic Peninsula) were followed throughout the course of a year in zooplankton samples collected once every three weeks from February 1993 to January 1994. The sampling was conducted at two stations: A in the central part of Admiralty Bay (400–0 m) and B in the entrance to the Bay from the Bransfield Strait (400–0 m). The samples were taken using a WP-2 net (square mouth opening of 0.196 m^2 and 200 μm mesh) hauled vertically from the bottom to the surface. Changes in the age structures of the populations of three species *Alacia belgicae*, *Alacia hettacra* and *Metaconchoecia isocheira* were tracked. Their population structures differed. The changes in *A. belgicae* suggested that it reproduces year-round, whereas both *A. hettacra* and *M. isocheira* probably complete their life cycles within a year. The cycle in *A. hettacra* probably starts earlier in the year than that of *M. isocheira*. Populations of *A. belgicae* and *M. isocheira* were more advanced in their development at station A, than at station B, but *A. hettacra* was more advanced at the latter. Advection appears to play a role in maintaining the populations in the shelf waters. Comparisons between populations in the shelf area (Admiralty Bay) and in open ocean waters (Croker Passage) show that the *M. isocheira* population is older in shelf water whereas the age structure of *A. belgicae* population is not influenced by the locality.

Keywords Antarctica · Admiralty Bay · Pelagic Ostracoda · Halocyprids · Life cycle · Seasonal changes

Guest editors: R. Matzke-Karasz, K. Martens & M. Schudack
Ostracodology – Linking Bio- and Geosciences

K. Blachowiak-Samolyk (✉)
Ecology Department, Institute of Oceanology, Polish Academy of Sciences, Powstancow Warszawy 55, 81-712 Sopot, Poland
e-mail: kasiab@iopan.gda.pl

M. V. Angel
George Deacon Division, National Oceanography Centre, Southampton SO14 3ZH, UK

Introduction

Pelagic halocyprid Ostracoda are an important component of the mesozooplankton in Antarctic waters, locally occurring in high abundance (Boden, 1985; Fukuchi et al., 1985; Hopkins, 1985a; Boden & Parker, 1986; Foster, 1987, 1989; Blachowiak-Samolyk & Zmijewska, 1997). The importance of the role played by the pelagic Ostracoda in oceanic ecosystems, both in the Southern Ocean and other oceans has been not fully recognised. The geographical ranges of Southern Ocean Ostracoda and their vertical distributions are well described (Hillman, 1967; Deevey, 1978; Kock, 1992, 1993; Blachowiak-Samolyk & Zmijewska, 1995, 1997; Blachowiak-Samolyk, 1999; Blachowiak-Samolyk & Osowiecki, 2002; Chavtur & Kruk, 2003; Drapun, 2004; see also Blachowiak-Samolyk & Angel, 2004), but few studies have focused on their life cycles. Kock (1992, 1993) examined the population structure of halocyprids in the epipelagic zone near Antarctic Peninsula. Blachowiak-Samolyk (2001) reported on populations in deep water in the Croker Passage and Blachowiak-Samolyk & Osowiecki (2002) studied the influence of pack-ice on the age structure of the ostracods in the Scotia Sea.

Admiralty Bay is the site for an intensive international scientific programme, and the Antarctic Special Management Area is an appropriate basin for investigations of land-ocean interactions and their influence on the distributions and structure of coastal zone zooplankton communities. The inshore coastal areas are more productive and support more diverse pelagic communities than the open ocean waters, possibly because they are not subject to iron limitation (de Baar & Boyd, 2000). Previous zooplankton studies in Admiralty Bay have only examined copepods (Chojnacki & Weglenska, 1984; Zmijewska, 1993a) and krill (Kittel, 1980; Kittel & Presler, 1980; Rakusa-Suszczewski & Stepnik, 1980). Menshenina & Rakusa-Suszczewski (1992) presented a general overview of the seasonal cycles of the zooplankton communities based on a year-round investigation in the central basin. Freire et al. (1993) described the responses of the zooplankton communities of Admiralty Bay to the physical forcing of tides and winds. Kittel et al. (2001) examined the heterogeneity of zooplankton communities in the West Antarctic coastal ecosystem, and summarised the current state of knowledge of zooplankton in Admiralty Bay. Their studies were based on the same set of samples used in this study. However, the only previously published study of pelagic ostracods from Admiralty Bay was based on three samples collected in summer 1988/1989 (Blachowiak-Samolyk & Osowiecki, 2002).

These year-round samples from Admiralty Bay provide a unique opportunity to describe the annual life cycles of three dominant pelagic ostracod species - *Alacia belgicae, A. hettacra* and *Metaconchoecia isocheira.*

The aim of this paper is to verify the hypothesis that there are substantial differences between the populations of the three dominant halocyprid species inhabiting the central (station A) and outlet regions (station B) of Admiralty Bay. We also compare the age structures of the two species sampled from the shelf area of Admiralty Bay and open oceanic waters of the Croker Passage using published data (Blachowiak-Samolyk, 2001).

Study area

Admiralty Bay (area 120 km^2, maximum depth ~500 m) is the largest bay of King George Island. It opens into the Bransfield Strait via an eight km wide outlet. The Bay is fjordic in character and connects to a system of smaller inlets - the Ezcurra Inlet, the MacKellar Inlet and the Martel Inlet (Kittel et al., 2001). Pruszak (1980) and Lipski (1987) have described the physico-chemical environment of Admiralty Bay. The water in the Bay is well-mixed and lacks any well-defined halocline or thermocline (Szafranski & Lipski, 1982). However, close to the ice barriers, the upper 15–40 m of water column has relatively low salinity and temperature, and a higher oxygen content (Szafranski & Lipski, 1982). In summer surface salinities are low near the glaciers and freshwater inflows. Physical circulation in the Bay is driven by exchanges with the open water of the Bransfield Strait via deep inflows and surface outflows. The upper 100 m of the water column is exchanged every 1–2 weeks (Pruszak, 1980). In the centre of the Bay the seabed is rather even and depth increases steadily towards the outlet to >500 m. There are some ridges mostly at the entrances to the inlets. These are important features because they generate eddies in the inflows from the Bransfield Strait and cause of some localised upwelling.

Materials and methods

About 54 zooplankton samples were collected during the XVII Polish Antarctic Expedition of Polish Academy of Sciences to the *Arctowski* Station (1992–1994), using a WP-2 net (0.25 m^2 square mouth opening and 200 μm mesh). Zooplankton samples were collected at four locations within the Bay, with distinctive hydrological regimes (Fig. 1). Samples from two of these stations were selected for populations studies of the ostracods: station A in the central part of Admiralty Bay (62°08′ S; 58°26′ W) where the depth was 400 m and station B in the opening from the Bay into the Bransfield (62°12′30″ S; 58°22′ W, depth > 500 m) (Fig. 1).

The other two stations were sampled, H—a glacial lagoon in Herve Cove (62°11′ S; 58°32′20″ W) with depth 18 m (see Sicinski et al., 1996) and C, in Cardozo Cove inside Ezcurra Inlet (62°10′S; 58°35′ W), where the maximum depth was 150 m. The samples contained too few ostracods, and although they have been analysed the data provide no innovative insights.

At stations A and B 400–0 m vertical hauls were collected every three weeks, between February 1993 and January 1994 (except for July). Each haul represented the zooplankton filtered from a nominal 100 m^3. All samples were preserved immediately in 4% seawater formalin buffered with sodium borate.

In the laboratory, all specimens of Ostracoda were sorted and identified from the samples. Determinations were based on the papers of Müller (1912), Sars (1928), Hillman (1969), Deevey (1978, 1982), Angel (1981) and Kock (1992), and sorted into development stages according to Hartmann (1968). All the species belonged to the family Halocyprididae, and have a life cycle consisting of six juvenile instars and a single adult stage. According to Hartmann (1968) these are designated increasingly A-6, A-5,...,A-1 and Ad (adult). The sexes of the three eldest instars, A-2, A-1 and Ad were determined. The age structures of the two most abundant species were compared by calculating the "mean population stage" index for each season. The index $\overline{S}$ modified by Kock (1992)

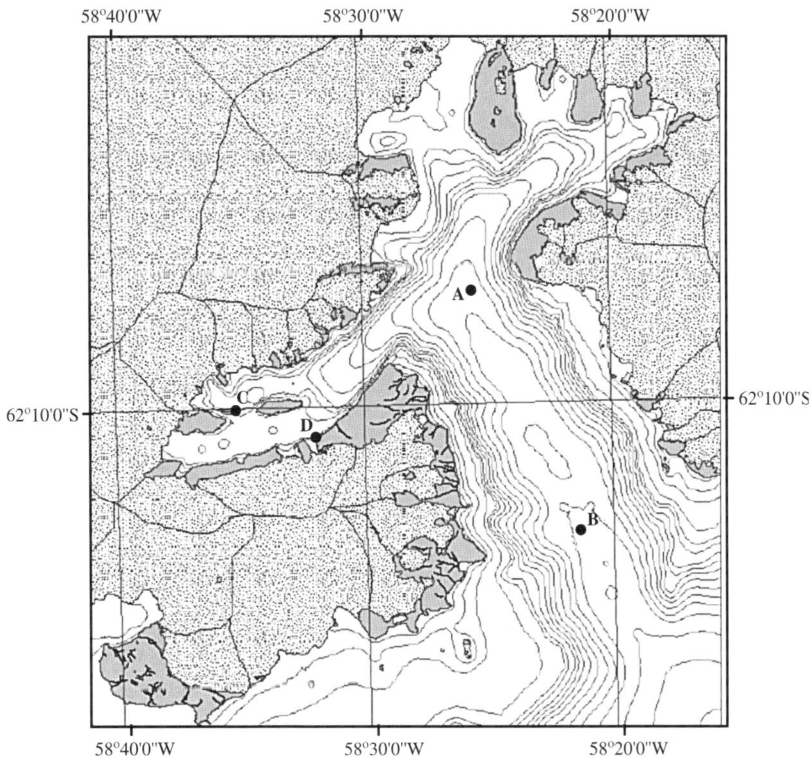

Fig. 1 Location of zooplankton sampling stations in Admiralty Bay

summarizes the relative abundances of developmental stages thus:

$$\overline{S} = \frac{N_{A-6} * 1 + N_{A-5} * 2 + N_{A-4} * 3 + N_{A-3} * 4 + N_{A-2} * 5 + N_{A-1} * 6 + N_{Ad} * 7}{N_{A-6} + N_{A-5} + N_{A-4} + N_{A-3} + N_{A-2} + N_{A-1} + N_{Ad}}$$

where $N_{A\text{-}6}$, $N_{A\text{-}5}$,..., N_{Ad} are the mean abundances of the discrete developmental stages. The "mean population stage" has a minimum of 1 when the whole population consists of the earliest A-6 instar and rises to a maximum of 7 when the population is entirely adult.

A multivariate analysis of the data matrix of indices for each species by season at the two stations has been conducted using the PRIMER package v.5 (Clarke, 1993). In addition Box Whiskers plots and the nonparametric Mann–Whitney *U*-test were carried out using the Statsoft software STATISTICA v. 6.

Results

The results focus principally on the population structures of three dominant species, *Alacia belgicae* (Müller, 1906), *A. hettacra* (Müller, 1906) and *Metaconchoecia isocheira* (Müller, 1906) (note Chavtur & Angel, in press are re-classifying the latter species in a monospecific genus) at stations A and B. All developmental stages except A-6 of *A. belgicae* and *A. hettacra* were sampled. Specimens of the earliest A-6 instar might have been extruded through the net's 200 μm mesh and so not retained. *Metaconchoecia isocheira* is even smaller and neither A-6 nor A-5 specimens were sampled. Herein instars A-6 to A-3 are referred to as the "younger" component of the population, and instars A-2 to Ad as the "older" component.

Seasonal changes:

(A) *Alacia belgicae* age structure

During autumn, winter and spring the population was dominated by the "younger" component (mainly A-3 instars) at both stations (Fig. 2). In summer, the "older" component of the population, consisting mainly of A-1 females dominated (~75%). Specimens of the A-5 instar were sampled at Station A, but the youngest instar sampled at station B was A-4.

(B) *Alacia hettacra* age structure

The *A. hettacra* population was dominated by the "older" component throughout the year (Fig. 2). In autumn, ~50% of the population at both A and B were A-2 females. The A-5 instar was only sampled at B in the autumn. In winter the proportions of the "older" component at A and B were respectively 75% and 70%. In spring only the older stages were present at station A and the species was absent from station B. During summer only developmental stages from A-3 to Ad occurred at station B (Fig. 2).

(C) *Metaconchoecia isocheira* age structure

The *M. isocheira* populations were dominated by adults (mainly females) in all seasons at both stations (Fig. 2). In autumn, the "older" component contributed 75% of the populations at both sites. In winter, the frequency of the pre-adult stage, A-1, increased at both stations. In spring and summer, the populations were dominated by adult females at both localities.

Dendrogram of double-root transformed abundances of each species' developmental stages in particular season distinguished two main groups: spring-summer and autumn-winter populations (Fig. 3). The only exception was spring (at station A) and summer (at station B) population structures of *A. belgicae*, which clustered together with autumn-winter populations of the three analysed species.

At the shallow station C (data not shown) the three species occurred seasonally—*M. isocheira* only during autumn and winter. The *A. belgicae* population was dominated by young instars, whereas the populations of *A. hettacra* and *M. isocheira* were predominantly older stages. Only single specimens of *A. belgicae* (from A-4 to Ad)

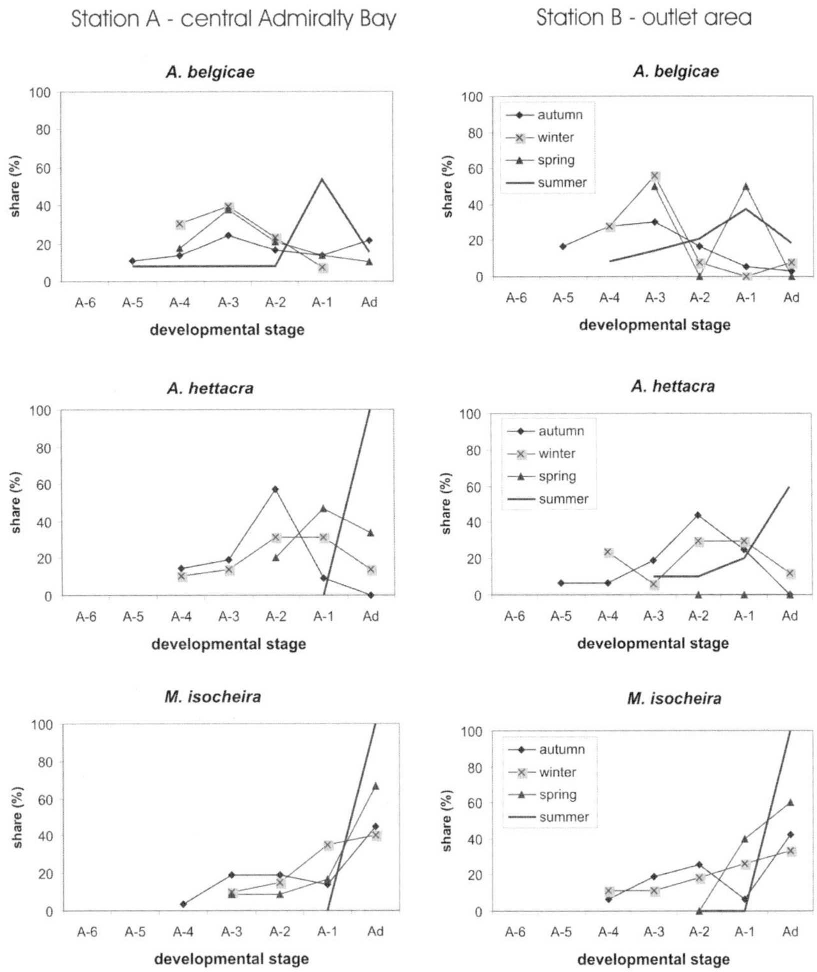

Fig. 2 Percentage composition of developmental stages in the populations of *Alacia belgicae*, *Alacia hettacra* and *Metaconchoecia isocheira* sampled at stations A and B in Admiralty Bay during various seasons

were sampled at station H in the winter-spring period.

"Mean population stage" analysis

The "mean population stage" indices calculated for the populations of *A. belgicae, A. hettacra* and *M. isocheira* at the two main sites during each season are shown in Table 1.

The Box Whiskers plot (Fig. 4) illustrates the differences between the $\overline{S}$ indices for the three species at the two sites during various seasons. The populations of *A. belgicae* and *M. isocheira* are older at the station A compared to those at B, whereas the opposite pattern emerges for *A. hettacra*.

The nonparametric Mann–Whitney *U*-test was used to verify whether "mean population stage" of the three species differs significantly between station A and station B. The results showed the differences are significant for all three species (*A. belgicae*: $P = 0.021$, $Z = -2.309$; *A. hettacra*: $P = 0.034$, $Z = -2.211$ and *M. isocheira*: $P = 0.029$, $Z = -2.165$).

An analogous Box Whiskers plot was calculated comparing data from Admiralty Bay (stations A & B) with those from the Croker Passage (station C) (Fig. 5). This analysis compares only

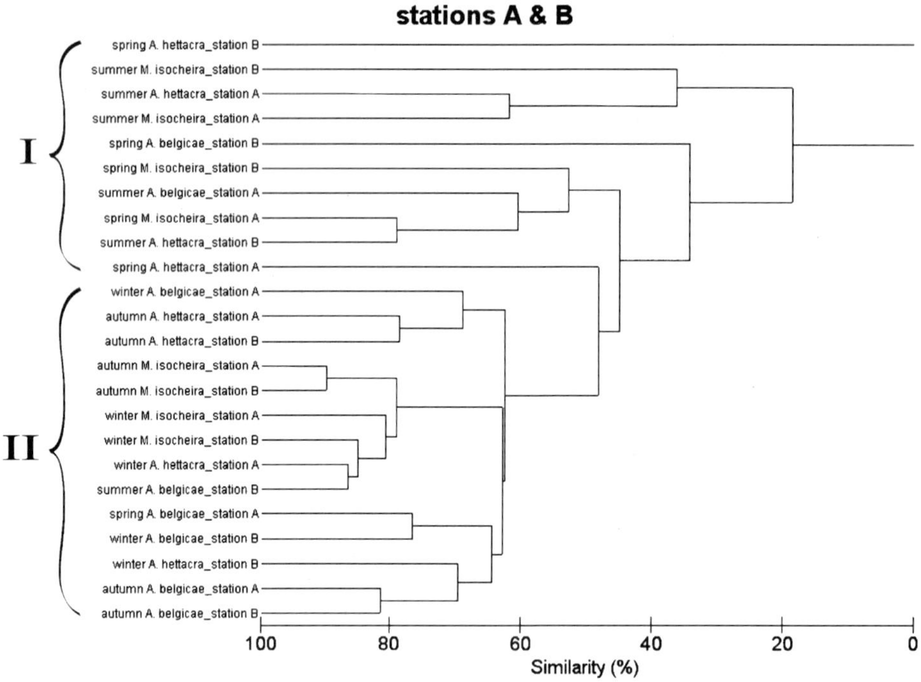

Fig. 3 Dendrogram (cluster) of double-root transformed data with two main groups: (**I**) spring-summer and (**II**) autumn-winter populations of *Alacia belgicae, Alacia hettacra* and *Metaconchoecia isocheira* at stations A and B of Admiralty Bay

A. belgicae and *M. isocheira*, which were described in the Croker Passage in autumn, winter and summer. It shows that the $\overline{S}$ indices increase towards shelf stations for *M. isocheira* whereas average $\overline{S}$ values were similar for *A. belgicae* in Admiralty Bay (station A) and the Croker Passage but lower at station B.

Table 1 "Mean population stage" for *A. belgicae, A. hettacra* and *M. isocheira* at stations A and B in Admiralty Bay

Investigated area	season	$\overline{S}$		
		A. belgicae	*A. hettacra*	*M. isocheira*
station A	autumn	4.7	4.6	5.8
	winter	4.1	5.2	6.1
	spring	4.6	6.1	6.4
	summer	5.4	–	7.0
station B	autumn	3.8	4.8	5.6
	winter	4.0	5.0	5.6
	spring	5.0	–	6.6
	summer	5.4	6.3	7.0

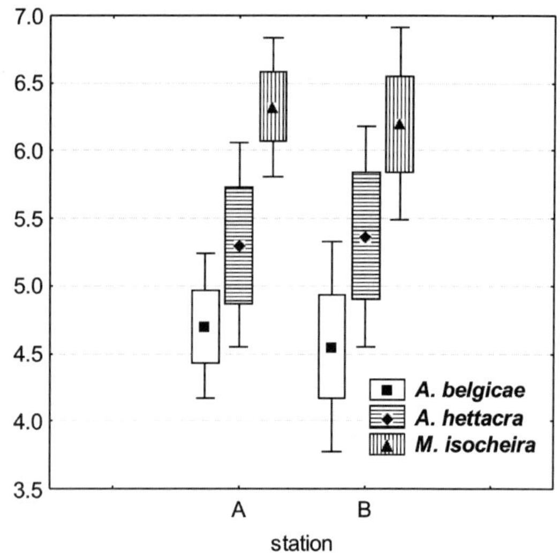

Fig. 4 Box Whiskers plot with differences of the $\overline{S}$ values for *A. belgicae*, *A. hettacra* and *M. isocheira* at the two sites (stations A and B) of Admiralty Bay. In this type of Box Whiskers plot, the smallest box in the plot represents the mean (central tendency) of the $\overline{S}$ values, while the dispersion (variability) among seasons is represented by ± 1 times the standard error (large box) and ± 1 times the standard deviation (SD) about the mean ("whiskers")

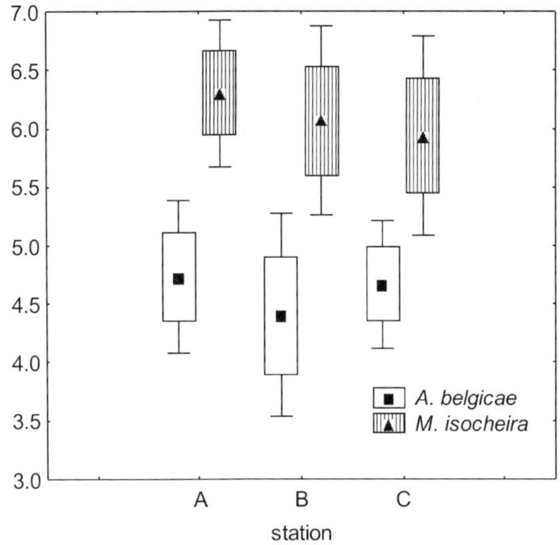

Fig. 5 Box Whiskers plot comparison between the $\overline{S}$ values for *A. belgicae* and *M. isocheira* at the two sites (stations A and B) of Admiralty Bay and Croker Passage (station C). For details see the explanation given for Fig. 4

Discussion

The only previous report of pelagic Ostracoda in Admiralty Bay was in general comments about zooplankton composition (Kittel et al., 2001). They found that the ostracods never attained very high abundances between February 1992 and January 1993, even so the group ranked third after Copepoda and Chaetognatha, in abundance and often in biomass. Frequencies of occurrence of both *A. belgicae* and *M. isocheira* oscillated between 92% and 93% in the upper 400 m at stations A and B, but were slightly lower for *A. hettacra* (71% and 53% at A and B respectively (Kittel et al., 2001). At station C, where the maximum depth was ~150 m, their frequencies were lower (*A. belgicae*, 57%, *A. hettacra* 38% and *M. isocheira* 36%). At the shallowest station H (depth ~20 m), only *A. belgicae* (38%) and *A. hettacra* (8%) occurred. Kock (1993) suggested that *A. belgicae* has a preference for the neritic zone, which is consistent with this species' high frequency of occurrences in Admiralty Bay. It is worth noting that this species was first described from samples collected through holes in pack ice during both the *Belgica* expedition (Müller, 1906) and the first *Discovery* expedition (Brady, 1907—identified as *C. inominata*). Plots of its geographical distribution based on all published data (Blachowiak-Samolyk & Angel, 2004) show that its range is closely associated with pack ice. Previous studies on distribution of *A. hettacra* (Siegel et al., 1992; Blachowiak-Samolyk & Zmijewska, 1995) have shown that it has a distinct preference for open oceanic waters, so its occurrence in rather low abundances in the shelf area of Admiralty Bay may be the result of advection of oceanic waters into the Bay. The higher frequency of *M. isocheira*, relative to *A. hettacra*, may result from its lack of any clear preference for a particular water mass (Deevey, 1982; Kock, 1993; Blachowiak-Samolyk, 1999). There are only about half the number of published records of *M. isocheira* compared to *A. hettacra* (see charts of Blachowiak-Samolyk & Angel, 2004) probably because of its much smaller size.

Only the upper 400-m of the water column could be sampled in Admiralty Bay, and this may have contributed to the three species' relatively low abundances and frequencies since they all have marked preferences for deeper water (Deevey, 1978, 1982). Kock (1992), sampling from 300–0 m, reported that the Ostracoda were inhabiting the lower epi- and upper mesopelagial. Whereas studies in the Bransfield Strait (Blachowiak-Samolyk, 1999) and the Croker Passage (Blachowiak-Samolyk & Zmijewska, 1997) have shown that all three species prefer mesopelagic depths (down to about 700 m).

A common characteristic of the bathymetric ranges of Ostracoda at high latitudes is their absence in the surface waters (Angel & Fasham, 1975, Moguilevsky & Angel, 1975). The rarity of Ostracoda at the two shallowest stations C and H in Admiralty Bay supports these observations. However, halocyprids do occur quite abundantly in the epipelagial and neuston at low latitudes (e.g. Moguilevsky & Angel, 1975).

Little attention has been paid to the life cycles of halocyprid ostracods in the Southern Ocean, because of a paucity of year-round sampling from a given area. Indeed very few such studies have been carried out anywhere. Ikeda & Imamura (1992) and Ikeda (1990, 1992) investigated the life cycle of a mesopelagic species *Conchoecia pseudodiscophora* in the Sea of Japan. Hartmann

(1968), Hillman (1969), Kock (1992), Blachowiak-Samolyk (1999, 2001) and Blachowiak-Samolyk & Osowiecki (2002) studied their developmental patterns in the Southern Ocean. Kock's (1992) study was the most thorough, and he based his interpretations on analyses of the ontogenetic migrations of Antarctic euphausiids, chaetognaths and copepods. He developed a model of life cycles of *Alacia belgicae* and *A. hettacra*, based on samples collected mainly from 200–0 m. The net he used had a relatively large mesh size (320 μm) so his samples did not collect the youngest instars (A-6, A-5). He observed that the mature stages, particularly adult females, were more abundant over deeper water. He hypothesized that, like many other high latitude species, the ostracods undertake both ontogenetic and seasonal vertical migrations. Angel (1977, 1979) observed the bathymetric ranges of adult females of several species North Atlantic halocyprids extended far deeper than other stages and postulated this could be the result of ontogenetic migrations.

Kock's (1992) model linked the distributions he observed with circulation in the vicinity of the Antarctic shelf. The maturing instars of *A. belgicae* and *A. hettacra* migrate up into shallower water in spring, where they are advected offshore by the general drift of the Antarctic Surface Waters. In autumn the process is reversed; the animals migrate down into deep water to overwinter as juvenile instars, where they are advected onshore in the deep circulation. In autumn in the Croker Passage the younger instars of *A. belgicae* and *M. isocheira* are almost entirely restricted to the deepest layers confirming the ontogenetic migrations of these species (Blachowiak-Samolyk, 2001). However, Kock's model may prove to be simplistic because it fails to take into account the presence of *A. belgicae* at shallow depths (<30 m) under pack-ice in winter (Müller, 1906; Brady, 1907).

The year-round age structure of *A. belgicae* populations in Admiralty Bay suggests it breeds throughout the year, hence the predominance of the younger instars during all seasons, except summer. During the autumn *A. belgicae* populations are dominated by younger instars both in Admiralty Bay (at station B - this study) and in the Croker Passage (Blachowiak-Samolyk, 2001). In winter the younger instars of *A. belgicae* continued to dominate in both areas. However the percentage of adults increased in the open waters (Blachowiak-Samolyk, 2001), but remained low in the shelf area (this study). In spring, at station A the population structure of *A. belgicae* was similar to that in winter, but the frequency of adults increased. This is consistent with earlier observations in the centre of the Bransfield Strait in spring 1986 (Blachowiak-Samolyk, 1999), and also near the Antarctic Peninsula in spring 1987 (Kock, 1992). The previous conclusion that *A. belgicae* reproduces year-round was based on observations in the open waters of the Croker Passage (Blachowiak-Samolyk, 2001), and is confirmed for the shelf zone of Admiralty Bay by the constant relatively low "mean population stage" indices observed throughout the year (Table 1 and Fig. 5) as well as by clustering spring-summer population of *A. belgicae* together with autumn-winter populations of all three analysed species (Fig. 3).

Zmijewska (1993b) concluded that the omnivorous copepod *Metridia gerlachei* also reproduces continuously, and its trophic flexibility (Hopkins, 1985b) enables it to maintain large populations in Antarctic waters. Hopkins showed that in the Croker Passage (1985b) the ostracods are mostly omnivorous, detritus-feeding animals making them independent of any single food source.

Relatively little is known about the population structure of *A. hettacra*. Throughout the year its age structure is dominated by the older instars confirming the observations of Kock (1993) in the 1989/1990 summer near the Antarctic Peninsula and in the Scotia Sea in the upper 300 m layer (Blachowiak-Samolyk & Osowiecki, 2002). As the distributional charts for this species suggest (Blachowiak-Samolyk & Angel, 2004), *A. hettacra* occurs throughout the oceanic regions of the Southern Ocean. So its presence in the coastal waters of Admiralty Bay probably results from, and hence indicates, advection of open ocean waters into the Bay (c.f. Pruszak, 1980). Thus its absence from station B in the spring may indicate there had been no recent influx of oceanic water into the Bay. Kock's (1992) model also implies that its seasonal absence from Admiralty Bay may result from the interaction between its life cycle and exchanges of water between the Bay and the open ocean. Box Whiskers plots (Fig. 4) also imply that *A. hettacra* might be influenced by Admiralty Bay (local) circulation more than the other two species. The high "mean population stage" of 6.3 observed in summer suggests that it was breeding at that time and that its life-cycle is essentially annual. A similar population index of 6.2 was observed both in the vicinity of Admiralty Bay in summer 1988/1989 (Blachowiak-Samolyk & Osowiecki, 2002) and near the Antarctic Peninsula in summer 1987 (Kock, 1992). However, in the same season in the Scotia Sea its index was 4.9 (Blachowiak-Samolyk & Osowiecki, 2002). Possibly maturation is slower in the open oceanic waters compared to shelf regions. If such a delay does occur, it could lead to separation of the breeding populations and hence become a mechanism for speciation. The lowest values of *A. hettacra* population indices (Table 1) were observed in autumn in Admiralty Bay, indicating recruitment by a new generation spawned in the summer. However, until information of the developmental rates for the various instars is available (c.f. those obtained by laboratory studies for *D. pseudodiscopora* by Ikeda (1992)), it will be difficult to interpret these data.

The populations of *M. isocheira* in Admiralty Bay were dominated by adults throughout the year. The dominance of adults in summer was in marked contrast to the two *Alacia* species. Similar population structures were observed in the upper 300 m of the Scotia Sea in summer 1988/1989 (Blachowiak-Samolyk & Osowiecki, 2002) and near the Antarctic Peninsula (Kock, 1992). In summer in the open waters of the Croker Passage, adults still dominated the populations but juvenile instars were also present (Blachowiak-Samolyk, 2001). The occurrence of A-4 and A-3 instars in autumn-winter in Admiralty Bay, suggests the species is reproducing there in summer. Along the Antarctic Peninsula Kock (1992) observed that adults were predominant at 500–300 m, but that stage A-2 were dominant at 700–500 m, implying a pattern of ontogenetic migration. The dominance of adult *M. isocheira* during winter in Admiralty Bay was consistent with wintertime observations at the 425–175 m depth interval in the Croker Passage (Blachowiak-Samolyk, 2001). Adults and A-1 individuals of *M. isocheira* reached their highest proportions in Admiralty Bay. In the spring of 1986 A-5 instars of *M. isocheira* were predominant at the centre of the Bransfield Strait (Blachowiak-Samolyk, 1999); at that time the stratified sampling showed that the mature component of the *M. isocheira* populations was at 1000–500 m, and the juvenile instars were in deeper waters (Blachowiak-Samolyk, 1999). Kock (1992) made similar observations off the Antarctic Peninsula. The mean population stage indices for *M. isocheira* in Admiralty Bay were highest in summer, implying that reproduction began during this season. The indices were similar in autumn and winter in Admiralty Bay, so either development is very slow or it had ceased possibly because the halocyprids entered a dormant state analogous to diapause in copepods. The mean population stage analysis implies that *M. isocheira* completes its life cycle within one year, and onset of its reproduction is delayed until summer in the Bay, whereas in the open waters of the Croker Passage it is initiated in spring (Blachowiak-Samolyk, 2001). The sequence of reproduction seems to be confirmed also by higher average $\overline{S}$ indices for *M. isocheira* at the shelf stations in comparison with open waters (Fig. 5).

Statistical analysis (the nonparametric Mann–Whitney *U*-test) demonstrates that the three halocyprids populations' indices differ significantly between the inner region (station A) and outlet (station B) of Admiralty Bay. On the other hand, Box Whiskers plots (Fig. 4) illustrate similar trends in the *A. belgicae* and *M. isocheira* populations (older populations at station A), contrary to *A. hettacra*.

Conclusions

1 The population structures of the three halocyprid species *A. belgicae*, *A. hettacra* and *M. isocheira* in Admiralty Bay differ between the two sites.
2 *Alacia belgicae* breeds continuously throughout the year, and its longevity probably exceeds one year.

3 *Alacia hettacra* and *M. isocheira* have similar life cycles that are most likely completed within one year, but reproduction of *A. hettacra* seems to be initiated in autumn, earlier than in *M. isocheira*.
4 Local factors play a role in breeding. At station A reproduction is initiated earlier in *A. belgicae* and *M. isocheira* than at station B, but advection also plays a role.
5 The seasonal absence of *A. hettacra* in spring in the outlet area towards the Bransfield Strait, and in summer in the central part of Admiralty Bay, implies that it is an expatriate species in shelf water.
6 Population development indices $\overline{S}$ indicate that in the shelf area of Admiralty Bay *M. isocheira* populations are older than in the open waters of the Croker Passage, whereas *A. belgicae* populations are similar (i.e. independent of locality).

Acknowledgements The authors would like to thank Prof. W. Kittel from Lodz University and Prof. M.I. Zmijewska from the University of Gdansk for their sharing plankton samples. Ph.D. student Monika Kedra carried out the original map for us. Special thanks are given to Dr. B. Urban-Malinga from the Sea Fisheries Institute in Gdynia for her very helpful statistical advice.

References

Angel, M. V., 1977. Studies on Atlantic Halocyprid Ostracods: vertical distributions of the species in the top 1000 m in the vicinity of 44° N, 13° W. Journal of the Marine Biological Association of the United Kingdom 57: 239–252.

Angel, M. V., 1979. Studies on Atlantic halocyprid ostracods: their vertical distributions and community structure in the central gyre region along latitude 30°N off Africa to Bermuda. Progress in Oceanography 8: 3–124.

Angel, M. V., 1981. Ostracoda. In Boltovskoy, D. (eds), Atlas del zooplankton del Atlantico suboccidental y metodos de trabajo con el zooplancton marino, Mar del Plata. INIDEP, Argentina, 543–585.

Angel M. V. & M. J. R. Fasham, 1975. Analysis of the vertical and geographic distribution of the abundant species of planktonic Ostracods in the North-East Atlantic. Journal of the Marine Biological Association of the United Kingdom 55: 709–737.

Baar de, H. J. W. & P. W. Boyd, 2000. The role of iron in plankton ecology, carbon dioxide transfer of the global oceans. In Ducklow, H. W., J. G. Field & R. Hanson (eds) The changing ocean carbon cycle: a mid-term synthesis of the joint global ocean flux study. Cambridge University Press, Cambridge, 61–140.

Blachowiak-Samolyk, K., 1999. Distribution and population structure of pelagic Ostracoda near the Antarctic Peninsula in spring 1986 (BIOMASS III, October-November 1986). Polish Archives of Hydrobiology 46: 9–25.

Blachowiak-Samolyk, K., 2001. Comparative studies on population structures of *Alacia belgicae* and *Metaconchoecia isocheira* (Ostracoda) in the Croker Passage during various seasons. Polar Biology 24: 222–230.

Blachowiak-Samolyk, K. & M. V. Angel, 2004. http://ocean.iopan.gda.pl/ostracoda.

Blachowiak-Samolyk, K. & A. Osowiecki, 2002. Distribution and population structure of pelagic Ostracoda near the sea-ice edge in the Scotia Sea and the King George Island (December 1988-January 1989). Polish Polar Research 23: 135–152.

Blachowiak-Samolyk, K. & M. I. Zmijewska, 1995. Horizontal and vertical distribution of Ostracoda in the Drake Passage and Bransfield Strait (BIOMASS-SIBEX, December 1983-January 1984). Polish Polar Research 16: 41–54.

Blachowiak-Samolyk, K & M. I. Zmijewska, 1997. Planktonic Ostracoda in Croker Passage (Antarctic Peninsula) during two austral seasons: summer 1985/1986 and winter 1989. Polish Polar Research 18: 79–87.

Boden, B. P., 1985. The South African SIBEX I Cruise to the Prydz Bay region, 1984: VIII. The plankton encountered during the survey. South African Journal of Antarctic Research 15: 28–33.

Boden, B. P. & L. D. Parker, 1986. The plankton of the Prince Edward Islands. Polar Biology 5: 81–93.

Brady, G. S., 1907. Crustacea. V- Ostracoda. Natural History Report of the British Antarctic Discovery Expedition 3: 1–9.

Chavtur, V. G. & N. V. Kruk, 2003. Latitudinal distribution of pelagic ostracods (Halocyprididae) in the Australian-New Zealand sector of the Southern Ocean. Biologiya Morya, Vladivostok 29: 166–172.

Chavtur, V. G. & M. V. Angel. Tribe Metaconchoecini: a systematic revision of pelagic ostracods (Ostracoda: Halocyprididae) formerly in the genus Metaconchoecia and the description of a new species. Zootaxa (in press).

Chojnacki, J. & T. Weglenska, 1984. Periodicity of composition, abundance and vertical distribution of summer zooplankton (1977/1978) in Ezcurra Inlet, Admiralty Bay (King George Island). Journal of Plankton Research 6: 997–1017.

Clarke, K. R., 1993. Non-parametric multivariate analysis of changes in community structure. Australian Journal of Ecology 18: 117–143.

Deevey, G. B., 1978. A taxonomic and distributional study of the planktonic Ostracods collected on three cruises of the Eltanin in the South Pacific and the Antarctic region of the South Pacific. In Kornicker, L.S. (ed.), Biology of the Antarctic Seas VIII. Antarctic Research Series 28: 43–70.

Deevey, G. B., 1982. A faunistic study of the planktonic Ostracoda (Myodocopa, Halocyprididae) collected on eleven cruises of the Eltanin between New Zealand, Australia, the Ross Sea, and the South Indian Ocean. In: Kornicker, L. S. (ed.), Biology of the Antarctic Seas X. Antarctic Research Series 32: 131–167.

Drapun, I. E., 2004. Distribution of pelagic ostracods (Halocyprididae) in the western part of the Bransfield Strait, Antarctica. Morski Ekologicznij Journal 3: 25–36. (in Russian).

Foster, B. A., 1987. Composition and abundance of zooplankton under the spring sea ice of McMurdo Sound, Antarctica. Polar Biology 8: 17–35.

Foster, B. A., 1989. Time and depth comparison of sub-ice zooplankton in McMurdo Sound, Antarctica. Polar Biology 9: 431–435.

Freire A. S., M. J. C. Coelho & S. L. C. Bonecker, 1993. Short term spatial-temporal distribution patterns of zooplankton in Admiralty Bay (Antarctica). Polar Biology 13: 433–439.

Fukuchi, M., A. Tanimura & H. Ohtsuka, 1985. Zooplankton community conditions under sea ice near Syowa Station, Antarctica. Bulletin of Marine Science 37: 518–528.

Hartmann, G., 1968. Ostracoda (3. Lieferung). In Gruner, H.-E. (ed.), Dr. H. G. Bronns Klassen und Ordnungen des Tierreichs. Band 5 (Arthropoda), Abteilung I (Crustacea), Buch 2, Teil IV (Ostracoda) Akademische Verlagsgesellschaft, Leipzig, 409–568.

Hillman, N. S., 1967. Ecology of Antarctic pelagic Ostracoda. Antarctic Journal of the United States 2: 199–200.

Hillman, N. S., 1969. Ontogenetic studies of Antarctic pelagic Ostracoda. Antarctic Journal of the United States 4: 189–190.

Hopkins, T. L., 1985a. The zooplankton community of Croker Passage, Antarctic Peninsula. Polar Biology 4: 161–170.

Hopkins, T. L., 1985b. Food web of an Antarctic midwater ecosystem. Marine Biology 89: 197–212.

Ikeda, T., 1990. Ecological and biological features of a mesopelagic ostracod, *Conchoecia pseudodiscophora*, in the Japan Sea. Marine Biology 107: 453–461.

Ikeda, T., 1992. Laboratory observations on spawning, fecundity and early development of a mesopelagic ostracod, *Conchoecia pseudodiscophora*, from the Japan Sea. Marine Biology 112: 313–318.

Ikeda, T. & A. Imamura, 1992. Population structure and life cycle of the mesopelagic ostracod, *Conchoecia pseudodiscophora*, in Toyama Bay, southern Japan Sea. Marine Biology 113: 595–601.

Kittel, W., 1980. Population studies on *Euphausia superba* Dana, 1852 (Euphausiacea, Crustacea) in waters of the Admiralty Bay during austral summer of 1978. Polish Archives of Hydrobiology 27: 267 272.

Kittel, W. & P. Presler, 1980. Morphology of the post-larval developmental stages of *Euphausia crystallorophias* Holt et Tattersall, 1906 (Euphausiacea, Crustacea). Polish Archives of Hydrobiology 27: 259–265.

Kittel, W., J. Sicinski, M. I. Zmijewska, L. Bielecka & K. Blachowiak-Samolyk, 2001. Antarctic neritic zooplankton community (Admiralty Bay, King George Island, South Shetlands Islands). Polish Polar Research 22: 11–33.

Kock, R., 1992. Ostracods in the epipelagial zone off the Antarctic Peninsula - a contribution to the systematics and to their distribution and population structure with regard to seasonality. Berichte zur Polarforschungen 106: 1–209.

Kock, R., 1993. Planktonic ostracods along the Antarctic Peninsula during the 1989/90 summer season. Polar Biology 13: 495–499.

Lipski, M., 1987 Variations of physical conditions and chlorophyll "*a*" contents in Admiralty Bay (King George Island, South Shetlands Islands, 1979). Polish Polar Research 4: 307–322.

Menshenina, L. & S. Rakusa-Suszczewski, 1992. Zooplankton changes during the year in Admiralty Bay (February 1990-January 1991). Polish Archives of Hydrobiology 39: 49–58.

Moguilevsky, A. & M. V. Angel, 1975. Halocyprid ostracods in Atlantic neuston. Marine Biology 32: 295–302.

Müller, G. W., 1906. In Expédition antarctique Belge. Résultats du voyage du S. Y. Belgica en 1897–1898–1899 sous le commandement de A. de Gerlache de Gomery. Rapports scientifiques publiés aux frais du Gouvernement Belge, sous le Direction de la Commission de la Belgica. Zoologie, Ostracoden: 1–7.

Müller G. W., 1912. Ostracoda. In Schulze, F. E. (ed.), Das Tierreich, 31. Lieferung. Friedländer & Sohn, Berlin.

Pruszak, Z., 1980. Currents circulation in the waters of Admiralty Bay (region of *Arctowski* station on King George Island). Polish Polar Research 1: 55–74.

Rakusa-Suszczewski, S. & R. Stepnik, 1980. Three species of krill from Admiralty Bay (King George Island, South Shetlands) in summer 1978/79. Polish Archives of Hydrobiology 27: 273–284.

Sars, G. O., 1928. An account of the Crustacea of Norway, with short descriptions and figures of all the species, 9, Ostracoda 1–277. Bergen Museum, Bergen.

Sicinski, J., O. Rozycki & W. Kittel, 1996. Zoobenthos and zooplankton of Herve Cove, King George Island, South Shetlands Islands, Antarctic. Polish Polar Research 17: 221–238.

Siegel, V., A. Skibowski & U. Harm, 1992. Community structure of the epipelagic zooplankton community under the sea-ice of the northern Weddell Sea. Polar Biology 12: 15–24.

Szafranski, R. & M. Lipski, 1982 Characteristic of water temperature and salinity at Admiralty Bay (King George Island, South Shetlands Islands, Antarctic) during the austral summer 1978/1979. Polish Polar Research 3: 7–24.

Zmijewska, M. I., 1993a. Copepoda. In Rakusa-Suszczewski, S. (ed.), The Maritime Antarctic Coastal Ecosystem of Admiralty Bay. Department of Antarctic Biology, Polish Academy of Sciences, Warsaw: 64–66.

Zmijewska, M. I., 1993b. Seasonal and spatial variations in the population structure and life histories of the Antarctic copepod species *Calanoides acutus*, *Calanus propinquus*, *Rhincalanus gigas*, *Metridia gerlachei* and *Euchaeta antarctica* (Calanoida) in Croker Passage (Antarctic Peninsula). Oceanology 35: 73–100.

Hydrobiologia (2007) 585:79–87
DOI 10.1007/s10750-007-0630-9

Water quality and diversity of the Recent ostracod fauna in lowland springs from Lombardy (northern Italy)

Valentina Pieri · Claudia Caserini · Stefano Gomarasca · Koen Martens · Giampaolo Rossetti

Abstract The Po river plain (northern Italy) is delimited to two mountain ridges, the Alps and the Apennines. It hosts peculiar lowland man-modified springs, locally known as "fontanili", which originate from natural resurgences occurring along the alluvial fans of the main watercourses, namely in the transition zone from the higher to lower plain which is characterized by changes in slope profile and sediment granulometry. These habitats usually show low variation in hydrological, hydrochemical and thermal conditions throughout the year. Twenty-eight springs, located in the provinces of Lodi and Cremona (Lombardy) in the alpine sub-catchment of the Po river were sampled in summer and autumn of 2004. Twenty-three of them were typical alluvial-fan springs, while the remaining five were terrace springs. The two groups of springs showed marked differences in their hydrochemical and hydrological characteristics. Sixteen ostracod species in three families (Candonidae, Ilyocyprididae, and Cyprididae) were identified. The most frequent species were *Cypria ophtalmica* (19 sites), *Herpetocypris reptans* (16), and *Prionocypris zenkeri* (13). Five species were found only once: *Chlamydotheca incisa*, *Scottia pseudobrowniana*, *Pseudocandona compressa* and *Candona neglecta*. Up to 6 taxa were recorded from a single site and the average number of taxa for each site was c. 3. The associations among ostracod taxa and their occurrence in relation to environmental factors were examined. Finally, the results of this survey were compared with a similar study previously conducted in 31 alluvial fan springs of the apennine sub-catchment of the Po river.

Keywords Lowland springs · Italy · Water chemistry · Freshwater ostracods · Taxonomy · Distribution

Guest editors: R. Matzke-Karasz, K. Martens & M. Schudack
Ostracodology – Linking Bio- and Geosciences

Electronic supplementary material The online version of this article (doi:10.1007/s10750-007-0630-9) and accessible for authorized users.

V. Pieri · C. Caserini · G. Rossetti (✉)
Department of Environmental Sciences, University of Parma, Parco Area delle Scienze 33A, 43100 Parma, Italy
e-mail: giampaolo.rossetti@unipr.it

S. Gomarasca
Department of Biology, University of Milano, Via Celoria 26, 20133 Milan, Italy

K. Martens
Freshwater Biology, Royal Belgian Institute of Natural Sciences, Vautierstraat 29, 1000 Brussels, Belgium

K. Martens
Department of Biology, University of Ghent, K.L. Ledeganckstraat 35, 9000 Ghent, Belgium

Introduction

The "fontanili" are man-modified springs characteristic of the lowland plains of northern Italy (Minelli, 2001). These springs typically occur along lines corresponding to the alluvial or paleoalluvial fans of rivers flowing from the Alps and the northern Apennines, at the boundary between higher and lower plains. The change in the valley slope profile favours the deposition of clay and sand layers; the presence of these less permeable sediments forces the groundwater coming from deep aquifers upwards. To facilitate the water outflow, small basins are artificially excavated and perforated metal tubes are usually inserted into the ground; in addition, periodic removal of bottom sediment and plant biomasses is required to guarantee the hydraulic effectiveness of springs. Another type of springs is also common in this area. They occurr in eroded river terraces and are in direct contact with the underlying water table.

Previously, these springs were used as sources of drinkable water and for crop irrigation. They were properly managed and gave origin to an important network formed by a vast system of lentic waters, streams and channels which sustained a rich freshwater biodiversity (Bracco et al., 2001). Changes in the agricultural practices and in the captation methods, associated with the expansion of transportation infrastructures and economic activities, have progressively led to the disappearance of many springs and a reduction in the ecological connectivity between lowland aquatic habitats, with negative consequences on their biological communities. In addition, pollution of agricultural origin is a growing concern for spring water quality. Nevertheless, only in few cases, protection areas have been established to safeguard springs of particular conservation interest.

Here, we report on the results of a monitoring programme on water quality and ostracod fauna of 23 alluvial-fan and five terrace springs in the Po river plain of Lombardy (northern Italy). The major objectives were to analyse the associations among ostracod taxa and to assess possible relationships between their occurrence and environmental parameters. The data are compared with those obtained from a previous investigation carried out in 31 springs located south of the present study area (Rossetti et al., 2005).

Materials and methods

Study area

All the springs analysed in the present study are located in the provinces of Lodi and Cremona (Fig. 1). The sites were selected on the basis of a cartographic search using both historical cadastral records (1904) and the 1:10000 scale Technical Regional Maps produced by the Lombardy Region in 1999. The actual presence of active springs was then verified during field surveys. A

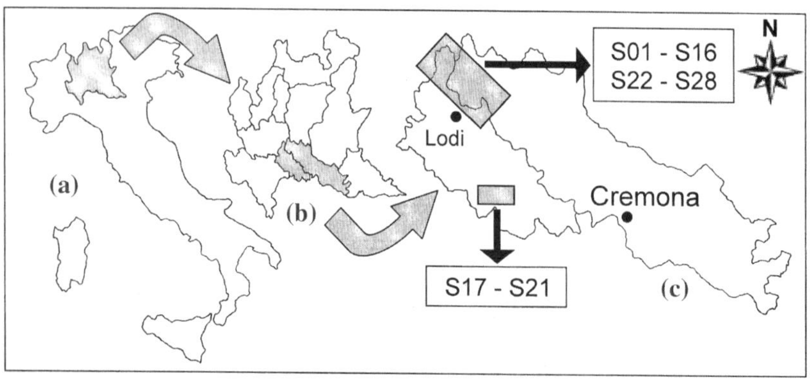

Fig. 1 Maps of Italy (**a**), Lombardy (**b**) and the provinces of Lodi and Cremona (**c**). Location of sampling sites is also shown (see Table 1 for codes)

total of 28 springs, considered to be representative for the area, were eventually included in this study (Table 1); 23 of them are typical alluvial-fan springs (S01–S16, S22-S28), and the remaining five are terrace springs (S17–S21). Nine of the studied springs (S12, S17–S20, S22–S25) are subjected to protection measures under different conservation programmes.

In the study area, alluvial-fan springs are clustered in the northern part, while the terrace springs occur in the southern part along the Adda river, one of the major tributaries of the Po river.

Alluvial-fan springs are fed by deep confined aquifers of alluvial valleys which guarantee a relatively constant flow, transparent waters and little seasonal fluctuation in water temperature. In the discharging part, they can either be small ponds or heads of channels with several abstraction points; the water depth rarely exceeds 1 m. Different springs may join together to form streams with large water flows. Terrace springs, on the other hand, are supplied by groundwater intersecting the land surface in eroded river banks; aquifers are mainly recharged by the areal precipitation and are particularly susceptible to contamination by percolation of polluted surface waters. These springs appear as small channels or ditches, and their outflows are usually modest and variable throughout the year.

The resurgence points have no direct connection to other surface waters, except for S05 which receives waters from an adjacent river (Addetta) through a small drain. S23 has concrete banks, while all the other springs have vegetated banks.

Sampling and analytical methods

Each spring was sampled twice, in summer (late June-early September) and autumn (November) 2004. Water temperature was measured in situ using a digital thermometer (ILC). Water samples were collected close to the discharge points and

Table 1 Geographic location of the studied springs

Spring	Code	Lat N	Long E	Municipality	Province
Quattro Ponti	S01	45°27′26″	09°24′56″	Lavagna	Lodi
Sorgente a T	S02	45°27′34″	09°25′45″	Lavagna	Lodi
Cavo Marocco	S03	45°27′01″	09°25′48″	Lavagna	Lodi
Casolate	S04	45°23′51″	09°26′41″	Zelo Buon Persico	Lodi
Lavagna	S05	45°27′53″	09°27′17″	Lavagna	Lodi
El Rio	S06	45°21′39″	09°30′33″	Boffalora d'Adda	Lodi
Fontana 2	S07	45°21′17″	09°30′16″	Boffalora d'Adda	Lodi
Fontana 1	S08	45°21′20″	09°30′14″	Boffalora d'Adda	Lodi
Fontana 0	S09	45°21′23″	09°30′13″	Boffalora d'Adda	Lodi
Galluppina	S10	45°21′29″	09°29′57″	Boffalora d'Adda	Lodi
Comazzo	S11	45°26′53″	09°28′26″	Comazzo	Lodi
Merlo Giovane	S12	45°26′22″	09°30′43″	Rivolta d'Adda	Cremona
Falconetta nord	S13	45°25′47″	09°30′06″	Rivolta d'Adda	Cremona
Nord Cascina Maleo	S14	45°25′55″	09°30′06″	Rivolta d'Adda	Cremona
Falconetta 2	S15	45°25′53″	09°30′02″	Rivolta d'Adda	Cremona
Falconetta sud	S16	45°25′33″	09°30′17″	Rivolta d'Adda	Cremona
Fontanile 1	S17	45°08′46″	09°38′56″	Somaglia	Lodi
Fontanile 2	S18	45°08′49″	09°39′23″	Somaglia	Lodi
Fontanile 3	S19	45°08′49″	09°39′33″	Somaglia	Lodi
Fontanile 4	S20	45°08′49″	09°39′33″	Somaglia	Lodi
Madonna del Fontanone	S21	45°09′32″	09°35′29″	Ospedaletto Lodigiano	Lodi
Francavalla	S22	45°21′12″	09°31′55″	Dovera	Cremona
Alipranda	S23	45°21′52″	09°31′16″	Dovera	Cremona
Cimitero	S24	45°22′20″	09°32′23″	Dovera	Cremona
El Rì	S25	45°22′58″	09°31′41″	Dovera	Cremona
Mozzanica	S26	45°22′51″	09°28′40″	Spino d'Adda	Cremona
Fontanina	S27	45°23′06″	09°29′25″	Spino d'Adda	Cremona
Meraviglia	S28	45°20′33″	09°31′57″	Lodi	Lodi

were kept refrigerated until analysed in the laboratory. Hydrochemical and physical variables and parameters were determined as follows: pH by potentiometry (TIM 90, Radiometer); total alkalinity by potentiometric end-point titration at pH 4.5 and 4.2 (TIM 90, Radiometer) and linearization according to Rodier (1978); electric conductivity at 20°C by conductometry (CDM 83, Radiometer); chemical oxygen demand by titration with ammonium iron(II) sulphate (A.P.H.A., 1998); nitrate (Rodier, 1978), ammonia (A.P.H.A., 1998), nitrite (A.P.H.A., 1998), soluble reactive phosphorus (Valderrama, 1981), dissolved reactive silica (Golterman et al., 1978), and chlorophyll-*a* (Golterman et al., 1978) by spectrophotometry (Beckman DU 65). The full data set of hydrochemical data is available as online appendix to the article.

Ostracod samples were collected using a 250 μm net equipped with a long handle pulled close to the sediment and swept through the vegetation. Living samples were transferred to the laboratory, where specimens were sorted under a binocular microscope and then fixed in ethanol. The specific allocation of the collected material was based on adult specimens; in few cases (*Herpetocypris* sp. and *Pseudocandona* sp.) the specific identification remained uncertain due to the presence of immature stages only. Both soft parts (dissected in glycerine and stored in sealed slides) and valves (stored dry in micropal slides) were checked for species identification, using Fox (1965) for *Chlamydotheca incisa* and Meisch (2000) for the remaining species.

Data analysis

The ordination of abiotic and biotic data was carried out through Principal Component Analysis (PCA) and Canonical Correspondence Analysis (CCA) using the statistical package CANOCO version 4.5 (ter Braak & Šmilauer, 2002), and Cluster Analysis (CA) using PAST ver. 1.06 (Hammer et al., 2001). PCA was performed on the correlation matrix of measured physical and chemical water variables after log (x + 1) transformation (except for pH) to normalize the distributions. The concentration of dissolved inorganic nitrogen (DIN) was the sum of nitrate, nitrite, and ammonia. The same matrix and the ostracod absence/presence data were used to assess possible relationships between ostracod distribution and environmental data by CCA. Monte Carlo permutation tests (9999 permutations) were used to assess the significance of the canonical axes and of the environmental variables that were selected in the forward selection procedure. Similarity between species assemblages was assessed by Unweighted pair-group average CA using the Raup-Crick index (Raup & Crick, 1979) for absence/presence data with a Monte Carlo randomization procedure (200 replicates).

Results

Water temperature ranged between 14.4 and 20.0°C in summer and between 12.3°C and 19.0°C in autumn; the difference between summer and autumn values in each spring was usually less than 2°C, except for S04, S13, S14, and S27 where the differences were slightly larger (from 4.2°C to 4.6°C). Conductivity generally varied between 300 μS cm^{-1} and 500 μS cm^{-1}, although in the southern springs values were rather higher especially in autumn, with a maximum of 908 μS cm^{-1} recorded in S17. The lowest values were measured in S05 (206 and 238 μS cm^{-1} in summer and in autumn, respectively), indicating a dilution effect due to the inflow of waters from the river Addetta. In the study area pH values showed a relatively low variability in both space and time, ranging from 6.85 (S13) to 7.47 (S02) in summer and from 6.80 (S27) to 7.25 (S05) in autumn. Also total alkalinity values were typical of well-buffered waters, with a minimum of 2.06 meq l^{-1} in S05 (summer) and a maximum of 4.80 meq l^{-1} in S20 (autumn); marked seasonal differences were chiefly observed in the terrace springs, with higher values in autumn. Soluble reactive phosphorus content was highly variable among sites; higher concentrations were generally measured in summer (up to a maximum of 38 μg l^{-1} in S06); in autumn, values were under the detection limit of 5 μg l^{-1} in 19 out of 28 springs, and a maximum of 30 μg l^{-1} was found in S06. Even though there was a considerable variability between sites, nitrate was the most abundant form of dissolved inorganic nitrogen in all springs; concentrations were ≤11 mg l^{-1}, except in S21 with 21.1 mg l^{-1} in summer and 22.2 mg l^{-1} in autumn. Nitrite concentrations ranged between below the detection limit (5 µg l^{-1}) and 16 µg l^{-1}, with two isolated autumn peaks of 55 and 27 µg l^{-1} in S17 and S20. Ammonia, with few exceptions, showed higher concentrations in summer samples (up to a maximum of 171 µg l^{-1} in S23), while in autumn values remained below 12 µg l^{-1} in all springs except for S04 (31 µg l^{-1}) and S17 (43 µg l^{-1}). Dissolved reactive silica had concentrations comprised between 0.6 and 9.2 mg l^{-1}; the highest values were observed in the terrace springs. In most of the sites, chemical oxygen demand concentrations were ≤20 mg l^{-1}; values ≥40 mg l^{-1} in at least one sample were measured in S01, S03, and S13. Chlorophyll-*a* concentrations, apart from S10 (6.2 µg l^{-1} in autumn), were always ≤2.2 µg l^{-1}.

PCA was used to summarize variation patterns in physical and chemical characteristics of waters among the studied springs (Fig. 2). The first two axes account for 72,7% of total cumulative variance. Total alkalinity and conductivity have the highest correlation with the first principal component ($r = -0.87$ and $r = -0.81$), soluble reactive phosphorus and chemical oxygen demand with the second principal component ($r = 0.67$ and $r = -0.59$, respectively). Most of the alluvial-fan spring samples are grouped around the origin of the axes and in general no clear seasonal pattern is evident. Also summer samples from terrace springs tend to partially overlap with this main cluster, while samples collected in autumn, characterized by more mineralised waters and higher concentrations of dissolved inorganic nitrogen, are restricted to the upper left quadrant. A third group of samples is located in an opposite position, consisting of the samples collected in S05 and S15 (summer and autumn) and S13 (only autumn) which have a low-ionic content. Finally, the summer sample from S16 occupies a rather isolated position in the upper right quadrant due to its low conductivity and total alkalinity values associated with high soluble reactive phosphorus concentration (Fig. 2).

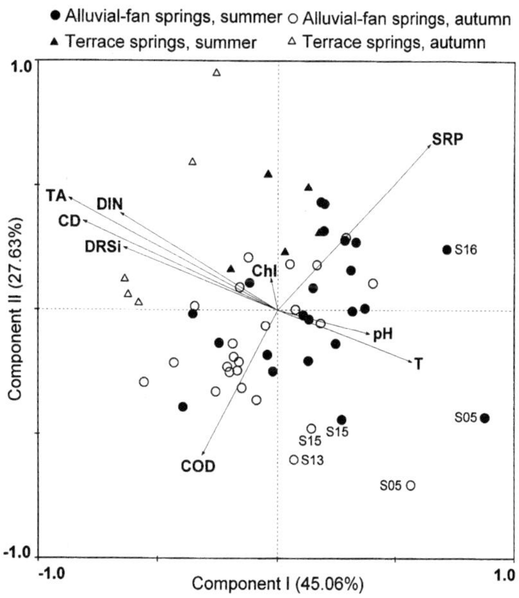

Fig. 2 PCA biplot of springs and environmental variables in the plane defined by the first two axes. To make the graph more readible, only six samples are labelled. T: temperature; CD: conductivity; TA: total alkalinity; SRP: soluble reactive phosphorus; DIN: dissolved inorganic nitrogen; DRSi: dissolved reactive silica; COD: chemical oxygen demand; Chl: chlorophyll *a*

Sixteen ostracod taxa in three families (Candonidae, Ilyocyprididae, and Cyprididae) were identified (Table 2). The most frequent species were *Cypria ophtalmica* (19 sites), followed by *Herpetocypris reptans* (16), and *Prionocypris zenkeri* (13). Four taxa (*Scottia pseudobrowniana*, *Pseudocandona compressa*, *Candona neglecta* and *Herpetocypris* sp.) were recorded in one sampling site only. Nine taxa were present in both seasons, while *Ilyocypris bradyi*, *Chlamydotheca incisa*, *Heterocycpris reptans*, *Pseudocandona compressa*, and *Candona neglecta* were collected only in summer, *Herpetocypris* sp. and *Pseudocandona* sp. only in autumn. A maximum of six taxa were recorded from a single site (S12), while in three springs (S01, S18, and S20) only one species was found; the average number of species per site was 3.32 (Fig. 3). The highest ostracod diversity was recorded in the Comazzo area, where eleven species were identified (Tables 1 and 2).

In the CCA, the Monte Carlo permutation test showed that all the canonical axes are not

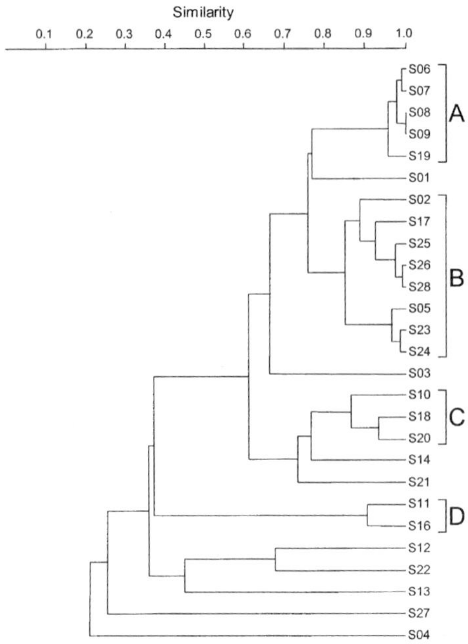

	Pzenk	Copht	Ibrad	Iiner	Cinci	Herpr	Herpb	Cvidu	Claev	Hetre	Spseu	Pcomp	Cnegl	Ccand	Psspp	Hersp	N
S06	1	1	0	0	0	1	0	0	1	0	0	0	0	0	0	0	4
S07	1	1	0	0	0	1	0	0	1	0	0	0	0	0	0	0	4
S08	1	1	0	1	0	1	0	0	1	0	0	0	0	0	0	0	5
S09	1	1	0	1	0	1	0	0	1	0	0	0	0	0	0	0	5
S19	0	1	0	0	0	1	0	0	1	0	0	0	0	0	0	0	3
S01	0	0	0	0	0	1	0	0	0	0	0	0	0	0	0	0	1
S02	1	1	0	0	0	1	0	0	0	0	0	0	0	1	0	0	4
S17	1	1	0	0	0	0	0	1	0	0	0	0	0	0	0	0	3
S25	1	1	0	0	0	1	0	0	0	0	0	0	0	0	0	0	3
S26	1	1	0	0	0	1	0	1	0	0	0	0	0	0	0	0	4
S28	1	1	0	0	0	1	0	1	0	0	0	0	0	0	0	0	4
S05	0	1	0	0	0	1	0	1	0	0	0	0	0	0	1	0	4
S23	0	1	0	0	0	1	0	1	0	0	0	0	0	0	0	0	3
S24	0	1	0	0	0	1	0	1	0	0	0	0	0	0	0	0	3
S03	0	1	1	1	0	1	0	0	0	0	0	0	0	0	0	0	4
S10	0	1	0	0	0	0	1	0	0	0	0	0	0	0	0	0	2
S18	0	1	0	0	0	0	0	0	0	0	0	0	0	0	0	0	1
S20	0	1	0	0	0	0	0	0	0	0	0	0	0	0	0	0	1
S14	0	1	0	0	0	0	0	1	1	0	0	0	0	0	0	0	3
S21	0	1	0	1	0	0	1	0	1	0	1	0	0	0	0	0	5
S11	1	0	0	1	0	0	0	0	1	0	0	0	0	0	0	1	4
S16	0	0	0	1	0	0	0	0	1	0	0	0	0	0	0	0	2
S12	1	0	0	0	0	1	1	1	1	1	0	0	0	0	0	0	6
S22	1	0	0	0	0	0	1	1	0	0	0	1	0	0	0	0	4
S13	0	0	0	0	0	0	0	1	0	0	0	0	0	1	0	0	2
S27	0	0	0	0	0	1	1	0	0	1	0	0	1	0	0	0	4
S04	1	0	1	0	1	0	0	1	0	0	0	0	0	0	1	0	5
Σ	13	19	2	6	1	16	5	11	10	2	1	1	1	2	2	1	

Fig. 3 Dendrogram obtained from the CA showing the similarity level between ostracod communities. Presence and absence (indicated as 1 and 0, respectively) of ostracod taxa in each spring is also indicated. Pzenk: *Prionocypris zenkeri*; Copht: *Cypria ophtalmica*; Ibrad: *Ilyocypris bradyi*; Iiner: *Ilyocypris inermis*; Cinci: *Chlamydotheca incisa*; Herpr: *Herpetocypris reptans*; Herpb: *Herpetocypris brevicaudata*; Cvidu: *Cypridopsis vidua*; Claev: *Cyclocypris laevis*; Hetre: *Heterocypris reptans*; Spseu: *Scottia pseudobrowniana*; Pcomp: *Pseudocandona compressa*; Cnegl: *Candona neglecta*; Ccand: *Candona candida*; Psesp: *Pseudocandona* sp.; Hersp: *Herpetocypris* sp.; Σ: number of total record for each taxon; N: number of taxa in each spring

significant ($P > 0.05$), indicating that the available data do not support a causal relationship between species occurrence and hydrochemical features.

CA was performed using the data matrix of presence/absence of taxa to examine patterns in ostracod assemblage composition. At a similarity value of approximately 0.8, four major groups were evidenced (Fig. 3). The first cluster (A) consists of five springs characterized by a relatively high number of taxa (3–5) and by the presence of *Cypria ophtalmica, Herpetocypris reptans* and *Cyclocypris laevis*. Four of these springs (S06, S07, S08 and S09) are located in the area of Boffalora d'Adda (Table 1). The second cluster (B) is formed by eight sites in which three or four taxa were found. Here, *Cypria ophthalmica* is alternatively associated with *Herpetocypris reptans, Cypridopsis vidua* and *Prionocypris zenkeri*, i.e., the species which are more common in the study area. Springs forming this cluster are spread across different geographic areas (Table 1); one of them (S17) is a terrace spring. The third (C) and fourth (D) cluster are each composed of a limited number of springs: cluster C includes two terrace springs (S18 and S20) and an alluvial fan spring (S10) which show a low-ostracod diversity; cluster D contains only two springs (S11 and S16) characterized by the simultaneous presence of *Ilyocypris inermis* and *Cyclocypris laevis*.

Discussion and conclusions

Compared to the total number of lowland springs included in historical cadastral registers, only few of them are still active in the study area. This accelerated loss is mainly due to inappropriate management of water bodies or even to their active destruction to meet the needs of large-scale agricultural practices. Another issue of concern is the water quality of the remaining springs.

Table 2 Comparison between ostracod taxa found in lowland springs of the southern (Rossetti et al., 2005) and the northern (this study) subcatchment of the Po river

	Rossetti et al., 2005	This study
Family Candonidae Kaufmann, 1900		
Candona candida (O.F. Müller, 1776)		X
Candona neglecta Sars, 1887		X
Pseudocandona sp.		X
Pseudocandona compressa (Koch,1838)	X[a]	X
Cypria ophtalmica (Jurine, 1820)	X	X
Cyclocypris ovum (Jurine, 1820)	X	
Cyclocypris laevis (O.F. Müller, 1776)	X	X
Family Ilyocyprididae Kaufmann, 1900		
Ilyocypris decipiens Masi, 1905	X	
Ilyocypris inermis Kaufmann, 1900	X	X
Ilyocypris bradyi Sars, 1890	X	X
Family Notodromadidae Kaufmann, 1900		
Notodromas persica Gurney, 1921	X	
Family Cyprididae Baird, 1845		
Prionocypris zenkeri (Chyzer & Toth, 1858)	X	X
Herpetocypris sp.		X
Herpetocypris brevicaudata Kaufmann, 1900	X	X
Herpetocypris reptans (Baird, 1835)		X
Heterocypris reptans (Kaufmann, 1900)		X
Scottia pseudobrowniana Kempf, 1971		X
Cypridopsis vidua (O.F. Müller, 1776)	X	X
Potamocypris fulva (Brady, 1868)	X	
Chlamydotheca incisa (Clauss, 1812)		X
Number of taxa	12	16

[a] *Pseudocandona* gr. *compressa*

Indeed, the results of this study reveal that waters are severely impacted by chemicals leaching into groundwater from the recharge zone or locally washed off from crop fields. In particular, high nitrate concentrations measured in several springs (with peaks over 20 mg l^{-1}) confirm the progressive contamination of aquifers in northern Italy mainly caused by the excessive use of fertilizers (Russo & Zavatti, 2001). Farming (including livestock) in lowlands is probably responsible for the high levels of chemical oxygen demand frequently observed in the studied springs. Obviously, springs situated in protected areas (usually restricted to a very limited surface) are also affected by these kinds of impact. Other variables reflecting the trophic state of spring waters (e.g., soluble reactive phosphorus and chlorophyll *a*) show, with few exceptions, low concentrations, most likely because of the development of dense stands of aquatic macrophytes and/or algal mats. In general, there was no clear pattern in seasonal variation of hydrochemical characteristics of alluvial-fan springs. In terrace areas, on the contrary, autumn samples showed notably higher conductivity and total alkalinity values than those collected in summer, as the superficial groundwater feeding the springs is directly influenced by the rainfall regime and by the agricultural cycles.

The studied sites usually host relatively simple ostracod communities, as it is expected in spring habitats (see the literature revised by Rossetti et al., 2005). Most of the species identified in this study are known for their high tolerance to different hydrochemical and ecological factors and have a wide geographic distribution or are cosmopolitan. Nevertheless, the records of *Scottia pseudobrowniana* in S21 and *Chlamydotheca incisa* in S04 are of particular interest. The subfamily Scottiinae consists of (semi-) terrestrial species; *Scottia* is the only genus within the tribe Scottiini and has a Holoarctic distribution (Martens et al., 2004); *S. pseudobrowniana* is the most common representative of the genus and, although not rare, it is known for a limited number of localities (Meisch, 2000). To our knowledge, in Italy this species was previously found only in a spring of the central-eastern Alps (Meisch, 2006), while it has never been reported for the Mediterranean area (Meisch, 2000). The specimens collected in S21 were very scarce and they were exclusively retrieved from the sediment of the spring channel within few meters from the resurgence point. All the attempts to recover animals from leaf litter and mosses (i.e., semi-terrestrial habitats) along the channel banks failed, as well as from other sampling points located downstream.

Chlamydotheca incisa, a South American species, has repeatedly been found in Italy in ricefields (Rossi et al., 2003) and in small water bodies of the Po river plain (Rossetti, unpubl.), although it has not been reported yet from other European countries (http://www.faunaeur.org). Its occurrence in contrasting habitats from several locations is a paradigmatic example of a species of "ospiti esteri" (*sensu* McKenzie & Moroni, 1986) which has successfully colonized the Italian waters.

The overall prevalence of euryecious ostracods in the studied springs probably accounts for the lack of strong relationships between species occurrence and physical and chemical variables. For example, apart from *Scottia pseudobrowniana*, all other taxa found in terrace springs were also present in alluvial-fan springs. No significant differences in species richness between protected and unprotected springs were observed. As evidenced by CA, similarities in community composition do not necessarily reflect the geographic proximity between springs, with the exception of the area of Boffalora d'Adda (Table 1, Fig. 2).

A previous study performed in 2001 in 31 alluvial-fan springs of the Apennine sub-catchment of the Po river (Rossetti et al., 2005) revealed the presence of 12 taxa in 9 genera; eight species and 7 genera were found in both areas (Table 2). In Lombardy, 5 species with ≥10 records were found, while in the other group of springs *Cypria ophtalmica* was overwhelmingly the most common species. The average number of taxa per site was 3.32 in the first case, and only 1.97 in the second case. Although such divergent figures can be partly due to a different sampling effort (the springs situated on the Apennine side were sampled only once), the comparison of the results obtained from the two studies seems to indicate that the springs of the province of Lodi and Cremona possibly support a more diversified ostracod fauna.

Acknowledgements Fabrizio Santantonio and Paola Rosa (Amministrazione Provinciale di Lodi), Fausto Cremascoli (Consorzio di Bonifica Muzza Basso Lodigiano, Lodi), and Pierluigi Viaroli (University of Parma) are acknowledged for their support in different stages of this study. Two referees are thanked for constructive suggestions and comments. Ostracod were partly identified during a visit of VP to KM founded by the EU project SYNTHESYS. Ricardo Pinto (University of São Paulo) provided information on the distribution of *Chlamydotheca incisa*. KM and GR acknowledge the support of the EU Project SEXASEX (MRTN-CT–2004-512492).

References

A.P.H.A., A.W.W.A., W.P.C.F., 1998. Standard methods for examination of water and wastewater. 20th edition, A.P.H.A., Washington.

Bracco, F., S. Paradisi, G. Sburlino & F. Stoch, 2001. Degrado e conservazione. In Minelli, A. (ed.), Risorgive e fontanili. Acque sorgenti di pianura dell'Italia Settentrionale Quaderni habitat, Ministero dell'Ambiente, Museo Friulano di Storia Naturale, Udine: 123–139.

Fox, H. M., 1965. Ostracod Crustacea from ricefields in Italy. Memorie dell'Istituto Italiano di Idrobiologia 18: 205–214.

Golterman, H. L., R. S. Clymo & M. A. M. Ohnstad, 1978. Methods for physical and chemical analysis of fresh water. I.B.P. Handbook No. 8, Blackwell, Oxford, 214 pp.

Hammer, Ø., D. A. T. Harper & P. D. Ryan, 2001. PAST: Paleontological Statistics Software Package for education and data analysis. Palaeontologia Electronica 4: 9.

Martens, K., P. De Deckker & G. Rossetti, 2004. On a new terrestrial genus and species of Scottiinae (Crustacea, Ostracoda) from Australia, with a discussion on the phylogeny and the zoogeography of the subfamily. Zoologischer Anzeiger 243: 21–36.

McKenzie, K. G. & A. Moroni, 1986. Man as an agent of crustacean passive dispersal via useful plants – exemplified by Ostracoda *ospiti esteri* of the Italian ricefields ecosystem – and implications arising therefrom. Journal of Crustacean Biology 6: 181–198.

Meisch, C., 2000. Freshwater Ostracoda of Western and Central Europe. Spektrum Akademischer Verlag, Heidelberg, 522 pp.

Meisch, C., 2006. Ostracoda (Muschelkrebse). In Sambugar, B., G. Dessi, A. Sapelza, A. Stenico, B. Thaler & A. Veneri (eds), Südtiroler Quellfauna. Biologisches Landeslabor. Autonome Provinz Bozen, Südtirol, Italien, 315–317.

Minelli, A. (ed.), 2001. Risorgive e fontanili. Acque sorgenti di pianura dell'Italia Settentrionale. Quaderni habitat, Ministero dell'Ambiente, Museo Friulano di Storia Naturale, Udine, 154 pp.

Raup, D. & R. E. Crick, 1979. Measurement of faunal similarity in paleontology. Journal of Paleontology 53: 1213–1227.

Rodier, J., 1978. L'analyse de l'eau. Dunod, Orleans, 1136 pp.

Rossetti, G., V. Pieri & K. Martens, 2005. Recent ostracods (Crustacea, Ostracoda) found in lowland springs of the provinces of Parma and Piacenza (Northern Italy). Hydrobiologia 542: 287–296.

Rossi, V., G. Benassi, M. Veneri, C. Bellavere, P. Menozzi, A. Moroni & K. G. McKenzie, 2003. Ostracoda of the Italian ricefields thirty years on: new synthesis and hypothesis. Journal of Limnology 62: 1–8.

Russo, E. & A. Zavatti (eds), 2001. Nitrati. Acqua e suolo da salvaguardare. Prevenzione dell'inquinamento dei sistemi idrogeologici. I quaderni di ARPA. Agenzia regionale prevenzione e ambiente dell'Emilia Romagna, 191 pp.

ter Braak, C. J. F. & P. Šmilauer, 2002. CANOCO Reference Manual and CanoDraw for Windows User's Guide: Software for Canonical Community Ordination (version 4.5). Microcomputer Power, Ithaca, 499 pp.

Valderrama, J. C., 1981. The simultaneous analysis of total nitrogen and total phosphorus in natural waters. Marine Chemistry 10: 109–122.

Hydrobiologia (2007) 585:89–98
DOI 10.1007/s10750-007-0631-8

OSTRACODA (ISO15)

Factors affecting spatial and temporal distribution of Ostracoda assemblages in different macrophyte habitats of a shallow lake (Lake Fehér, Hungary)

Anita Kiss

Abstract Spatial and temporal distribution, abundance and habitat preferences of ostracod assemblages were studied in various habitats of Lake Fehér in Hungary. Of the 20 species *Fabaeformiscandona hyalina* and *Cypridopsis hartwigi* are new to Hungary. Significant spatial differences were recorded in the composition and abundance of the Ostracoda assemblages in the different habitats. The most diverse and abundant assemblages were found in the reed-belt, however, the mean abundance of the species differed in the different macrophyte communities of the reed-belt. The species richness and abundance were considerably lower in disturbed sites than in the other habitats. Most of the 20 species showed distinct habitat preferences and ecological tolerances. Especially the presence and coverage of the vegetation and the dissolved oxygen content influenced the distribution and abundance of the species.

Keywords Spatial and temporal patterns · Habitat preference · Shallow lakes · Hanság · Hungary

Guest editors: R. Matzke-Karasz, K. Martens & M. Schudack
Ostracodology – Linking Bio- and Geosciences

A. Kiss (✉)
Hungarian Danube Research Station, Alkotmány u. 2-4, Vacratot 2163, Hungary
e-mail: kissa@botanika.hu

Introduction

Although the ecological tolerance of freshwater ostracods is generally broad in terms of the physical, chemical and ecological variables of different aquatic habitats, each species show different responses and tolerances to abiotic and biotic factors. Consequently ostracod distribution might be correlated with water quality and the types of habitats (Mezquita et al., 2001; Külköylüoglu, 2003). Ostracod species respond sensitively to environmental changes and if the species tolerances are known, it might be possible to use them as environmental indicators (Forester, 1991; Külköylüoglu, 2004). However, there is little information about the habitat requirements and tolerance levels of many ostracod species (Meisch, 2000). Correlations have been found between species' distribution and abundance and different environmental factors such as altitude and habitat type (Malmquist et al., 1997), oxygen content (Dole-Olivier et al., 1997), temperature (Horne, 1983; Forester, 1991), water level (Forester, 1991) and ionic composition (Baltanás et al., 1990).

Detailed surveys of microcrustacean assemblages (Cladocera, Ostracoda, Copepoda) were carried out in different habitats of a shallow lake to investigate the spatial and temporal patterns of assemblages and the relationship between the composition of assemblages and the examined habitat variables. The results on Ostracoda assemblages are summarized in this paper.

Materials and methods

Study site

Lake Fehér is situated in the north-western part of Hungary, in a strictly protected area of the Fertő-Hanság National Park. The Fertő-Hanság area is one of the most important wetlands in Central Europe. The examined Lake Fehér (47°41′ N, 17°21′ E) is small (area: 2.69 km^2, open water: 0.25 km^2) and shallow (mean depth: 50 cm, maximum depth: 110 cm). The littoral zone of the lake is characterised by dense beds of emergent macrophytes (*Phragmites australis* (Cav.) Trin. and *Typha angustifolia* L.) and the open water was entirely covered by dense stands of *Najas marina* L. in 1999 and 2000.

Sampling and sample processing

The study was carried out from March 1998 to August 2001. Samples were collected at monthly intervals for most of the period, but at biweekly intervals in 1999. The following habitats were sampled: (1) open water—water column and sediment surface, (2) *Najas marina* beds, (3) edge of the emergent macrophyte zone—water column and sediment surface, (4) Lemno-Utricularietum (in the reed-belt)—macrovegetation and sediment surface, (5) Phragmitetum communis—macrovegetation and sediment surface, (6) Typhetum angustifoliae—macrovegetation and sediment surface, (7) Hydrocharicetum morsus-ranae (in the reed-belt)—macrovegetation and sediment surface, (8) *Drepanocladus* water moss beds (in the reed-belt), (9) Die-back reed-belt (habitats disturbed by the damage of reed harvesting), (10) Channel (narrow channel in the reed-belt), (11) Flooded bog meadow.

Measurements of physical and chemical variables (temperature, pH, conductivity, dissolved oxygen and oxygen saturation) were taken in situ using a WTW Multiline-P4 portable meter. The composition and coverage of vegetation was estimated with a 1 m^2 quadrat according to the Braun-Blanquet (1951) method.

Ostracods were collected from the open water of the lake (open water samples), from the macrophyte stands (plant samples) and from the sediment surface (sediment samples). In the open water 50 l of water were filtered from the water column with a 70 μm mesh net and then preserved in 5% formaldehyde. Five-litre samples were collected among the different macrophyte beds by one-liter plastic box sampler (Frisch & Wohltmann, 2005). Samples were taken in three replicates from the different macrophyte communities. On the sediment surface, a 70 μm mesh hand dredge net with a sampling area of 0.3 m^2 was used and the net was dragged over the bottom surface to a depth of 3–5 cm. The subfossil ostracod valves were collected in the open water with an Ekman grab sampler. The upper 10 cm of the sediment layer was collected and after washing preserved in 5% formaldehyde.

Statistical analyses

During the sampling period 535 monthly and biweekly samples were collected. In the statistical analyses 228 samples which included ostracods were used. Principal Component Analysis (PCA) was used for the estimation of the influence of environmental variables on ostracod assemblages. Non-metric multidimensional scaling (NMDS) based on the Horn similarity index was used to compare the composition of the assemblages in the different habitats. The PCA and NMDS analyses were done using the SYN-TAX 5.1. program package (Podani, 1997). Spearman's correlation (Nucosa program package, Tótmérész, 1995) with log transformed data was used to show the relationship between the species abundances and abiotic variables and the macrophyte cover measured in situ. To test for differences of environmental variables between habitats, one-way analysis of variance (ANOVA) was carried out using the Statistica 6.0 software package. The relationship between ostracod densities and habitats was tested also by one-way ANOVA and F-test from ANOVA.

Results

Environmental variables

There were significant differences between the environmental variables measured in the open water and in the reed-belt (ANOVA, F = 22.4, $P < 0.01$). In the reed-belt, the water depth, temperature, pH, dissolved oxygen content and oxygen saturation were lower, while the conductivity was higher, than in the open-water (F = 9.5, $P < 0.01$). In the reed-belt the coverage of macrophytes Phragmitetum and Typhetum were higher than the Lemno-Utricularietum and Hydrocharicetum. In the different submerged macrophyte beds, the dissolved oxygen content, temperature and pH were notably lower inside the beds than on their margins (ANOVA, F = 6.8, $P < 0.05$). The sampling sites of the reed-belt usually dried up for a few weeks in summer and early autumn. Because of the low water level and oxygen content the reed-belt was probably unsuitable for fish except in spring. The ranges of the most important parameters of the examined habitats are summarized in Table 1.

Ostracoda assemblages

Twenty ostracod species were found during this survey (Table 2). Significant spatial differences were recorded in the composition and the abundance of the Ostracoda assemblages in the different habitats. In the MDS analysis (Fig. 1) the characteristic assemblages of the reed-belt with *Candona weltneri*, *Cyclocypris* spp., *Notodromas monacha* and *Pseudocandona compressa* were clearly separated (A). Besides this group, *Physocypria kraepelini* and the rare *Candona candida* and *Fabaeformiscandona hyalina* as representative members of the fauna on the sediment surface of the reed-belt edge (B), as well as species of *Cypridopsis* with low density and similar habitat preference in Lake Fehér (C), also formed distinct clusters. The most segregated species on the graph was *Candona neglecta*, occurring only on the sediment surface in the open water.

On the biplot of the PCA analyses three groups were separated (Fig. 2). The coverage of the vegetation was the main factor which influenced the distribution of the species included in the characteristic assemblages of the reed-belt (dominated by *Cyclocypris ovum*) (A). The samples of the open water and the edge of the reed-belt with the dominance of *Physocypria kraepelini* belong to group B. In the open water and on the edge of the reed-belt, water temperature was the main factor which influenced the distribution of the species. The C group is a mixed cluster with open water, reed-belt edge, channel and *Najas marina* samples. In this group, the depth and the oxygen saturation were the most important factors which

Table 1 The range of examined parameters in the different habitats of the lake

	WD	Temp.	pH	Cond.	Oxygen	O. Sat.	Macro
OW	28–110	0.9–30.6	7.6–10.3	335–54	4.5–15.2	49.8–130	0
NM	40–85	20–26.1	8.2–10.0	225–614	5.4–20.3	62.6–242	75–100
E	8–91	0.7–31.3	7.3–10.1	317–771	0.7–15.5	7.9–110	0–30
L	0–67	1.3–21	5.7–8.1	367–1012	0.1–10.8	1.2–100.5	30–55
P	0–58	2.1–22.2	6.9–7.8	420–1060	0.2–6.5	1.4–48.7	60–100
T	0–52	1.9–22.5	7.1–7.8	443–2410	0.0–5.9	0.0–38.4	65–100
HY	0–54	2.7–24.3	6.9–7.7	432–1263	1.5–11.5	22–120.4	40–55
DR	0–32	1.8–25	7.2–8.3	510–2520	2.3–11.3	30.4–115	30–60
D	0–41	1.3–26.9	7.4–8.6	554–2010	0.5–5.4	1.6–36.7	0–10
CH	12–48	1.2–21	7.5–8.6	423–850	2.4–6.8	29.7–50.2	0–20
M	0–45	7.6–22	5.8–7.1	663–3014	6.7–14.5	50.7–147	20–30

Abbreviations and units: Wlevel—water depth (cm), Temp.—water temperature (°C), Cond.—conductivity (μS cm^{-1}), Oxygen—dissolved oxygen content (mg dm^{-1}), O. Sat.—oxygen saturation (%), Macro—macrovegetation coverage (%), OW—open water, NM—*Najas marina* beds, E—edge of the emergent macrophyte zone, L—*Lemno-Utricularietum*, P—*Phragmitetum communis*, T—*Typhetum angustifoliae*, HY—*Hydrocharicetum morsus-ranae*, DR—*Drepanocladus* water moss beds, D—die-back reed-belt, CH—channel, M—flooded bog meadow

Table 2 The number of collected individuals, the relative abundance and the sex-ratio of the species as well as the range of water depth, temperature, pH, conductivity, dissolved oxygen content and the percentage cover of macrovegetation

	Abbr.	n	%	Sex ratio	WD	Temp.	pH	Cond.	Oxygen	Macro
Candona candida (O. F. Müller)	Ccan	21	0.08	1:2	70–82	17.6–18.2	7.71–7.95	674–680	2.55–4.56	0 and 40
Candona neglecta Sars	Cneg	11	0.042	–	55–97	9.4–19.3	7.23–7.81	488–652	3.64–8.76	0
Candona weltneri Hartwig	Cwel	1038	3.94	1:2.25	2–102	1.1–25.4	5.57–8.63	284–1059	0–10.79	0–90
Fabaeformiscandona fabaeformis (Fischer)	Ffab	26	0.1	1:17	30–50	5.1–21	5.57–7.68	494–853	0.37–9.37	30–80
Fabaeformiscandona balatonica (Daday)	Fbal	27	0.1	1:2.14	19–105	4.4–10.0	7.2–8.03	356–799	6.33–8.57	0 and 20
Fabaeformiscandona fragilis (Hartwig)	Ffra	117	0.44	1:1.28	14–88	1.1–25.4	6.98–8.91	362–1059	0.37–12.35	0–90
Fabaeformiscandona hyalina Brady et Rob.	Fhya	1	0.004	–	35	5.3	7.81	493	10.33	0
Fabaeformiscandona protzi (Hartwig)	Fpro	48	0.18	1:1.33	5–85	0.9–25.3	7.00–8.22	443–1010	0.37–14.51	0–100
Pseudocandona rostrata (Brady et Norman)	Pros	18	0.07	–	5–43	1.5–13.1	7.33–7.44	587–1010	0.11–1.69	30 and 60
Pseudocandona compressa (Koch)	Pcom	1159	4.4	1:2.14	2–97	0.9–25.3	5.57–10.07	245–2410	0–14.81	0–100
Candonopsis kingsleii Brady et Robertson	Ckin	121	0.46	1:2.07	15–82	0.9–23.2	5.57–8.09	563–959	0–6.43	0–90
Cypria ophtalmica (Jurine)	Coph	339	1.29	1:1.94	2–102	3.4–26.4	5.7–8.15	405–1012	0.11–10.12	0–100
Physocypria kraepelini G. W. Müller	Pkra	1053	4.0	1:2.04	13–105	0.9–29.8	6.72–10.44	241–799	1.19–19.12	0–100
Cyclocypris globosa (Sars)	Cglo	544	2.07	–	8–67	1.3–26.1	6.84–8.05	472–1059	0–10.79	0–100
Cyclocypris laevis (O. F. Müller)	Clae	761	3.0	–	2–100	1.3–25.3	6.84–8.46	310–1059	0–13.11	0–100
Cyclocypris ovum (Jurine)	Covu	17123	65.06	–	2–100	1.3–26.4	5.57–10.23	241–2410	0–17.61	0–100
Notodromas monacha (O. F. Müller)	Nmon	3826	14.54	1:2	5–60	13.1–26.1	5.57–8.15	361–1012	0–10.12	0–100
Cypridopsis vidua (O. F. Müller)	Cvid	81	0.3	–	22–47	8.5–19.9	6.98–7.59	462–784	0.62–3.46	0–80
Cypridopsis elongata (Kaufmann)	Celo	1	0.004	–	49	17.2	8.21	460	10.61	0
Cypridopsis hartwigi G.W. Müller	Char	2	0.007	–	30	8.6	7.26	513	0.72	80

The abbreviations of the environmental parameters and the units see in Table 1. Other abbreviations: n—number of specimens, %—relative abundance, NM—no males

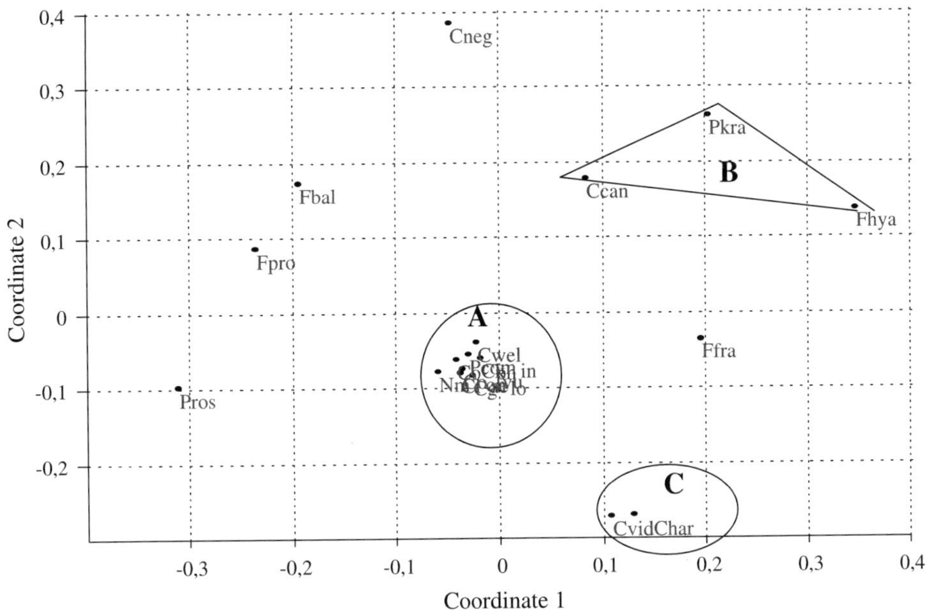

Fig. 1 NMDS ordination for ostracod species. Labels for species correspond to those used in Table 2 and 3. Circles refer to groups of species as explained in the text

affected the composition and distribution of these assemblages.

In the Correlation Analyses the distribution and abundance of the species were particularly influenced by the presence and coverage of the macrophytes, the temperature and the dissolved oxygen content (Table 3). The summer species of the reed-belt and the dominant species of the *Typhetum angustifoliae* habitat (the dissolved oxygen content was the lowest in this habitat)

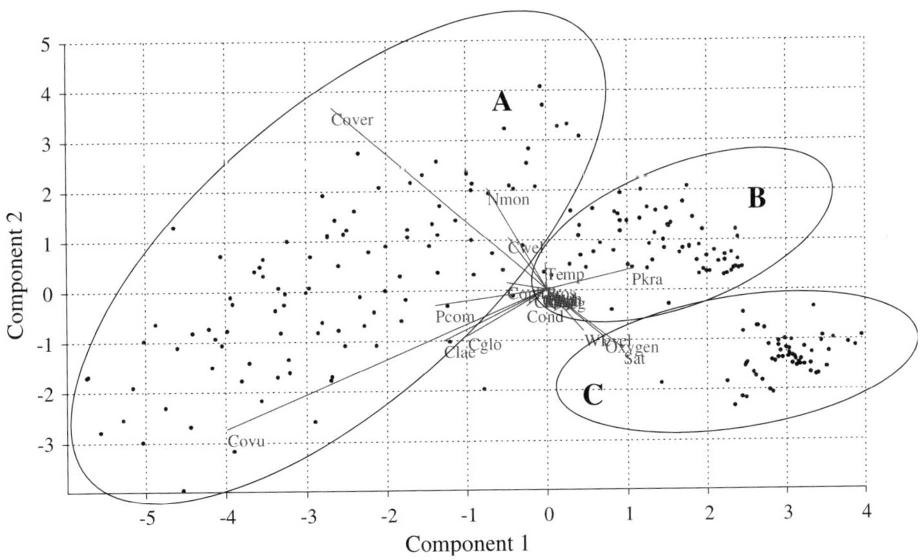

Fig. 2 PCA ordination biplot of ostracod species and their assemblages against the first two principal components for environmental variables. Abbreviations: Cover—coverage of macrovegetation, Wlevel—depth, Oxygen Sat—oxygen saturation, Temp—temperature, Cond—conductivity. Labels for species correspond to those used in Table 2 and 3. Ellipses refer to groups of species as explained in the text

Table 3 The values of the Spearman's correlation coefficients between the abundance of the species and the examined abiotic and biotic parameters as well as the spatio-temporal distribution of the species

	WD	Temp.	pH	Cond.	Oxygen	Macro	Period	Habitat
Ccan	0.192	**–0.309**	–0.146	0.162	0.090	**0.293**	IV, VI	LS, **ES**
Cneg	0.210	–0.183	–0.072	0.053	0.125	**–0.261**	IV, X	OS
Cwel	**–0.259**	–0.016	0.122	0.103	**–0.302**	**0.517**	I-XII	OS, EW, ES, LM, **LS**, PM, PS, TM, TS, HM, HS, DR,
Ffab	–0.063	–0.109	–0.044	0.134	–0.056	**0.296**	IV, VI, XI,	**PS**, ES,
Fbal	0.206	**–0.429**	0.124	0.146	**0.283**	**0.263**	III, XI	LM, **LS**, ES,
Ffra	0.143	–0.233	–0.094	0.173	0.193	0.197	I-XII	OS, **ES**, LM, **LS**, PM, PS, TM, TS,
Fhya	–	–	–	–	–	–	XI	ES
Fpro	0.139	**–0.272**	–0.063	0.173	0.201	**0.396**	I-V, IX-XII	OS, LM, **LS**, PM, PS, TM, TS, EW, ES,
Pros	0.190	–0.214	–0.176	0.200	0.019	**0.563**	X, XII	LM, PM,
Pcom	**–0.269**	0.154	0.176	0.081	**–0.251**	**0.423**	I-XII	all, **PM, PS, TM, TS**, M,
Ckin	0.140	–0.193	0.156	0.167	**–0.294**	**0.915**	I-V, X, XII	ES, LS, PS, TM, **TS**,
Coph	–0.197	0.127	0.154	0.045	**–0.312**	**0.247**	I-XII	ES, LM, LS, PM, **PS**, TM, TS, HM, HS, DR, **D**, M,
Pkra	0.193	**–0.267**	0.195	0.097	0.209	–0.163	I-XII	OW, OS, EW, **ES**, NM,
Cglo	–0.141	0.102	**–0.304**	0.197	–0.137	**0.620**	I-VII X-XII	LM, LS, PM, PS, **TM**, TS, HM, HS, DR, D, M,
Clae	0.157	–0.184	–0.167	0.163	–0.195	**0.712**	I-VII X-XII	EW, LM, LS, PM, PS, TM, TS, HM, HS, DR, D, M,
Covu	0.119	–0.209	–0.193	0.141	**–0.283**	**0.859**	I-XII	all, **TM, PM**,
Nmon	**–0.287**	**0.341**	0.152	0.095	**–0.263**	**0.412**	IV-X	**LM**, LS, PM, PS, TM, TS, HM, HS, DR, D, M,
Cvid	0.147	–0.176	–0.097	0.122	–0.021	**0.718**	V, VII, X	ES, LM, LS, **PM**, PS, TM, TS,
Celo	–	–	–	–	–	–	IX	OS
Char	–	–	–	–	–	–	X	PM

The significant values ($P < 0.01$) were designed in bold. Abbreviations: open water—water column (OW) and sediment surface (OS), *Najas marina* beds (NM), edge of the emergent macrophyte zone—water column (EW) and sediment surface (ES), *Lemno-Utricularietum*—macrovegetation (LM) and sediment surface (LS), *Phragmitetum communis*—macrovegetation (PM) and sediment surface (PS), *Typhetum angustifoliae*—macrovegetation (TM) and sediment surface (TS), *Hydrocharicetum morsus-ranae*—macrovegetation (HM) and sediment surface (HS), *Drepanocladus* water moss beds (in the reed-belt) (DR), die-back reed-belt (D), channel (Ch) and flooded bog meadow (M). The preferred habitats are designed in bold. The abbreviations of the habitat parameters as in Table 1, whereas the abbreviations of species as in Table 2 can tolerate hypoxic conditions (below 3 mg dm^{-3}) and sometimes near zero oxygen level.

Ostracoda abundance and species richness increased inshore from the open water to the *Typhetum*. In the open water habitats 97 individuals of 8 species (*Candona neglecta*, *Candona weltneri*, *Fabaeformiscandona fragilis*, *F. protzi*, *Pseudocandona compressa*, *Cyclocypris ovum*, *Physocypria kraepelini*, *Cypridopsis elongata*) were collected. The most frequent species was *Physocypria kraepelini* (relative abundance of 40.3%), whereas *Candona neglecta* and *Cypridopsis elongata* were found only in this habitat of the lake.

Interestingly, the ostracod species richness and abundance was low in the homogenous *Najas marina* beds, as only 155 specimens of three species (*Cyclocypris ovum*, *Physocypria kraepelini*, *Pseudocandona compressa*) were detected in this habitat.

In the reed-belt edge, 1154 individuals of 14 species were recorded. The most frequent species were *Physocypria kraepelini* (78.4%), *Fabaeformiscandona fragilis* (4.9%) and *Cyclocypris ovum* (4.8%). The preferred habitat of *Physocypria kraepelini* was the sediment surface of the reed-belt edge ($P < 0.01$, t-test, ANOVA) and the average density of this species on the sediment surface was higher in autumn and winter (maximum: 466 ind. m^{-2} in November 2000) than in spring and summer ($P < 0.05$, t-test ANOVA).

In the reed-belt, diverse and abundant assemblages occurred–13,625 individuals of 16 species were recorded. The mean abundance of species differed in the examined macrophyte communities of the reed-belt (Table 4) and the species richness and the total abundance of the species were considerably lower in disturbed, die-back reed-belt sites than in the other habitats of the reed-belt ($P < 0.01$). The dominant species were *Cyclocypris ovum* (72.7% relative abundance), *Notodromas monacha* (8.5%), *Candona weltneri* (5.8%) and *Pseudocandona compressa* (5.1%). *Pseudocandona rostrata*, *Notodromas monacha* and *Cypridopsis hartwigi* occurred only in the reed-belt, whereas *Physocypria kraepelini* was replaced by *Cypria ophtalmica* in this habitat. Most of the reed-belt species showed distinct habitat preferences. The density of *Candona weltneri*, *Pseudocandona compressa* and *Candonopsis kingsleii* was notably higher on the sediment surface than in the vegetation while *Cyclocypris globosa*, *Cyclocypris laevis*, *Cyclocypris ovum* and *Notodromas monacha* preferred the vegetation ($P < 0.01$). The density of *Cyclocypris* species increased with increasing coverage of vegetation ($P < 0.05$) while *Cypria ophtalmica* and *Notodromas monacha* preferred the Lemno-Utricularietum and the Hydrocharicetum morsus-ranae with lower density of macrovegetation. Interestingly the total abundance of *Cypria ophtalmica* was the highest in the degraded die-back reed-belt habitat. The species of the reed-belt tolerated hypoxic conditions and a significant fraction of the individuals occurred below 3 mg dm^{-3} dissolved oxygen content ($P < 0.01$). Most species occurred between pH 7.00–8.00, however, *Cyclocypris ovum* was found even at pH 10.2.

Table 4 The average density (±SD) of the most frequent species among the vegetation (ind. dm^{-3}) and on the surface of the sediment (ind. m^{-2}) in the submerged *Lemno-Utricularietum* and the emergent *Phragmiteum communis* and *Typhetum angustifoliae* habitats

	Lemno-Utricularietum		*Phragmitetum communis*		*Typhetum angustifoliae*	
	Water	Sediment	Water	Sediment	Water	Sediment
Candona weltneri	1.7 ± 3.1	**98.5 ± 160.4**	0.2 ± 0.3	16.9 ± 32.6	0.1 ± 0.3	1.5 ± 3.9
Pseudocandona compressa	0.5 ± 0.7	10.4 ± 13.4	1.8 ± 2.5	38 ± 51.4	1.5 ± 2.1	38.6 ± 118.6
Candonpsis kingsleii	0	0.6 ± 8.7	0	0.5 ± 1.8	0.04 ± 0.15	**7.9 ± 26.9**
Cypria ophtalmica	0.2 ± 0.5	8.5 ± 17	0.1 ± 0.2	6.9 ± 9.2	0.2 ± 0.4	6.1 ± 16.8
Cyclocypris globosa	0.3 ± 0.3	2.9 ± 6.7	0.8 ± 1.1	6.9 ± 14	1.1 ± 2.5	3.8 ± 8.7
Cyclocypris laevis	0.4 ± 0.6	2.3 ± 4.2	1.4 ± 2.2	11 ± 17.1	1.6 ± 2.7	14.6 ± 31.4
Cyclocypris ovum	15.3 ± 33.8	90.8 ± 197.6	24.9 ± 42.6	199.7 ± 170	**46.4 ± 61.1**	**306.6 ± 580**
Notodromas monacha	**5.5 ± 9.1**	19.2 ± 31.2	2.1 ± 3.7	6.9 ± 15.6	0.5 ± 1.1	0.2 ± 0.9

The significant values ($P < 0.01$, one-way ANOVA) were designed in bold

Contrary to this, *Cyclocypris globosa* did not tolerate high pH values; above pH 7.7 this species disappeared from the reed-belt.

In the 52 Ekman grab sediment samples only two living specimens of *Physocypria kraepelini* were found. A significant part of the subfossil fauna belonged to the representatives of the genera *Candona*, *Fabaeformiscandona* and *Physocypria*. Subfossil valves of *Darwinula stevensoni*, *Ilyocypris monstrifica*, *Limnocythere sanctipatricii* and *Metacypris cordata* were found in the samples in very low densities; however, these species should not be considered as representatives of the living ostracod assemblages studied during the sampling period.

Discussion

In Lake Fehér, the Ostracoda assemblages primarily colonized the reed-belt, where mainly the extremely shallow water level, the drying period in summer and the low dissolved oxygen content influenced the presence and distribution of the species. These environmental conditions can explain the very diverse Candonidae fauna (16 out of the 20 species) since species of *Candona*, *Fabaeformiscandona* and *Pseudocandona* are known to tolerate the periodical desiccation of their habitat (Horne, 1993), while the frequent *Notodromas monacha* is also a desiccation resistant species (Tétart, 1971).

There were significant differences in the composition and density of assemblages between the habitats and most species showed distinctive habitat preferences. The abundance of the species in the different macrophyte communities also differed, similar to the reports by Paterson (1993) and Benzie (1989). Most of the individuals in the open water and reed-belt edge habitats were found on the sediment surface and in the sediment. However, in the sediment living animals were recorded only in the upper 1–2 cm layer, supposedly because of low oxygen content. According to Löffler (1972) sediment is frequently anoxic below 1–2 cm sediment depth in eutrophic shallow lakes.

The Ostracoda species richness and density changed horizontally and increased from the open water to the reed-belt. These results conform well with Benzie's (1989) observations, who determined that Ostracoda diversity and species number increase from open water to the submerged and emergent macrophyte associations, being correlated with the increased quantity of detritus. In Lake Fehér, the average of species number and abundance increased with increased coverage of vegetation, so that the most diverse assemblages were formed in the homogenous *Phragmites* stands with 80–100% coverage of vegetation. These results are similar to numerous observations (e.g. Irvine et al., 1990; Paterson, 1993) which confirm that the density of several microcrustacean groups, especially phytophilous species, is correlated with the biomass of vegetation. In Lake Fehér the species richness and density minima were recorded in the die-back reed-belt. In this disturbed habitat, fragmented and degraded by human activity, Ostracoda assemblages with low species numbers and densities parallel the destruction of the habitat. According to Smock & Stoneburger (1980) in harvested reed-belts the decreasing species richness is the best indicator of habitat loss and water quality degradation for many freshwater animal groups.

In Lake Fehér the occurrence of several species is correlated with the examined abiotic and biotic variables. The presence and coverage of the macrophytes were the most important factors which effected the distribution and abundance of the different species. In case of the abiotic variables, the conductivity did not correlate significantly with the abundance of species. The pH also did not significantly affect the distribution of species, except for *Cyclocypris globosa* which prefers moorland and peaty areas with acidic conditions (Meisch, 2000). The abundance of stenothermal species (*Candona candida*, *Fabaeformiscandona balatonica*, *Notodromas monacha*) showed correlation with the temperature. In case of water depth, contrary to Tabacchi & Marmonier's (1994) observations, only three species showed weak correlation. Most of the reed-belt species can tolerate low oxygen content (0–1 mg dm^{-3}) and during diurnal or acyclic micromigration the individuals can detect the microhabitats with higher oxygen level. An alternative oxygen pool for these species might be the extended aerenchyma of some plant species (Douglas & Healy, 1991). In the case of *Cypria ophtalmica*, *Cyclocypris ovum*, *Cyclocypris laevis* and *Pseudocandona compressa*, laboratory observations proved that these species can tolerate near-zero oxygen levels for a short time (Delorme, 1991). During the diurnal investigations of Lake Fehér these species also tolerated zero oxygen levels from 5 to 9 a.m. in the *Typhetum* habitat. Contrary to this, the low density of one of the most common ostracods (*Cypridopsis vidua*) might be explained by the low dissolved oxygen content of the reed-belt. *Cypridopsis vidua*, despite its wide ecological tolerance, has a low tolerance for poorly oxygenated waters and according to Roca & Danielopol (1991) its dissolved oxygen minimum is around 5.0 mg dm^{-3}. Similar to Benzie's (1989) observations, densities of the phytophilous *Cypridopsis vidua* were very low in *Phragmites* stands because of the low oxygen content.

Conclusion

Species richness and abundance as well as the composition of assemblages varied significantly between the examined habitats. Most of the 20 recorded species showed distinct habitat preferences and ecological tolerances. In particular temperature, dissolved oxygen content and the presence and coverage of the vegetation influenced the distribution and the abundance of the ostracod species. These examinations confirm that ostracods can be used as bioindicators and based on dominant and characteristic species one can separate and describe different habitats and water types.

Acknowledgements This project was supported by the MTA/KvVM Program and the NKFP (Széchenyi Research Project). I deeply appreciate Finn Viehberg and the anonymous reviewer for their valuable comments.

References

Baltanás, A., C. Montes & P. Martıno, 1990. Distribution patterns of ostracods in Iberian saline lakes. Influence of ecological factors. Hydrobiologia 197: 207–220.

Benzie, J. A. H., 1989. The distribution and habitat preference of ostracods (Crustacea: Ostracoda) in a coastal sand-dune lake, Loch of Strathbeg, north-east Scotland. Freshwater Biology 22: 309–321.

Braun-Blanquet, J., 1951. Pflanzensoziologie. Grundzüge der Vegetationskunde. 2. Auflage, Springer-Verlag, Wien.

Delorme, L. D., 1991. Ostracoda. In Thorp, J. H. & A. P. Covich (eds), Ecology and classification of North American freshwater invertebrates. Academic Press, San Diego, 691–722.

Dole-Olivier, M.-J., P. Marmonier & J. L. Beffy, 1997. Response of invertebrates to lotic disturbance: is the hyporheic zone a patchy refugium? Freshwater Biology 37: 257–276.

Douglas, D. J. & B. Healy, 1991. The freshwater ostracods of two quagmires in Co. Louth, Ireland. Verhandlungen der internationalen Vereinigung für theoretische und angewandte Limnologie 24: 1522–1525.

Forester, R. M., 1991. Ostracode assemblages from springs in the Western United States: inplications for paleohydrology. Memoirs of the Entomological Society of Canada 155: 181–201.

Frisch, D. & A. Wohltmann, 2005. The bag-sampler: A simple device for collecting zooplankton in shallow vegetated ponds. International Review of Hydrobiology 90: 596–602.

Horne, D. J., 1983. Life-cycles of podocopid Ostracoda—a review (with particular reference to marine and brackish-water species). In Maddocks, R. F. (ed.), Applications of Ostracoda. Proceedings of the Eighth International Symposium on Ostracoda, July 26–29, 1982. Department of Geosciences, University of Houston, 238–249.

Horne, F. R., 1993. Survival strategy to escape desiccation in a freshwater ostracod. Crustaceana 65: 53–61.

Irvine, K., H. Balls & B. Moss, 1990. The Entomostracan and Rotifer Communities Associated with Submerged Plants in the Norfolk Broadland—Effects of Plant Biomass and Species Composition. Internationale Revue der Gesamten Hydrobiologie 75: 121–141.

Külköylüoglu, O., 2003. Ecology of freshwater Ostracoda (Crustacea) from lakes and reservoirs in Bolu, Turkey. Journal of Freshwater Ecology 18: 343–347.

Külköylüoglu, O., 2004. On the usage of ostracods (Crustacea) as bioindicator species in different aquatic habitats in the Bolu region, Turkey. Ecological Indicators 4: 139–147.

Löffler, H., 1972. The distribution of subfossil ostracoda and diatoms in prealpine lakes. Verhandlungen der internationalen Vereinigung für theoretische und angewandte Limnologie 18: 1039–1050.

Malmquist , B., C. Meisch & A. Nilsson, 1997. Distribution patterns of freshwater Ostracoda (Crustacea) in the Canary Islands with regards to habitat use and biogeography. Hydrobiologia 347: 159–170.

Meisch, C., 2000. Freshwater ostracoda of Western and Central Europe. In Schwoerbel, P. & Zwick (eds), Süßwasserfauna von Mitteleuropa 8/3. Spektrum Akademischer Verlag, Heidelberg, Berlin

Mezquita, F., H. I. Griffiths, M. I. Dominguez & M. A. Lozana-Quilis, 2001. Ostracoda (Crustacea) as ecological indicators: A case study from Iberian

Mediterranean brooks. Archiv für Hydrobiologie 150: 545–560.
Paterson, M. J., 1993. The distribution of microcrustacea in the littoral zone of a freshwater lake. Hydrobiologia 263: 173–183.
Podani, J., 1997. SYN-TAX 5.1: A new version for PC and Macintosh computers. Coenoses 12: 149–152.
Roca, J. R. & D. L. Danielopol, 1991. Exploration of interstitial habitats by the phytophilous Ostracod *Cypridopsis vidua* (O. F. Müller): experimental evidence. Annales Limnologie 27: 243–252.
Smock, L. A. & D. L. Stoneburger, 1980. The response of macroinvertebrates to aquatic macrophyte decomposition. Oikos 35: 397–403.
Tabacchi, E. & P. Marmonier, 1994. Dynamics of the interstitial ostracod assemblage of a pond in the Adour alluvial plain. Archiv für Hydrobiologie 131: 321–340.
Tétart, J., 1971. Etude de quelques populations d'Ostracodes dans des milieux astatiques de la vallée de l'Isère. Travaux de Laboratoire Hydrobiologique Grenoble 62: 75–130.
Tótmérész, B., 1995. Comparative analysis of different methods of diversity orderings. Journal of Vegetation Science 6: 283–290.

Hydrobiologia (2007) 585:99–118
DOI 10.1007/s10750-007-0632-7

OSTRACODA (ISO15)

Groundwater Ostracods from the arid Pilbara region of northwestern Australia: distribution and water chemistry

Jessica M. Reeves · Patrick De Deckker · Stuart A. Halse

Abstract An attempt has been made at a comprehensive study of the diversity and distribution of subterranean ostracods in the Pilbara region, northwestern Australia. The area is a "hot spot" for subterranean biodiversity, some of which is currently under threat from extensive mining operations. Both bore and well sites were targeted, totalling 445 sites, to obtain a thorough coverage of the 200,000 km^2. In addition, physical and hydrochemical measurements were obtained for all of the samples (temperature, conductivity, dissolved oxygen, pH, Eh, turbidity, nutrients, major ions). Ostracods were retrieved from approximately 47% of the samples and 56% of the sites. Twenty-one genera and around 110 species of ostracods have been identified. Of these, 72 are new species and a further 10 are currently in open nomenclature, due to the lack of suitable material for formal taxonomic description. The Candoninae are particularly well represented with 12 genera; some, such as *Areacandona* and *Deminutiocandona*, with 25 and 10 species respectively. Most sites (80%) were dominated by only one or two species, with up to six species at some sites. Population density varied from 1–370 individuals/sample. The most abundant and diverse sites occur in fresh, bicarbonate-rich aquifers utilised for water extraction, such as Pannawonica (Robe River), Cane River and Millstream. There is a clear distinction between taxa at the genus level from coastal and low-lying alluvial sites, and upland sites (>300 m altitude). Beyond this, the majority of species are confined within a surface water catchment, or in many cases, a specific aquifer. There are, however, some morphological similarities of the carapaces between different species within similar hydrogeologic settings. Ornate and ridged-valved species are common in the Mg–HCO_3 waters of the Newman and Marillana Creek areas, whereas smooth-shelled, tapered forms are prevalent in alluvial aquifers. The more saline, Na–Cl rich aquifers at the edge of Great Sandy Desert have a particularly distinctive fauna, including one almost triangular species. The distribution of the stygobitic ostracod species in relation to the hydrogeology and water chemistry is discussed.

Guest editors: R. Matzke-Karasz, K. Martens & M. Schudack
Ostracodology – Linking Bio- and Geosciences

Electronic supplementary material The online version of this article (doi:10.1007/s10750-007-0632-7) and accessible for authorized users.

J. M. Reeves (✉) · P. De Deckker
Department of Earth and Marine Sciences,
The Australian National University, Canberra,
ACT 0200, Australia
e-mail: jessica@ems.anu.edu.au

S. A. Halse
Department of Conservation and Land Management,
Woodvale, WA 6026, Australia

Keywords Stygofauna · Pilbara · Groundwater · Hydrochemistry · Candoninae · Biogeography

Introduction

The Pilbara region in northwestern Australia is best known for its arid landscape and iron-ore industry. The expansion of mining operations to below the watertable poses several potential problems for the ecology of groundwater, including increased salinisation and contamination of aquifers as well as destruction of habitat. The scarcity of surface water resources in this semi-arid region has lead to groundwater being utilised for town water supplies, mining operations and watering of stock on pastoral leases. In consideration of the threats to groundwater-dwelling organisms, or stygofauna, posed by mining operations and the exploitation of groundwater, the Department of Conservation and Land Management (CALM) is undertaking a survey of stygofauna in the region to provide a framework for the assessment and conservation of groundwater biodiversity. Ostracods have been targeted as a key group for survey, owing to their diversity, abundance and preservation potential in bicarbonate-rich waters, as well as their known response to hydrochemistry (e.g. Forester, 1983, 1986; Radke et al., 2003).

Stygobitic Ostracods

Stygobitic fauna (sensu Gibert et al., 1994) are defined as animals that complete their entire life cycle within subterranean environments. Such animals are known to have slower metabolisms, longer ontogenies and subsequently longer life spans than their surface water counterparts (Culver, 1982). Groundwater-dwelling ostracods share some features that are known to be beneficial to subterranean life, referred to as troglomorphic characters by Christiansen (1962). They are blind, have un-pigmented valves and are comparatively small. Carapace shapes vary from highly elongated to triangular, even with dorsal protuberances (Danielopol & Hartmann, 1986). The majority of taxa within this study are stygobitic. Further studies in the Pilbara region will focus on stygophilic animals, those that live only part of their life in groundwater, such as in hyporheic and spring discharge sites.

Stygofaunal research in Western Australia began in the 1990s with a diverse and unique fauna being found that showed Tethyan affinities at Cape Range Peninsula (Humphreys, 1993a, b, c, 2000; Bradbury & Williams, 1996a; Danielopol et al., 2000; Jaume & Humphreys, 2001) and Barrow Island (Bradbury & Williams, 1996b; Humphreys 2000, 2001a), and Gondwanan affinities in the adjacent arid, cratonic parts of the Pilbara (Poore & Humphreys, 1998; Eberhard & Humphreys, unpublished; Humphreys, unpublished). The Yilgarn to the south (Humphreys, 1999, 2001b; Watts & Humphreys, 1999; De Laurentis et al., 2001) and the tropical Kimberley to the north (Wilson & Ponder, 1992) (Fig. 1) are also rich in stygofauna. Prior to the present study, 23 groundwater ostracod species, 20 candonids and 3 darwinulids, were known to occur in the Pilbara region, all of which were described only recently (Martens & Rossetti, 2002; Halse et al., 2002; Karanovic, 2003, 2005; Karanovic & Marmonier,

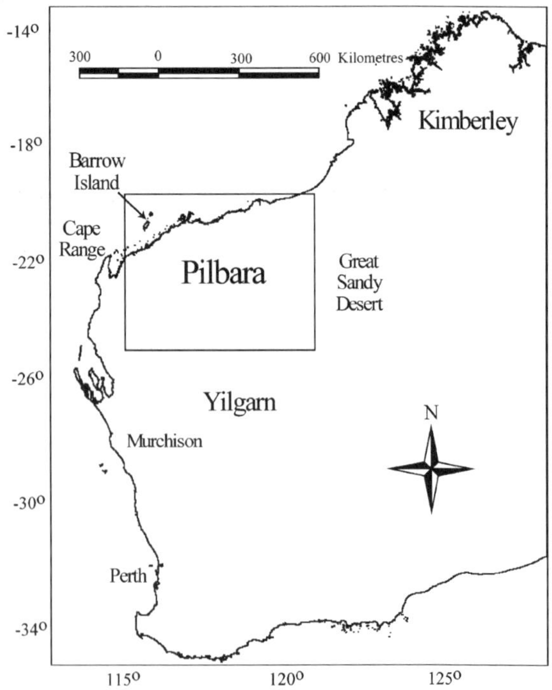

Fig. 1 Locality map of Western Australia. The area in the box represents the Pilbara region

2003). This compares well with the diversity of fauna in the adjacent Murchison (five species, one genus) and Cape Range (one species) regions (Danielopol et al., 2000; Karanovic & Marmonier, 2003).

The groundwater of the Pilbara region provides a refugium for aquatic invertebrates in this arid environment, with a high degree of subterranean biodiversity and endemicity (Humphreys, 1999, 2001, unpublished). The waters are typically rich in bicarbonate, and therefore ostracods are particularly well represented, because of the suitability of such waters for readily forming calcite valves. This study looks at the distribution of the ostracod species in relation to the physical constraints of the aquifers and the hydrochemistry in the Pilbara region.

Study area

The Pilbara region (20–24° S, 115–122° E) of northern Western Australia, covering ~200,000 km^2, is hot and dry. Although climatically regarded as semi-arid, with annual evaporation outweighing precipitation 10:1, the Pilbara is located at the tropical fringe. Seasonality is distinct with hot summers (25–36°C mean summer minimum and maximum) and mild winters (12–27°C mean winter minimum and maximum). Rainfall is erratic and localised, occurring predominantly in the summer months during thunderstorms and cyclonic events (averaging 200–350 mm annually, decreasing inland). Winter rainfall is sometimes significant, particularly in southern areas. There is little permanent surface water and all rivers are ephemeral; however, groundwater is plentiful and mostly fresh.

The region may be divided broadly into three physiographic types: low ranges, wide floodplains and a coastal zone (Fig. 2). The ranges form part of the Pilbara craton which has been emergent since the Palaeozoic. They comprise the Early Proterozoic—Archaean metasedimentary Hamersley Range in the central Pilbara, reaching around 900 m asl, with peaks around 1250 m asl, and the predominantly volcanic Chichester Range to the north, with a more subdued topography of around 600 m asl (Trendall, 1990). These units overlie the Archaean greenstones and granites, which outcrop to the north-east of the region. The regolith comprises a fine red blanket over much of the region, resulting in a very thin vadose zone. The Fortescue and

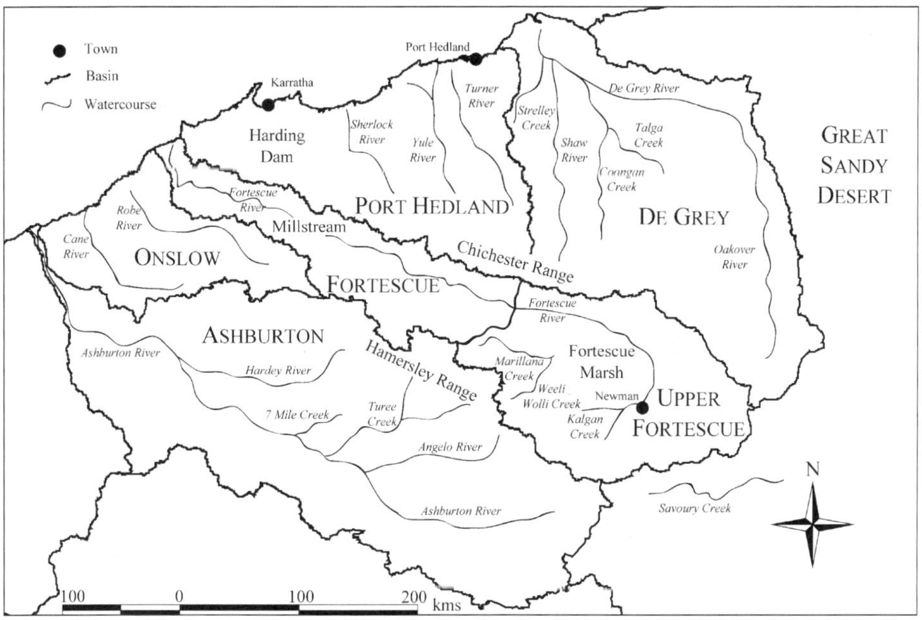

Fig. 2 Map of the Pilbara region, showing key localities within the study. The darker lines represent surface water drainage basins, the finer lines represent major drainage features

Ashburton Rivers form extensive floodplains, draining either side of the Hamersley Ranges. The Robe, Yule and De Grey Rivers extend as broad deltas from the highlands toward the Indian Ocean. Several other minor rivers also traverse this path. The coastal zone comprises broad, flat hummock and tussock grasslands, with scattered woodlands, on cracking clays or sandy soils. Minor Tertiary limestone outcrops occur across the plain.

The current drainage system of the Pilbara is thought to have developed through the Late Cretaceous—Early Tertiary (Beard, 1973, 1998). The calcretes, common to arid regions of Australia, formed from palaeochannels that dried up in the Palaeocene, when the climate switched from humid to arid (Bowler, 1976). Carbonate precipitation is active in the spring discharge regions, such as Millstream and Weeli Wolli.

There are three significant aquifer-types in the Pilbara region (Fig. 3): (1) unconsolidated sedimentary aquifers, including recent valley-fill alluvium and colluvium, and coastal deposits; (2) chemically-deposited calcretes and pisolites within Tertiary drainage channels; and (3) fractured-rock dolomite, banded-iron formations and granite, which form local aquifers (Johnson & Wright, 2001). The Wittenoom Dolomite forms an extensive aquifer, skirting the base of the Hamersley Ranges, commonly with cavernous karst development (Balleau, 1972). There is a noted relationship between the host rock aquifers and the resultant hydrochemistry. Groundwater in the region is typically fresh to low salinity (200–1500 mg l^{-1}) and bicarbonate-dominated, although Na–Cl-rich waters are common in both the coastal and arid eastern margins (Fig. 4). Isotopic analysis ($\delta^{18}O$, δD) of a selection of groundwater samples taken along a transect from the coast to ~300 km inland reveals that the majority of recharge is resultant from cyclonic rains (to be presented elsewhere). There is also a component of seepage through the major waterways to the alluvial aquifers during peak flow times.

The major rivers, including the Ashburton, Fortescue and De Grey systems, all have their

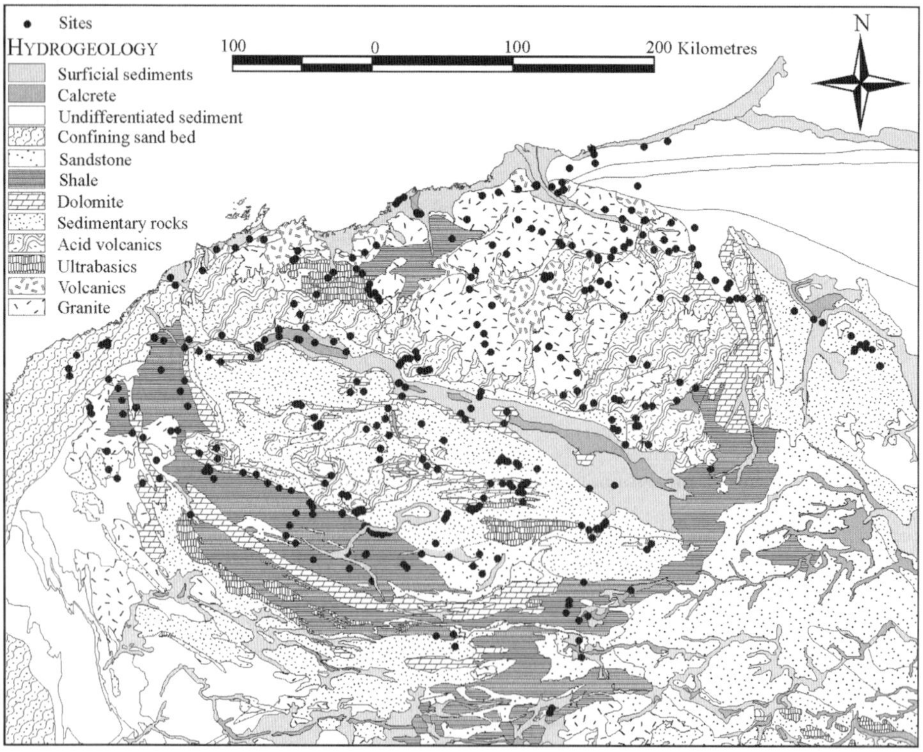

Fig. 3 Hydrogeological map of the Pilbara region, showing the sampling sites of this study (base map courtesy of WA WRC)

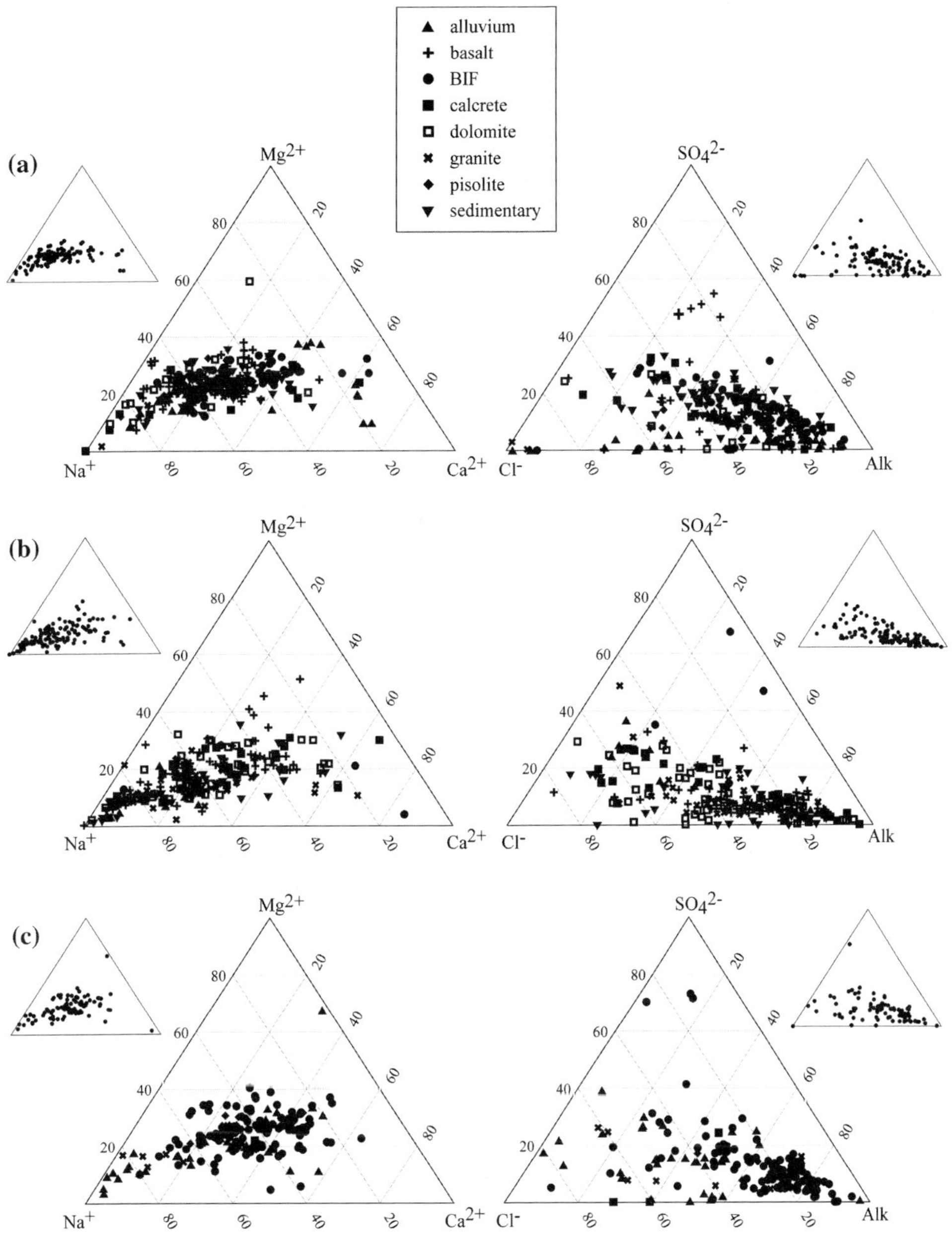

Fig. 4 Ternary plots for major cation (*left*) and anion (*right*) relative concentrations of the groundwaters from (**a**) the Fortescue, (**b**) Ashburton and Onslow and (**c**) De Grey, Port Hedland Coastal and Great Sandy Desert drainage basins. The symbols refer to the hydrogeology of the aquifers from which the samples were taken. The small, inset diagrams refer only to samples containing Ostracods

headwaters in the Proterozoic metasediments and Archaean granites of the Hamersley and Chichester Ranges. Surface waters are largely restricted to drainage lines, with river pools sustained by local bank storage or local water table. Springs are fed by local aquifers, particularly in the karstic areas. Yield from the aquifers is greatest in the calcrete Millstream region and in the Wittenoom Dolomite, producing up to 5,000 Gl/day (Johnson & Wright, 2001). The general direction of flow is from the headwaters in the ranges toward the coast.

There is a groundwater divide within the western Fortescue; the diversion of the lower

Fortescue away from the Robe River is a recent phenomenon, most likely since the Last Glacial Maximum (Barnett & Commander, 1985). Upstream of Mulga Downs, water flows toward the Fortescue Marsh (Barnett & Commander, 1985).

Sampling strategy

Sampling sites were selected to provide broad geographic coverage and to encompass the range of geologic, topographic, physiographic and hydrochemical environments across the Pilbara region. Shallow and deep aquifers were targeted; hyporheic and spring discharge sites were not included in this study. The majority of samples were taken from bores (85%), with the remainder from wells. The sampling area spans latitudes of 19–25° S and longitude of 115–122° E. Samples were taken from a range of altitudes, from the coast to 726 m asl, with a mean height of around 290 m asl. The bore construction differed between sites, in terms of diameter (50–400 mm), casing material (P.V.C or steel) and slotted interval, i.e. the depth along which the bore is open to the aquifer. Several sampled bores were from borefields; clusters of bores used for water extraction for mining and domestic purposes. Wells were all constructed from concrete or wood and between 700–2000 mm in diameter. Environmental attributes recorded for each site include latitude, longitude, altitude, bore/well construction details (including, where available, the depth and geology at the slotted interval), surface geology, vegetation, land-use, and impacts. The covering of the bore/well was also noted. Each bore/well was sampled once in late autumn–early winter (April–July, 2003–2004) and once in late winter–early spring (August–October, 2002–2004) to capture possible seasonal influences on species occurrence (extremely hot summers preclude field sampling at that time of year, with daily temperatures exceeding 40°C).

Materials and methods

Environmental and hydrochemical sampling

Standing water level (SWL, in metres below ground level) and the maximum depth of each bore were measured to the nearest 0.05 m with a Richter Electronic Depth Gauge or weighted Lufkin tape measure. Temperature, pH, Eh, electrical conductivity, dissolved oxygen and turbidity were measured at –1 m SWL using a calibrated Yeo-Kal 611 water quality analyser. Water samples for laboratory analysis (undertaken at Chemistry Centre, Perth, Western Australia) were collected from –1 m SWL using a sterile bailer (Clearwater PVC disposable 38 × 914 mm), and stored in sterile, acidified, 250 ml plastic bottles. One 250 ml water sample was filtered through a 0.45 μm membrane and frozen for analysis of nutrients (total soluble N, total soluble P). Highly turbid samples were pre-filtered though a glass-fibre filter using a hand vacuum pump (Millipore Sterifil Aseptic 47 mm OM041). A second 250 ml water sample was refrigerated for laboratory determination of solute concentrations (Na^+, Ca^{2+}, Mg^{2+}, K^+, Cl^-, HCO_3^-, CO_3^{2-}, SO_4^{2-}, NO_3^-, SiO_2, Fe^{2+}/Fe^{3+}, Mn^{2+}, Sr^{2+}), alkalinity, hardness, colour, turbidity, pH, and total dissolved solids. Laboratory methods followed APHA (1995). Saturation indices and activity coefficients were calculated using PHREEQC (Pankhurst & Appelo, 1999).

Stygofaunal sampling

Bores and wells were sampled for all stygofauna using a plankton net of suitable diameter (47 mm, 97 mm, 147 mm, or 197 mm) to match the bore/well diameter. The net, with a weighted McCartney vial attached, was lowered to the base of the bore/well, then agitated up and down (±1 m, 6 times) to disturb the bottom sediment. Six hauls of the entire water column were made, the first three hauls used a 150 μm net to capture macrofauna, the second three hauls used a 50 μm net to capture microfauna such as rotifers. To minimise loss of fauna through bow-wave effects during hauling, the McCartney vial had the bottom removed and replaced with 50 μm mesh. The entire net haul sample was transferred to a labelled 120 ml polycarbonate container and preserved in 100% ethanol. To maximise preservation for possible DNA analysis, the ethanol was replaced after a few hours by decanting the sample through a 50 μm net and refilling the sample bottle with fresh 100% ethanol. To eliminate the possibility of faunal contamination between sites, the nets were sterilised by washing in a decontaminant (5% solution of Decon 90), then rinsed in distilled water and air-dried.

Ostracod sorting and identification

Preliminary subdivision of fauna was undertaken at CALM's Woodvale laboratory. Prior to sorting, samples were first separated into three size fractions by sieving through 250, 90, and 53 μm Endicott sieves. Ostracods were separated under Leica MZ dissecting microscopes and sent to the senior author and Dr I. Karanovic for further description and counting. Data from both whole specimens with soft parts intact and valves alone were recorded, with the type of record and state of preservation noted. Identification to species level was undertaken using a Leitz binocular microscope. The distinction of new species was based on adult specimens, utilising where possible, soft part and carapace morphology.

Images of each taxon were obtained using a Wild M400 photomicroscope with digital imagery and Cambridge S260 Scanning Electron Microscope (SEM) at the ANU Electron Microscopy Unit. Examples of each new species have been sent to Dr I. Karanovic at the Western Australian Museum for full taxonomic description, to be presented in a forthcoming monograph, and deposition of voucher specimens within the museum's collection.

Statistical analysis

The relationship between ostracod species and ecological variables was investigated using canonical correspondence analysis (CCA) in CANOCO version 4.5 (ter Braak & Smilauer, 2002). Environmental variables that showed strong covariance were omitted from the analysis, leaving a total of fourteen parameters. Scaling was focussed on the inter-species distance, using the bi-plot scaling method and no transformations were made to the species data. Only the 209 samples without missing values were included in the analysis. Samples that had a very strong influence on the analysis and were recognised to be outliers in one or more environmental parameters were classed as supplementary. A total of 89 species, including 81 candonids and 8 *Gomphodella* species, and 1002 occurrences were incorporated into the analysis. The model was evaluated using the Monte-Carlo test with 499 permutations. The significance and the explanatory power of variance for each of the environmental variables were determined by manual forward selection of individual variables. Species abundance data were used in all analyses.

Results

Species composition and distribution

Thus far, 111 species of ostracods have been recovered from bores and wells, 73 of which are new species. Eighteen taxa are referred to here as "cf", thus reflecting a close affinity to an already described species, six have been assigned to genera, but not yet species and four are of undetermined genera. In most of these cases, representatives with well-preserved soft parts have not yet been recovered. Only 29 species have previously been described, 14 of these are known only from the Pilbara. The other 15 known species are not restricted to groundwater habitats and have a broader distribution. The distribution and number of occurrences of each of the candonid species recovered are tabulated in Electronic supplementary material.

In total, 56% sites of the 448 sites and 47% of the 751 samples contained ostracod fauna. Of the sites with fauna that were sampled more than once, 63% had fauna on all occasions, 18% only in the wet season and 19% only in the dry season. Many of the samples (60%) with only seasonal occurrence of ostracods had very few animals present, often with only a few valves. In samples with ostracods present on all occasions, the assemblage did not appreciably change between seasons, but in some cases there was a difference in abundance.

Species richness was usually very low (mean = 1.75), the majority of samples (55%) are comprised of only one species (Fig. 5a). Three

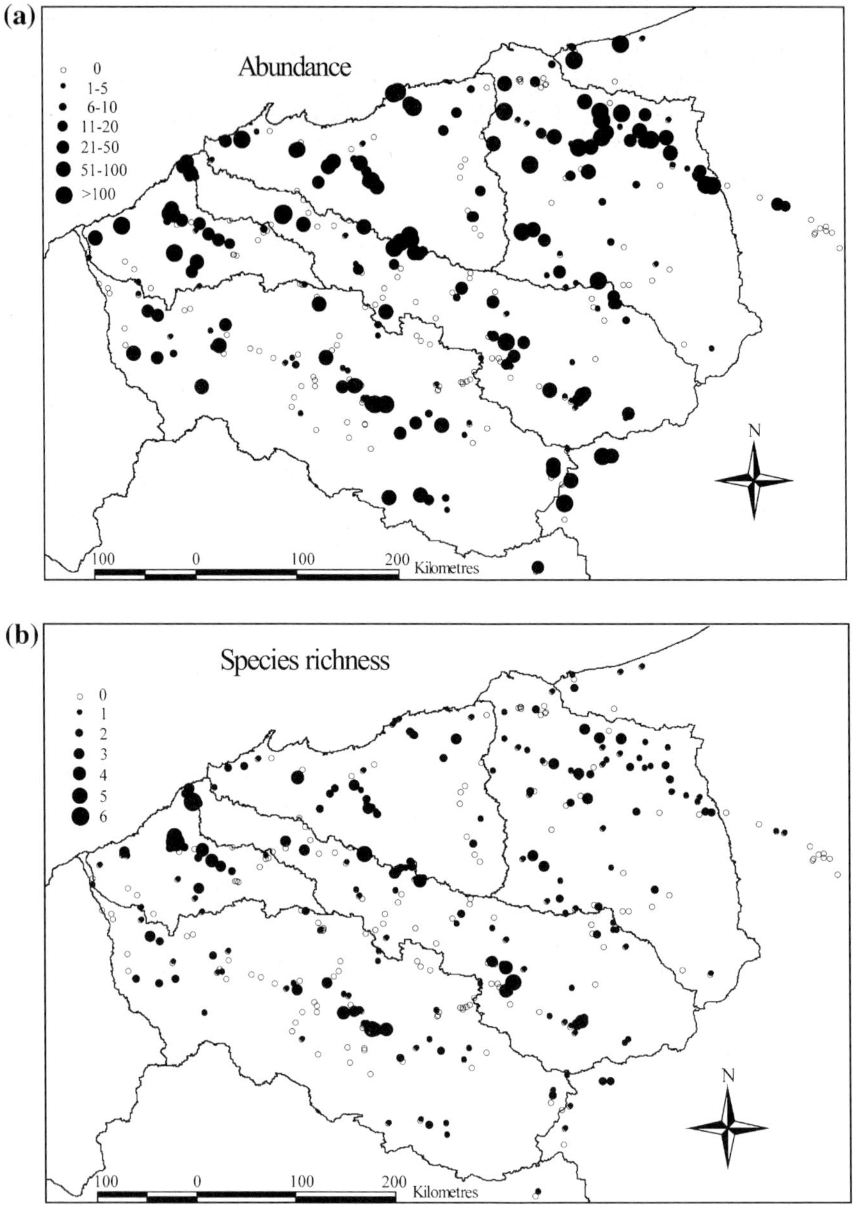

Fig. 5 Maps showing ostracod (**a**) abundances and (**b**) species richness across the Pilbara region. Open circles refer to sites with no recovered ostracods

samples had 6 species present; these were from the Robe River borefield and the Fortescue borefield. Other diverse sites were the Turee Creek and 7 Mile Creek borefields within the Ashburton basin, Weeli Wolli Creek and Marillana Creek in the upper Fortescue area, Cane River borefield in the Onslow basin, Tampathanna Pool, on the southern edge of the Chichester Ranges, and Harding Dam (see Fig. 2).

The abundance of ostracods was highly variable, with a mean number of 40 valves per sample (Fig. 5b), while 42% of samples were found to have less than 10 valves. The most abundant sites were dominated a single species of ostracod. Of the samples containing more than 100 individuals, six were from bore sites containing stygobitic species and four were well sites, with surface-water fauna. Most of the groundwater species

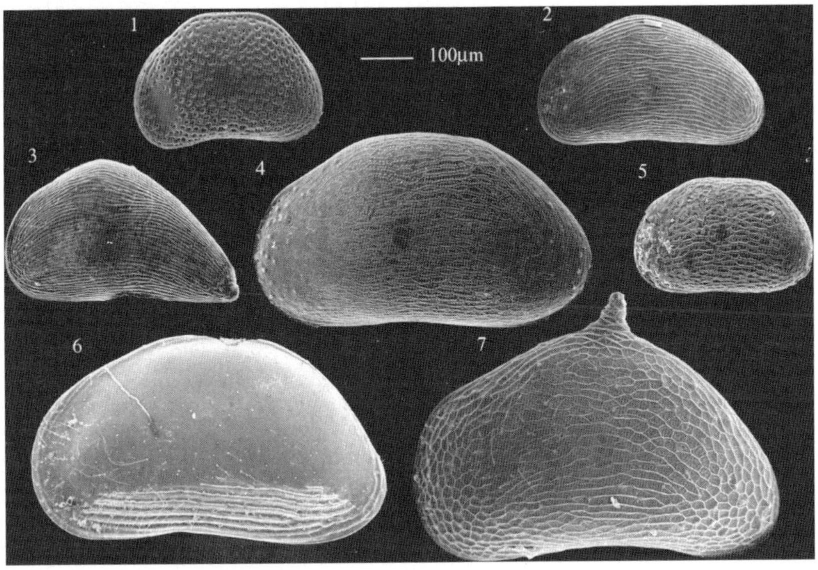

Fig. 6 SEM images of a selection of previously named stygobitic ostracod taxa from the Pilbara, identified in this study. All are left valves of adults; (**1**) *Candonopsis kimberleyi*, (**2**) *Notacandona modesta*, (**3**) *Pilbaracandona eberhardi*, (**4**) *Meridiescandona facies*, (**5**) *Meridiescandona lucerna* (**6**) *Humphreyscandona adorea*, (**7**) *Humphreyscandona fovea*

occurred in high abundance in localised areas, for example: *Humphreyscandona adorea* at Millstream, *Humphreyscandona woutersi* from the Robe River borefield, *Meridiescandona facies* from Marillana Creek and *Deminutiocandona* sp. 4 from the Cane River borefield.

The most broadly distributed fauna consists of surface-water taxa, most commonly found in samples from wells. Species such as *Cypretta seurati* were found in large numbers in such sites from each of the basins. The groundwater fauna was largely restricted to single drainage basins, and in many cases, aquifers. Forty-nine species were recorded from single sites; however, nine of these have been described previously from elsewhere.

By far, the most abundant group was the Candonidae. These included 64 new species and a further 15 that have been described before. Examples of previously described fauna also found in this study are presented in Fig. 6. The Pilbara candonids have been separated into 12 genera, four of which are considered to be new (I. Karanovic, in prep). All genera are represented by a number of species; *Areacandona* and *Deminutiocandona* are the most speciose, having 25 and 10 identified species respectively. Most of these genera are considered endemic to the Pilbara region, with only two species of *Candonopsis* having been previously recorded elsewhere. These include *C. tenuis*, which was described from eastern Australia (Brady, 1886; Sars, 1896) and *C. kimberleyi* (Karanovic & Marmonier, 2002), which was identified from the subterranean waters of the Kimberley region to the north of the Pilbara. All species identified in previous studies in the Pilbara were again collected in this study, with the exceptions of *Humphreyscandona pilbarae* (Karanovic & Marmonier, 2003), *Neocandona novitas*, *N. newmani*, *Areacandona arteria*, *A. mulgae* and *Origocandona gratia* (Karanovic, 2005). The bores from which these species were described were not re-sampled in the present study.

The groundwater ostracod fauna show clear distributional patterns, associated primarily with the extent of the surface water catchment or the aquifer (Fig. 7). Although there are a large number of *Areacandona* species, most were found within the low-lying coastal areas and alluvial aquifers of the Port Hedland, Robe and lower Fortescue basins. One species (*A.* sp. 25), considered to belong to the genus, occurs only to the east of the Oakover River, in the Great Sandy Desert. This is contrast to the previously known

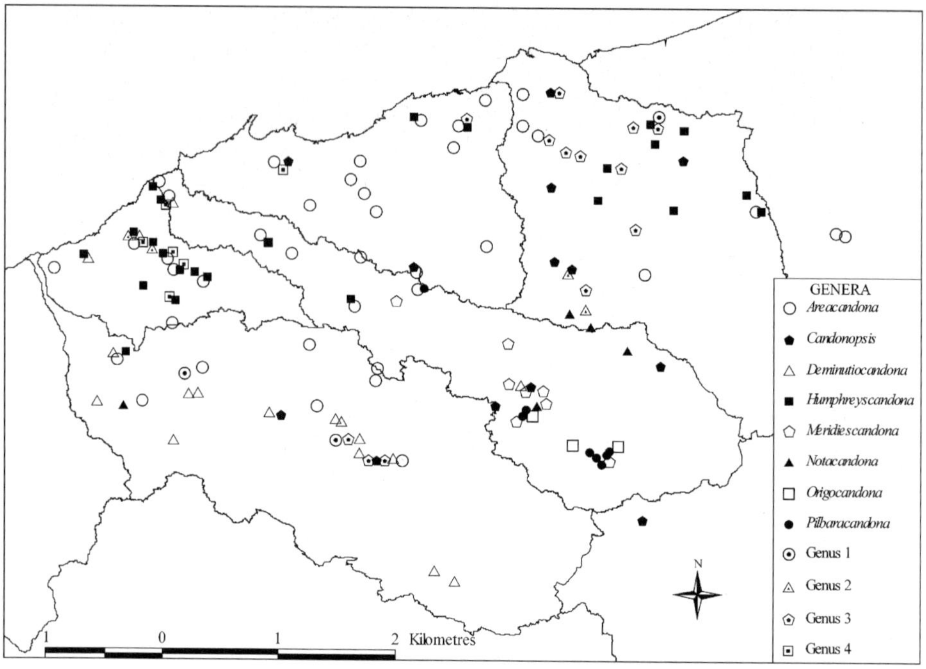

Fig. 7 Map showing the distribution of the genera of Candonidae across the Pilbara region

examples of the genus, *A. mulgae* and *A. arteria*, which were described from the Mulga Downs site in the central Fortescue (Karanovic, 2005). Each species of *Areacandona* occurred over a relatively small area, associated with one or two surface sub-catchments of tributaries flowing into the major rivers. Many species were restricted to single localities. More than one species of *Areacandona* was identified from several well sites. Most species of the genus have smooth, oblong valves, with rounded margins.

Humphreyscandona was associated mainly with the lower Fortescue and Robe River catchments, with each species being geographically distinct. The exceptions were *H.* sp. 2, which occurs in the De Grey/Oakover system and *H.* sp. 1 found around the upper Robe and Cane River region. These distributions were generally consistent with the known extent of the genus (Karanovic & Marmonier, 2003; Karanovic, 2005). *H. waldockae*, which was previously described from Mulga Downs, has now also been found near Port Hedland. Most of these species are comparatively large and well calcified, commonly with ridged or reticulated ornamentation of the valves concentrated around the periphery.

In contrast, *Deminutiocandona* was best represented both in diversity and abundance, in the Ashburton basin, with some species restricted to the Cane River borefield or Robe River borefield of the Onslow basin. Again, individual species showed narrow geographical distributions. Only one species, *D. mica*, has previously been described from the Weeli Wolli area (Karanovic, 2003). This species has been recovered from only one site in this study, and no other species of this genus have been found in Fortescue upland areas.

The unnamed candonid Genus 1, currently comprising two species, has a disparate distribution. Characteristically, species of this genus include valves that are rounded anteriorly and pointed posteriorly with an arched dorsum. The first species, with a heavily ridged carapace, was found in abundance in Eel Ck region of the De Grey basin. The second species, with smooth valves, was recovered from two bore sites in Ashburton basin.

A second unnamed candonid (Genus 2) is also broadly spread, although individual species occur in restricted regions. The genus is characterised by very small, smooth, oblong valves. Two species have been identified; one from the Robe borefield and a second from the upper Shaw River in the De Grey Basin. A third species, thought to belong to the genus, has been identified from a number of sites in the Harding Dam regions, although specimens with soft parts have not yet been recovered to confirm this.

The third open genus (Genus 3) is found largely in the De Grey Basin, from where six species have been differentiated, and a further species from the adjacent Port Hedland region. Species within the genus are characteristically small, with smooth carapaces, clearly differentiated from the other genera on the basis of soft part morphology. Most of the species are restricted to single localities or close regions, however valves of one species (Genus 3 sp. 1) have been located in the Turee Creek region of the Ashburton Basin, more than 200 km from the type locality in the headwaters of the Coongan River. Soft parts of these specimens have not been recovered to confirm identification.

Two species have been tentatively placed in the new candonid Genus 4, their taxonomic affinities are yet to be finalised. They both have smooth valves, tapered anteriorly, and are restricted to the lower Fortescue—Robe River systems.

There was a marked distinction within the Fortescue system between the fauna of the low-lying areas and those of the uplands of the Hamersley Ranges with almost no genera in common. *Meridiescandona* are dominant in the Marillana Creek—Weeli Wolli region, from where the genus was first described (Karanovic, 2003). All upland *Meridiescandona* species were comparatively large and had highly ornamented valves. Two species occurred in the central Fortescue region but these have smooth carapaces.

Pilbaracandona, with clearly ridged valves, also occurs in the uplands, concentrated around the Newman region. This region is particularly diverse, with four species identified. Two species of the genus *Origocandona* were previously described from the Newman region (Karanovic, 2005), only one of which, *O. inanitas*, was collected during this study. In addition, a new species from Kalgan Creek and Weeli Wolli Creek has been identified.

The genus *Notocandona* also showed disparate distribution, with representatives in the uplands

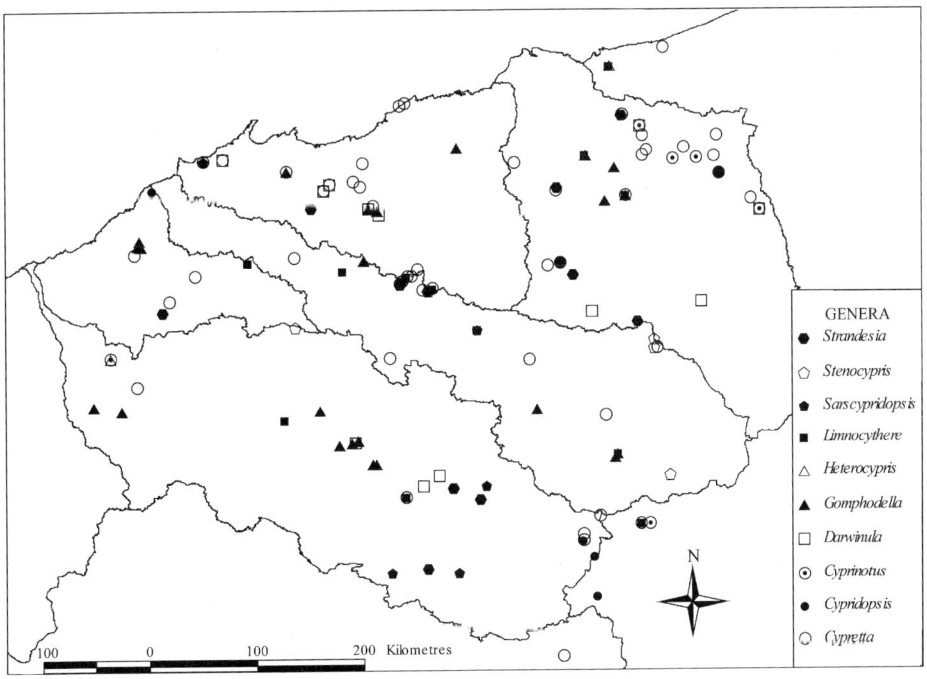

Fig. 8 Map showing the distribution of the genera of non-Candonidae taxa across the Pilbara region

of the Weeli Wolli region, the lower Ashburton, and north of the Chichester Range in the southern De Grey basin, with species unique to each region. Both *Notcandona boultoni* and *N. modesta* have been identified from their type locality, near Weeli Wolli Creek. As for other genera, the species from the upland Weeli Wolli sites have the highest degree of ornamentation, with well-developed ridges across the entire carapace.

Apart from the stygobitic Candonidae, several other stygoxene (sensu Gibert et al., 1994) taxa were recovered. The genus, *Gomphodella*, which is commonly found in springs, thus would best be referred to as crenophilic, was well represented with eight species. Most of the *Gomphodella* species were restricted to the Pilbara region, with several species found at only one locality (Fig. 8). The Ashburton basin was particularly well represented, with five species, each restricted to a separate sub-catchment. The Port Hedland basin had two species, also very localised. The species *Gomphodella hirsuta*, originally described from the Newman borefield (Karanovic, in press), showed a broader distribution, and has now also been identified from the Robe borefield, and two sites in the De Grey basin.

Darwinulids have been found at numerous localities, although have presently not been identified. Three species of Darwinulidae: *Penthesilenula brasilensis, Vestalenula matildae* and *Vestalenula marmonieri* have previously been described from Cape Range and the Pilbara region (Halse et al., 2002; Martens & Rosetti, 2002). In the present study, darwinulids were found in the Port Hedland, De Grey and Ashburton basins, mostly from well sites. There was no clear pattern to their distribution.

Several Cyprididae were found, particularly in the samples taken from uncovered wells, where epigean fauna may have colonised via passive dispersal (Fig. 8). These included *Cypretta* sp., *Cyprinotus kimberleyi*, *C. edwardi*, *Heterocypris kimberleyi*, *Strandesia kimberleyi* and three other unnamed species, *Stenocypris bolieki*, some fragments of *Bennelongia* sp., *Cypridopsis vidua*, *Sarscypridopsis ochracea* and *Limnocythere* sp.. These species showed no clear geographic distribution, and have previously been described from outside the Pilbara region, many in the Kimberley to the north (McKenzie, 1966). Valves of mixed marine origin, including *Xestoleberis*, *Paranesidea*, *Australimoosella* and *Paracytheroma*, were recovered from two coastal sites from within porous aquifers, indicating influence of the proximal seawater. These genera are common in the nearby marine waters (Hartmann, 1978). As these taxa were not the primary concern of this study, they are not considered further here.

Most groundwater species occurred primarily in isolation, as observed by the poor diversity in most sites, but a few associations were noted. These include: *Deminutiocandona* sp. 2, *D.* sp. 9 and *Candonopsis* sp. 2 from the Turee Creek borefield; *D.* sp. 1 and *Humphreyscandona woutersi*, with *Areacandona* sp. 2 also common from the Robe River borefield; and *H.* sp. 2 with either candonid Genus 1 sp. 1 in the Oakover system or *Areacandona* sp. 25 in the Great Sandy Desert. At this stage, it would be premature to assign sympatry to these assemblages and further sampling is required.

In many of the samples from bores sites, in particular, the recovered ostracods were poorly preserved, with soft parts missing. This may be a result of the sampling technique, or an indication that the ostracods were not actually living in the boreholes. For most species, a biocoenosis assemblage was recovered from only one site, or from several boreholes in close proximity. Other sites comprised few well-preserved individuals, empty or disarticulated valves. Species that were dominant in the well sites, such as *Cypretta* sp., *Stenocypris bolieki* and *Cyprinotus kimberleyensis*, were recovered as well preserved, life assemblages from many sites.

Physico-chemistry of the waters

A summary of the data collected from the sites, with and without ostracods, is presented in Electronic Supplementary Material—Appendix 2. The basal depth of bores and wells varied from 3–160 m and depth to groundwater varied from 0–88 m, increasing inland. Coastal and alluvial sites were generally less than 10 m SWL, whereas those on the Hamersley Range exceeded 20 m. Water temperature ranged from 17.3–35.5°C, with a mean of 29.6°C. The highest temperatures were recorded from the lower De Grey River, in the northeast corner of the study area. Temperature variations of up to 5°C were noted between seasons in some of the shallower sites. Turbidity was mostly low, with a mean of less than 6 NTU. Isolated sites with high turbidity were mostly wells in the lower Ashburton. High levels of total nitrogen and phosphorus were also recorded from these sites. Dissolved oxygen (DO) measurements were taken for all sites at the time of sampling. However, several sites selected for profiling of water chemistry were shown to have highly variable values of DO down the core. In some cases changed DO related to a halocline but in others there was no obvious explanation. Routinely collected DO values did not reflect whole of bore conditions reliably and were not used in analyses. Eh values mostly indicated oxidising conditions (mean = 334 mV, ranges = –558–837 mV). Exceptions to this were isolated wells within the lower Ashburton and De Grey basins.

The chemistry of groundwaters of the Pilbara is, at a first glance, relatively predictable (Fig. 4). However, host rock and the aquifer-type primarily determine the composition of the water, and chemical patterns were identified both between and within basins. Most drainage basins showed a clear gradation from characteristically hard, Mg–Ca–Alk-rich headwaters to more Na–Cl dominated lowland water. This was coincident with the change from banded-iron formations (BIF), granite and basaltic fractured rock aquifers in the uplands to the sedimentary and alluvial valley-fill aquifers downstream. Intermediary waters of the Na–Mg–HCO_3-type were also common, particularly in coastal and alluvial sites.

Salinity in Pilbara groundwaters is mostly low, with mean total dissolved solids (TDS) below 1 g l^{-1}, increasing toward the coast. The highest values were recorded at Fortescue Marsh and the remote inland areas of the De Grey basin, both with salinity in excess of 10 g l^{-1}. Some sites within alluvial systems also had increased salinity (~1.5–2 g l^{-1}), reflecting recharge by seepage of evaporated river waters. The freshest waters were recovered from the deeper waters of the fractured rock aquifers. In most cases, there was no appreciable seasonal difference in salinity, although some of the shallow, alluvial sites showed decreased salinity of up to 0.5 g l^{-1} in the wet season. Salinity was found to be generally higher in well sites (mean 1.2 g l^{-1}, range 0.2–5.2 g l^{-1}) than bores (mean 0.9 g l^{-1}, range 0.04–13 g l^{-1}).

Alkalinity varied across the Pilbara, according to the aquifer host-rock. Mean total alkalinity was 284 mg l^{-1}, with a range of 5–3180 mg l^{-1} and the highest values were recorded in the lower Ashburton and the upper De Grey basins, associated with the coastal plain and fractured Proterozoic volcanics, respectively. The lowest values were in the central Fortescue, the Millstream region and in some of the coastal sites, away from areas of active carbonate precipitation.

Carbonate saturation showed a similar distribution to alkalinity, with waters supersaturated with respect to both calcite and dolomite near the upper De Grey and Oakover Rivers, Hooley Creek, around Newman, the Angelo River, the lower Robe and western Port Hedland coastal basin. Undersaturated areas included the upper Robe, the mid-Fortescue and De Grey coastal basins, from where ostracod valves were only rarely recovered.

Chlorinity was highly variable throughout the Pilbara, ranging from 0.2–175 meq L^{-1}, with a mean of 8.8 meq L^{-1}. The pattern of chlorinity across the region was predictable, being highest near the coast and in the alluvial deposits of the mid Fortescue and the lower Ashburton. Wells in the arid inland parts of the upper De Grey, Great Sandy Desert and Ashburton basins also had high chlorinity, probably because of evaporative enrichment. Lowest values were recorded from the upper Fortescue catchments of Weeli Wolli and Marillana Creek.

Most of the Pilbara waters are considered hard to very hard; mean 424 mg l^{-1} and range of 10–6400 mg l^{-1}. There is notable variation in Mg/Ca ratio across the Pilbara; Ca^{2+} dominates waters in the calcrete areas of the De Grey and upper Ashburton basins, whereas Mg^{2+} dominates those in the dolomite-rich parts of the Fortescue, such as Millstream and Weeli Wolli.

With respect to pH, the majority of sites sampled were circum-neutral to moderately alkaline. An exception is the Marillana Creek catchment, where pH as low as 4.4 was recorded.

Oxidation of pyritic shale has been noted in some of the mining lease areas, resulting in acidic, sulphate-rich groundwaters (Johnson & Wright, 2001; Woodward-Clyde, unpublished). Surprisingly, ostracods were identified from these sites, with *Meridiescandona facies* being recovered in large numbers.

Ostracod distribution and water chemistry

The presence of ostracods in Pilbara bores and wells was predominantly determined by the pH ($P < 0.001$) and the carbonate saturation ($P = 0.001$) of the host waters (see Electronic supplementary material for other parameters). Samples with pH below 6 and with Eh values indicating reducing environments or total nitrogen concentrations in excess of 10 mg l^{-1}, rarely contained ostracods.

Among the samples with ostracods, there was a clear distinction between sites with surface water fauna, such as cypridids dominating well sites, and those with a candonid fauna dominant in most bore sites. There was also a significant secondary relationship with salinity and solute composition, with surface water species preferring the more saline waters of higher chlorinity, although the distribution of surface versus groundwater fauna was somewhat distorted by the sampling method, as most samples from the Great Sandy Desert, Oakover River and other remote areas were taken from wells, leading to a larger proportion of surface-water species being present.

The relationship between ostracod species distribution and environmental variables is explored through CCA analysis for samples with only stygobitic ostracods present (Fig. 9). The results of the analysis, incorporating 10 variables, are summarised in Tables 1 and 2. Both the first axis and the model are significant at the 99% level ($P < 0.01$). The first four axes of the CCA combined explained only 5.6% of the variance in species composition, but 54.3% of the variance in species–environment relations. This low explanatory power is due to the very large number of zeroes in the data set, with many species occurring at only one site.

Correlation coefficients for each of the environmental variables incorporated into the CCA

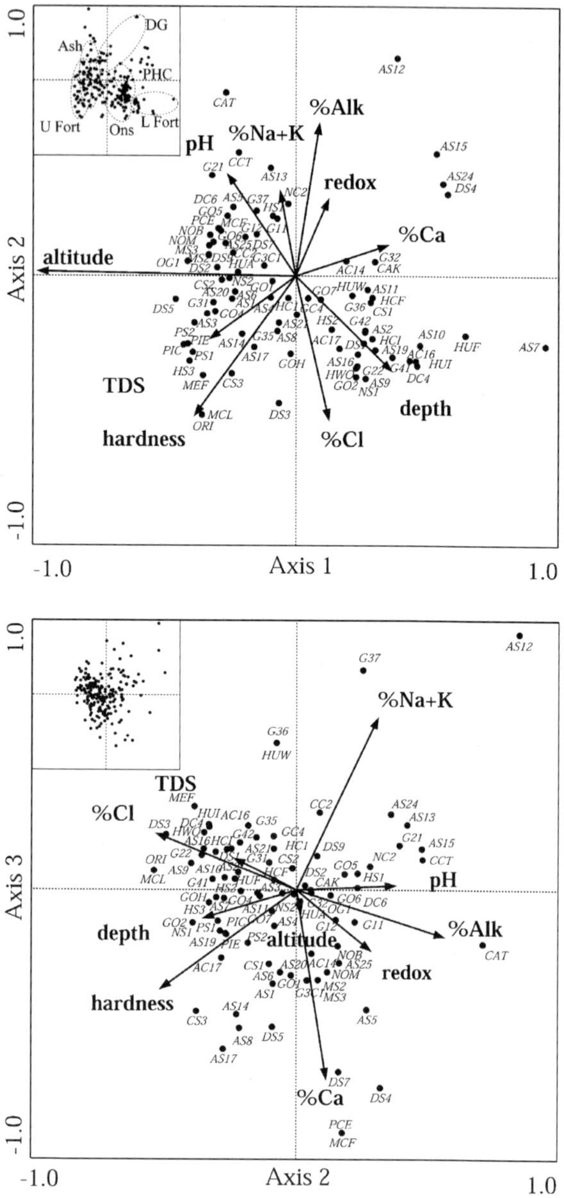

Fig. 9 CCA species–environment biplots for (**a**) axes 1&2 and (**b**) axes 2&3. *Arrows* and *heavy font* refer to environmental variables, species codes, as in Electronic Supplementary Material Appendix Table 3, are in italics. The small inset plots refer to the ordination of samples. See Table 1 for results of the CCA and text for further details. The codes for the surface water basin are Ash—Ashburton, DG—De Grey, PHC—Port Hedland Coastal, L Fort—Lower Fortescue, U Fort—Upper Fortescue

with the resulting first four axes are tabulated in Electronic supplementary material. Altitude (–0.95 correlation) was the by far the dominant

Table 1 Percentage of variance explained and significance of each of the environmental variables used in the CCA

Variable	Variance	*P*-value	*F*-value
Altitude	0.91	0.002	3.53
%Na + K	0.65	0.002	2.53
%Alk	0.67	0.002	2.62
%Ca	0.55	0.006	2.13
Hardness	0.55	0.002	2.17
Depth	0.53	0.004	2.12
TDS	0.50	0.002	2.03
%Cl	0.44	0.008	1.79
pH	0.42	0.002	1.72
Eh	0.42	0.002	1.69
Sum	5.64		

environmental factor determining the first axis. Depth to water (0.35) gave a notable positive correlation, which is related primarily to proximity to coast and altitude. Ostracods of the Fortescue system were classified into three main categories; (1) a lowland group, incorporating the coastal and lower floodplain taxa (*Areacandona, Humphreyscandona*), (2) an upland group (*Pilbaracandona, Origocandona, Meridiescandona*), focussed around the upper Fortescue and (3) an intermediary group. Very few genera were shared between the upland and lowland groups. Taxa from the midland areas show confluence at generic level but species are restricted to middle elevations. Coincident with changes in altitude are changes in temperature, depth to water and salinity. In addition, aquifer lithology differs from metasedimentary and volcanic rocks in the uplands to predominantly soft sediment, alluvial floodplain and coastal deposits in lowlands.

Position on the second CCA axis was determined by the relative concentrations of alkalinity (0.51) and chloride (–0.48), with hardness also significant (–0.46). This describes a solute evolution, from HCO_3^- rich waters, such as in the dolomite aquifers of the uplands and the spring discharge sites, to Cl^- being the dominant anion in the coastal and central floodplain deposits, where groundwaters are shallow and recharged in part by evaporated stream water. Perhaps associated with a clear relationship between solute evolution and an increasing proportion of surface water species, the stygofaunal ostracod species occurring in Cl^- rich water, such as *Candonopsis* and *Areacandona*, usually have smooth and more poorly calcified valves, indicating a bicarbonate-limited environment. In contrast, those in waters with higher relative bicarbonate, such as *Meridiescandona* and *Notacandona*, have more robust and ornate valves (see Fig. 6 for examples).

The third axis reflected the cations, determined by the relative concentration of $Na^+ + K^+$ (0.54) and Ca^{2+} (–0.59), which is closely related to host-rock lithology, as well as solute evolution. Sites with Ca^{2+}–Mg^{2+}-rich waters in the Newman and Ethel Gorge area, and in calcretes of the upper Ashburton and De Grey basins contained taxa such as *Pilbaracandona* that have large, heavily-ridged valves compared to the small, smooth valves of *Deminutiocandona* and Genus 3 in the more Na^+-dominated waters of the lower De Grey and Ashburton basins. Even the species of *Areacandona* and *Humphreyscandona* that were found in hard waters have heavily calcified valves.

Other significant environmental factors included Eh and pH. Although pH was relatively uniform across the Pilbara, rare sites had low pH (<5) associated with oxidation of pyrite. Only a few taxa, such as *M. facies* and one population of *A.* sp. 12, were found at these sites, although soft parts were not preserved. Eh was shown to be

Table 2 Summary of the results of the CCA performed on 209 active samples and 89 active species (see text for further details)

Axes	1	2	3	4	Total inertia
Eigenvalues	0.930	0.794	0.686	0.628	
Species–environment correlations	0.973	0.900	0.842	0.821	
Cumulative percentage variance					
Of species data	1.7	3.2	4.4	5.6	
Of species–environment relation	16.5	30.6	42.9	53.9	
Sum of all eigenvalues					54.278
Sum of all canonical eigenvalues					5.636

quite variable across the region, with ostracods preferring the oxidised sites. There is no clear relationship between species traits and redox potential, although sites in the lower Robe and De Grey Rivers registered a low Eh.

Discussion

Sampling methods

Preliminary sampling by Halse et al. (2002) of a series of five spring sites in the Pilbara revealed a stygophilic ostracod fauna. These species have been found to not be representative of those recovered from the deeper aquifer itself, contrary to the findings of Gibert et al. (1994). Of the four species identified, *Candonopsis tenuis* and *Limnocythere dorsosicula* have been described from sites across Australia and *Vestalenula marmonieri* from New Caledonia. Only *V. matildae* is thus far considered endemic to the Pilbara (Martens & Rosetti, 2002). Such species have been found in the current sampling program in wells but not bores. *Humphreyscandona adorea* is the only named species currently known from in the hyporheic zone and at depth (S.A. Halse, unpublished data). The hydrochemistry of the groundwater samples is significantly different to that of springs; where the average pH measured was 8.2, conductivity 1700 μS cm^{-1}, and dissolved oxygen in excess of 100%.

The validity of sampling from bores as representatives of aquifers has been questioned, due to potentially increasing DO and dissolved organic carbon, introducing metals to the system via bore casings and permitting mixing of both fauna and water types (Humphreys, 2001b). Although this cannot be categorically ruled out, it is most likely the abundance rather than the diversity of taxa present that would be affected, as supported by this study. There was no observable correlation between bore type and presence, abundance or diversity of ostracod fauna.

The sampling efficiency of the net haul method versus pumping was evaluated in five bores. Sampling by the net haul method collected 34% of the abundance (mean summed proportions for 5 bores) of what was collected by the pumping method (S. Eberhard, pers. commun.). Sampling in subsequent seasons showed variation in the abundance of fauna present in alluvial aquifers, but the taxa present were not greatly altered. This may be in part due to sampling discrepancies between seasons; however, the sites in the alluvial aquifers particularly, are subject to disturbance via scouring of the streambeds during peak flow times during monsoonal and cyclonic rainfall (Davies, 1996; Marmonier et al., 2000).

Pilbara species diversity

Prior to this study, there were published records of 332 species of all stygofaunal taxa from the Pilbara, the majority in the Fortescue basin (Eberhard et al., in press). Fifteen of the seventeen major taxonomic groups of stygofauna have been found in the Pilbara. Even in the preliminary studies, ostracods were particularly well represented, comprising 12.7% of all groundwater taxa, compared with around 3% worldwide (Eberhard et al., in press). The results of this study suggest ostracods represent about 30% of all stygofaunal species in the Pilbara but it should be recognised that ostracods have received more comprehensive examination than many other groups.

High degrees of endemicity are common in groundwater faunal distributions (Gibert et al., 1994). By mid-2004, the PASCALIS (Protocols for the Assessment and Conservation of Aquatic Life in the Subsurface) group had 1239 species of all stygobitic taxa from 11,000 distribution records in France, Italy, Belgium, Slovenia, Spain and the Canary Islands combined (Gibert, 2004). The Pilbara fauna, with more than 70 species of stygobitic ostracods alone, supports the notion of this region being a subterranean biodiversity "hot spot". Szczechura (1980) attributed the genetic isolation of ostracods, and subsequent speciation, to the ability of the group to prosper in a wide range of habitats and withstand or respond to environmental change.

Most previous investigation in the Pilbara was concentrated on groundwater primarily in calcrete deposits (Humphreys, 2001b). However, the alluvial aquifers of the Pilbara are some of the most extensive and contain abundant supplies of freshwater. Alluvial aquifers in Europe, such as in the Rhône, are known to be some of the most abundant and diverse habitats for stygofauna (Marmonier et al., 1993; Rouch & Danielopol, 1997) and also yield well in the Pilbara (Eberhard et al., 2004, in press). Fractured rock aquifers in the Pilbara have previously attracted almost no attention, but have been shown to host ostracods, and other stygofauna (Eberhard et al., 2004).

The regional groundwater ostracod fauna in the Pilbara is extremely large for the size of the region, despite individual sites usually yielding few species. Surrounding regions, such as the Kimberley (Humphreys 1999; Karanovic & Marmonier, 2002) and Murchison (Karanovic & Marmonier, 2002) have not been studied as intensively but clearly have much smaller faunas. Interestingly, there is far greater diversity among stygofaunal ostracods in the Pilbara than among their surface water counterparts (S.A. Halse unpublished data). This has previously been noted only for the stygobitic harpacticoid fauna of Brazil (Rouch & Danielopol, 1997).

Ostracod species distribution

As with most groundwater faunas, distributions in the Pilbara are usually restricted (see Strayer, 1994). Perhaps as a result of the localised distributions and discontinuity of many of the aquifers, particularly in the patchy calcrete regions, many species were collected at only one site. However, a high proportion of singletons are also a feature of surface water studies in Western Australia (Halse et al. 2000; Pinder et al. 2004), suggesting there may be a more general explanation than the nature of groundwater distributions.

Of the truly stygobitic ostracod fauna, no species are shared with the adjacent regions. Furthermore, no Tethys-associated ostracods have been recorded from the Pilbara, unlike within the Tertiary limestones on Cape Range (Danielopol et al., 2000). This region has water of marine origin intersecting with fresh groundwater, forming anchialine systems and was inundated with marine water during the Cainozoic (BMR, 1990).

All stygobitic species in the Pilbara are from freshwater lineages and have Gondwanan affinities (i.e. Candonidae, Darwinulidae). The Candonidae in particular, with their lack of pigmentation, blindness and exaggerated limbs, are considered to have been long adapted to subterranean life. In a climate as arid as Australia's, groundwater provides one of the few permanent freshwater environments suitable for the candonids. In fact, the only known epigean candonids, other than *Candonopsis*, found in Australia from either modern or fossil deposits, are from two dolomitic groundwater-fed swamp sites in Tasmania (De Deckker, 1982). Even *Candonopsis* has been located predominantly in temporary pools, which may also be groundwater fed. In the Pilbara, speciation of the ostracod fauna most likely occurred subsequent to the termination of large surficial conduits during the late Tertiary (van de Graaff et al., 1977), resulting in geographic isolation of fauna within the aquifers. The calcretes that were the focus of the early studies formed in the late Tertiary along the palaeodrainage channels (Bowler, 1976).

Biogeography

There is clear sub-regional differentiation in ostracod fauna, determined primarily by altitude, then surface drainage basin, then aquifer. Altitude does not only reflect height above sea-level, it also affects many secondary factors, such as depth to water, temperature, host-rock chemistry and aquifer type, which may all have an effect on the divergent morphology and taxonomy of ostracods.

The three broad stygofaunal groupings suggested by Humphreys (unpublished) were supported in this study within the Fortescue system: a lowlands group, dominated by *Areacandona* and *Humphreyscandona*, an uplands group comprising *Meridiescandona*, *Notocandona* and *Pilbaracandona*, and an intermediary group. The intermediate group of fauna, centred on the drainage divide between the upper and lower Fortescue, is not as well established in this study because many of the taxa found previously in the Mulga Downs region were not recovered. However, a small suite of ostracods, including species of *Meridiescandona* and *Areacandona*, were found restricted to this central Fortescue region.

There are also shared fauna between lower Fortescue and lower Robe catchments, which reflect the previous course of the Fortescue River that shifted only during the Late Pleistocene. The upper De Grey and Ashburton basins both comprise distinct fauna, each dominated by one genus.

Morphological response to environment

Although belonging to different genera, species can show similarities in morphology in the same or adjacent sites, most likely in response to the ecological/environmental variables. This convergent morphology is in support of Danielopol et al. (1994) who realised common traits in unrelated species from similar environments, both in groundwater and the deep sea and attributed them to adaptive responses. For example, the smooth and tapered forms of *Areacandona* and *Deminutiocandona* are found in alluvial aquifers; the large and well-calcified valves of *Humphreyscandona* in bicarbonate-rich waters and the ornate valves of *Meridiescandona* in Mg^{2+}-rich, lower pH environments. These morphological responses are seen at the generic level, with species of different genera at the same locality showing similar characteristics in the carapace.

Conclusion

This study is the first time that a systematic sampling program for stygobitic ostracods has been undertaken in Australia. A plethora of new species has been discovered and many are currently being described. The restriction of many taxa to single aquifers has great implications for their conservation and management in this economically significant region.

The distribution of species appears to be controlled primarily by historical events that lead to the formation and the extent of the host aquifer, with pre-adaptive colonisation and subsequent speciation. Within an aquifer, alkalinity, salinity (as% Na + K), and pH, together, govern the occurrence of taxa, as determined by canonical correspondence analysis. A combined knowledge of hydrology and hydrochemistry is required to assess the likelihood of ostracod occurrence within aquifers of the Pilbara region. The development of tolerance limits of a wide range of parameters for the known occurrences of each species will assist in the assessment of the effect of any likely impact to the subterranean ecosystem.

The results of the current study confirm and expand upon the predictions of Humphreys (unpublished) that (1) Pilbara stygofauna are restricted to the Pilbara; (2) there are distinct sub-regional patterns of taxonomic groupings; and (3) that not only all undisturbed calcretes, but nearly all local Pilbara aquifers are likely to have ostracods.

Acknowledgements The authors wish to thank M. Scanlon, J. Cocking, and H. Barron for tirelessly undertaking the fieldwork and sorting out the ostracods on which this paper is based. Dr I. Karanovic identified the ostracods in some of the samples on which this paper is based and provided advice on ostracod identification. Jenny McGuire at the Western Australian Chemistry Centre performed the analyses on water chemistry. JMR would also like to thank the Statistical Consulting Unit and the Electron Microscopy Unit of the ANU. Funding for this project was provided by Conservation and Land Management, WA awarded to PDD. We are grateful for comments of two anonymous reviewers that clarified some of the finer points of the manuscript.

References

APHA, 1995. Standard Methods for the Examination of Water and Wastewater. American Public Health Association, Washington.

Balleau, W. P., 1972. Outline of groundwater in the Fortescue River valley: Western Australia Geological Survey Record 1986/8.

Barnett, J. C. & D. P. Commander, 1985. Hydrogeology of the Western Fortescue Valley, Pilbara region, Western Australia. Western Australia Geological Survey Record 1986/8.

Beard, J. S., 1973. The elucidation of palaeodrainage patterns in Western Australia through vegetation mapping. Vegmap Publications, Perth. Vegetation Survey of Western Australia, Occasional Paper 1.

Beard, J. S., 1998. Position and development of the history of the central watershed of the Western Shield, Western Australia. Journal of the Royal Society of Western Australia 81: 157–164.

BMR Paleogeographic Group, 1990. Australia: Evolution of a Continent. Bureau of Mineral Resources, Australia.

Bowler, J. M., 1976. Aridity in Australia: age origins and expression in aeolian landforms and sediments. Earth-Science Reviews 12: 279–310.

Bradbury, J. H. & W. D. Williams, 1996a. Two new species of anchialine amphipod (Crustacea: Hadzzidae:

Liagoceradocus) from Western Australia. Records of the Western Australian Museum 17: 395–409.

Bradbury, J. H. & W. D. Williams, 1996b. Freshwater amphipods from Barrow Island, Western Australia. Records of the Western Australian Museum 48: 33–74.

Brady, G. S., 1886. Notes on the freshwater entomostraca from South Australia. Proceedings of the Scientific Meetings of the Zoological Society of London 1: 82–93.

Culver, D. C., 1982. Cave Life: Evolution and Ecology. Harvard University Press, Cambridge, MA, USA.

Christiansen, K., 1962. Proposition pour la classification des animaux cavernicoles. Spelunca 2: 76–78.

Danielopol, D. L., M. Creuze des Chatelliers, F. Mösslacher, P. Popisil & R. Popa, 1994. Adaptation of Crustacea to interstitial habitats: a practical agenda for ecological studies. In Gibert, J., D. L. Danielopol & J. A. Stanford (eds), Groundwater Ecology. Academic Press, San Diego, USA, 218–243.

Danielopol, D. L., A. Baltanas & W. F. Humphries, 2000. *Danielopolina kornickeri* sp. nov. (Ostracoda, Thaumatocypridoidea) from a Western Australian anchialine cave: morphology and evolution. Zoologica Scripta 29: 1–16.

Danielopol, D. L. & G. Hartmann, 1986. Ostracoda. Part I: stygobiont Ostracoda from inland subterranean waters. In Botosaneanu, L. (ed.), Stygofauna Mundi. A Faunistic Distributional and Ecological Synthesis of the World of Fauna Inhabiting Subterranean Waters. E.J Brill, Leiden, 265–267.

Davies, P. M., 1996. The influence of flow conditions on the structure of the aquatic fauna of streams of the Pilbara, Western Australia. In Noonan, D. (ed.), An Ecological Perspective on Cooper's Creek. Australian Conservation Foundation, Adelaide, 3.

De Deckker, P., 1982. Non-marine Ostracods from two Quaternary profiles at Pulbeena and Mowbray Swamps. Alcheringa 6: 249–274.

De Laurentis, P., G. L. Pesce & W. F. Humphreys, 2001. Copepods from ground waters of Western Australia, VI. Cyclopidae (Crustacea: Copepoda) from Yilgarn region and Swan coastal plain. Records of the Western Australian Museum 19: 243–257.

Eberhard, S. M., S. A. Halse, M. D. Scanlon, J. S. Cocking & H. J. Barron, 2005a. Assessment and conservation of aquatic life in the subsurface of the Pilbara region, Western Australia. In Gibert, J. (ed.), World Subterranean Biodiversity, Proceedings of an International Symposium held 8th–10th December 2004 in Villeurbanne, France, Université Claude Bernard of Lyon 1, PASCALIS European Research Programme, 61–68.

Eberhard, S. M., S. A. Halse & W. F. Humphreys, 2005b. Stygofauna in the Pilbara region, north-west Western Australia: a systematic review. Journal of the Royal Society of Western Australia 88: 167–176.

Forester, R. M., 1983. Relationship of two lacustrine ostracode species to solute composition and salinity: implications for palaeohydrology. Geology 11: 435–438.

Forester, R. M., 1986. Determination of the dissolved anion composition of ancient lakes from fossil ostracodes. Geology 14: 796–798.

Gibert, J., J. A. Stanford, M.-J. Dole-Olivier & J. V. Ward, 1994. Basic attributes of groundwater ecosystems and prospects for research. In Gibert, J., D. L. Danielopol & J. A. Stanford (eds), Groundwater Ecology. Academic Press, San Diego, USA, 8–40.

Gibert, J., 2004. How to assess and conserve groundwater biodiversity. The PASCALIS European project in the international context: main results. Programme and Abstracts of the Symposium of World Subterranean Biodiversity, Lyon, 14.

Halse, S. A., R. J. Shiel, A. W. Storey, D. H. D. Edward, I. Lansbury, D. J. Cale & M. S. Harvey, 2000. Aquatic invertebrates and waterbirds of wetlands and rivers of the southern Carnarvon Basin, Western Australia. Records of the Western Australian Museum Supplement 61: 217–265.

Halse, S. A., M. D. Scanlon & J. S. Cocking, 2002. Do springs provide a window to the groundwater fauna of the Australian arid zone? In International Association of Hydrogeologists (ed.), Balancing the Groundwater Budget, Proceedings of the International Groundwater Conference, Darwin, 2002, 1–12.

Hartmann, G., 1978. Die Ostracoden der Ordnung Podocopida G.W. Müller, 1894 der tropisch-subtropisch Westküste Australiens (zwischen Derby im Norden und Perth im Süden). In Hartmann, G., G. Hartmann-Schröder (eds), Zur Kenntnis des Eulitorals der australischen Küsten unter besonderer Berücksichtigung der Polychaeten und Ostracoden, Teil 1. Mitteilungen aus dem Hamburgischen Zoologischen Museum und Institut 75: 64–219.

Humphreys, W. F., 1993a. Patterns of genetic diversity within selected subterranean fauna of the Cape Range Peninsula, Western Australia: systematic and biogeographic implications. Records of the Western Australian Museum Supplement 45: 145–164.

Humphreys, W. F., 1993b. The significance of the subterranean fauna in biogeographical reconstruction: examples from Cape Range Peninsula, Western Australia. Records of the Western Australian Museum Supplement 45: 165–192.

Humphreys, W. F., 1993c. Stygofauna in semi-arid tropical Western Australia: a Tethyan connection? Mémoire de Biospéléologie 20: 111–116.

Humphreys, W. F., 1999. Relict stygofaunas living in sea salt, karst and calcrete habitats in arid northwestern Australia contain many ancient lineages In Ponder, W. & D. Lunney (eds), The Other 99%. The Conservation and Biodiversity of Invertebrates. Royal Zoological Society of New South Wales, Mosman, 219–227.

Humphreys, W. F., 2000. The hypogean fauna of the Cape Range Peninsula and Barrow Island, northwestern Australia. In Wilkens, H., D. C. Culver & W. F. Humphreys (eds), Ecosystems of the World,

Vol. 30 – Subterranean Ecoystems. Elsevier, Amsterdam, 581–601.

Humphreys, W. F., 2001a. The subterranean fauna of Barrow Island, Northwestern Australia, and its environment. Mémoire de Biospéléologie 26: 108–127.

Humphreys, W. F., 2001b. Groundwater calcrete aquifers in the Australian arid zone: the context to an unfolding plethora of stygal biodiversity. Records of the Western Australian Museum, Supplement 64: 63–83.

Jaume, D. & W. F. Humphreys, 2001. A new genus of epacteriscid calanoid copepod from an anchialine sinkhole of Northwestern Australia. Journal of Crustacean Biology 21: 157–169.

Johnson, S. L. & A. H. Wright, 2001. Central Pilbara Groundwater Study. Water and Rivers Commission, Perth.

Karanovic, I., 2003. Towards a revision of Candoninae (Crustacea: Ostracoda): description of two new genera from Australian groundwaters. Species Diversity 8: 353–383.

Karanovic, I., 2005. Towards a revision of Candoninae (Crustacea: Ostracoda): Australian representatives of the subfamily, with descriptions of three new genera and seven new species. New Zealand Journal of Marine and Freshwater Research 39: 29–75.

Karanovic, I. & P. Marmonier, 2002. On the genus, *Candonopsis* (Crustacea: Ostracoda: Candoninae) in Australia, with a key to the world recent species. Annuals of Limnology 38: 199–240.

Karanovic, I. & P. Marmonier, 2003. Three new genera and nine new species of the subfamily Candoninae (Crustacea: Ostracoda: Podocopida) from the Pilbara region (Western Australia). Beaufortia 53: 1–51.

Karanovic, I., On the genus *Gomphodella* De Deckker, 1981 (Crustacea, Ostracoda, Limnocytheridae) with descriptions of three new species from Australia and redescription of the type species. Species Diversity (in press).

Marmonier, P., P. Vervier, J. Gibert & M.-J. Dole-Olivier, 1993. Biodiversity in groundwaters. Trends in Ecology and Evolution 8: 392–395.

Marmonier, P., M. Creuzé des Châtelliers, M. J. Dole-Oliver, S. Plenet & J. Gibert, 2000. Rhône groundwater systems. In Wilkins, H., D. C. Culver & W. F. Humphreys (eds), Ecosystems of the World 30 – Subterranean Ecosystems. Elsevier, Amsterdam, 481–512.

Martens, K. & G. Rossetti, 2002. On the Darwinulidae (Crustacea, Ostracoda) from Oceania. Invertebrate Systematics 16: 195–208.

McKenzie, K. G., 1966. Freshwater Ostracoda from northwestern Australia. Australian Journal of Marine and Freshwater Research 17: 259–279.

Pankhurst, D. L. & C. A. J. Appelo, 1999. User's Guide to PHREEQC (version 2) – a computer program for speciation, batch-reaction, one-dimensional transport, and inverse geochemical calculations. US Geological Survey Water-Resources investigation report 99-4259.

Pinder, A. M., S. A. Halse, J. M. McRae & R. J. Shiel, 2004. Aquatic invertebrate assemblages of wetlands and rivers in the wheatbelt region of Western Australia. Records of the Western Australian Museum Supplement 67: 7–37.

Poore, G. C. B. & W. F. Humphreys, 1998. The first record of the Spelaeogriphacea (Crustacea) from Australasia: a new genus and species from an aquifer in the arid Pilbara of Western Australia. Crustaceana 71: 721–742.

Radke, L. C., S. Juggins, S. A. Halse, P. De Deckker & T. Finston, 2003. Chemical diversity in south-eastern Australian saline lakes II: biotic implications. Marine and Freshwater Research 54: 895–912.

Rouch, R. & D. L. Danielopol, 1997. Species richness of microcrustacea in subterranean freshwater habitats. Comparative analysis and approximate evaluation. Internationale Revue gesamten der Hydrobiologie 82: 121–145.

Sars, G. O., 1896. On fresh-water entomostraca from the neighbourhood of Sydney, partly raised from dried mud. Archiv for Mathematik og Naturvidenskab 18: 1–81.

Szczechura, J., 1980. Causes for speciation in Ostracods. Neues Jahrbuch für Geologie und Paläontologie. Monatshefte 7: 439–441.

Strayer, D. L., 1994. Limits to the biological distributions in groundwater. In Gibert, J., D. L. Danielopol & J. A. Stanford (eds), Groundwater Ecology. Academic Press, San Diego, 287–310.

ter Braak, C. J. F. & P. Smilauer, 2002. CANONO Reference manual and CanoDraw for Windows User's Guide: Software for Canonical Community Ordination (version 4.5). Microcomputer Power, Ithaca, NY, USA.

Trendall, A. F., 1990. Hamersley basin. In geology and mineral resources of Western Australia. Western Australian Geological Survey, Memoir 3: 163–191.

van de Graaff, W. J. E., R. W. A. Crowe, J. A. Bunting & M. J. Jackson, 1977. Relict early Cainozoic drainages in arid Western Australia. Zeitschrift für Geomorphologie 21: 379–400.

Watts, C. H. S. & W. F. Humphreys, 1999. Three new genera and five new species of Dytiscidae (Coleoptera) from underground waters in Australia. Records of the South Australian Museum 32: 121–142.

Wilson, G. D. F. & W. F. Ponder, 1992. Extraordinary new subterranean isopods (Peracarida: Crustacea) from the Kimberley region, Western Australia. Records of the Western Australian Museum 44: 279–298.

Hydrobiologia (2007) 585:119–133
DOI 10.1007/s10750-007-0633-6

Ecological requirements of Ostracoda (Crustacea) in a heavily polluted shallow lake, Lake Yeniçağa (Bolu, Turkey)

Okan Külköylüoğlu · Muzaffer Dügel · Mustafa Kılıç

Abstract Although the area of Lake Yeniçağa is a potential candidate for RAMSAR convention, several anthropogenic factors compromise its biological diversity. This is mostly due to nutrient-rich water released from both point and nonpoint sources. Thirteen ostracod taxa (*Candona neglecta, C. candida, Ilyocypris bradyi, Darwinula stevensoni, Cypridopsis vidua, Physocypria kraepelini, Cypria ophtalmica, Prionocypris zenkeri, Eucypris virens, Herpetocypris reptans, Pseudocandona compressa, Fabaeformiscandona fabaeformis Potamocypris* cf. *fulva*) were found during this study. *Potamocypris* cf. *fulva* is a new record for the Turkish freshwater ostracod fauna. The first nine of these species have broad geographic ranges, implying high tolerance levels to different environmental variables. Based on the estimated species optima and tolerance levels, two species exhibited higher effective number of occurrences (*C. neglecta*, and *D. stevensoni*, respectively) than the other species. Three species (*C. neglecta*, *D. stevensoni*, *I. bradyi*) did not show significant correlation with any environmental variable we used. Both Canonical correspondence (CCA) and Pearson correlation analyses showed that temperature was the most effective predictor of species occurrence, followed by electrical conductivity and redox potential. In contrast, pH and dissolved oxygen of water were the least effective predictors. Approximately 71% of the correlation between community composition and environmental variables was explained by the first axis of the CCA diagram, which had a relatively low (7.7%) cumulative variance of species. The lower (560 μg/l) and the upper (2030 μg/l) levels of ammonia (NH_3) exceeded the limits during winter season. The concentrations of total coliform and *Escherichia coli* bacteria were measured up to 10×10^7 cfu/ml and 10×10^3 cfu/ml, respectively. Results of physicochemical measurements, microbiological counts, and species data indicate that water quality of Lake Yeniçağa has been rapidly deteriorated by anthropogenic factors that are the main threat for not only the lake's aquatic diversity but also human health around the lake.

Keywords Ostracoda · Tolerance · Optimum estimates · CCA · Conservation · Eutrophication

Guest editors: R. Matzke-Karasz, K. Martens & M. Schudack
Ostracodology – Linking Bio- and Geosciences

O. Külköylüoğlu (✉) · M. Dügel
Department of Biology, Faculty of Arts and Science, Abant İzzet Baysal University, 14280 Gölköy, Bolu, Turkey
e-mail: kulkoyluoglu_o@ibu.edu.tr

M. Kılıç
Department of Biology, Zoology Building, Faculty of Science, İstanbul University, 34650 Beyazıt, Istanbul, Turkey

Introduction

Many aquatic habitats of Turkey are under the protection of both national environmental laws and international conventions. However, like many other lakes in the world such as in Europe (e.g. Griffiths et al., 2002), North America (e.g. McFadden et al., 2004), Australia (e.g. Chan et al., 2003) and Asia (e.g. Guo et al., 2004), lakes in Turkey are faced with severe environmental problems related to anthropogenic factors (Beklioğlu et al., 2003). Although such problems are known in the literature (Balık, 1995; Külköylüoğlu, 2004), there are few long-term studies (Külköylüoğlu & Dügel, 2004) focused on the subject in Turkey, where serious conservation efforts are still needed. It is, for example, known that eutrophication is the major problem in a shallow Lake Yeniçağa (Bolu), threating adversely not only the biological diversity but also creating risk for the public health around the lake (this study). Lake Yeniçağa continuously receives significant amounts of nutrient-rich water from five to six creeks, whereas lake water is carried out of the lake via Çağa creek (Kılınç, 2003; Saygı-Başbuğ & Demirkalp, 2004a, b). Additionally, untreated municipal wastewater and nutrient-rich water from a slaughterhouse located on the eastern coast of the lake increase the levels of pollution in the lake.

Such simultaneous activities may reduce water quality of the lake and disrupt the biodiversity in the area. On the other hand, the lake area is considered a candidate for the Ramsar Convention, which was signed in Ramsar (Iran) in 1971 to provide a wise use and sustainable conservation of wetlands along with both national action and international cooperation. Accordingly, the lake area is a candidate because it supports habitats for at least 183 wading bird species (Ertan et al., 1989), 345 plant taxa (Sümer, 2002), four fish species (*Cyprinus carpio*, *Tinca tinca*, *Leuciscus cephalus*, and *Carassius carassius*), and many zooplanktons that have been documented to date.

Additionally, Lake Yeniçağa has also economic importance from at least two ways. First, fishery activities done under the control of Yeniçağa Fishery Cooperation provide important economic contribution to its members. *C. carpio* comprises the majority (~63%) of the lake's total fish biomass, followed by *T. tinca* (22% of total fish biomass), *C. carassius* (10%), and *L. cephalus* (5%) (Kılıç, pers. commun.). Second, the environment of the lake has rich peat deposits whose thickness varies from 0 to 15 m (Kavuşan & Karaseyfioğlu, 1997; Kavuşan & Serdaroğlu, 1997). Accordingly, the peat-deposits currently mined nearby the lake are of high quality with its high water-holding capacity, low porosity, and low salt and conductivity levels. Peat soil is commonly used for plant growth in greenhouses; therefore, peat business is economically important for the local people. Thus, all these bring out the economic importance of the lake, and another reason for the need of its conservation for the posterity.

Apart from fish, several other taxonomic groups have been examined. For example, Kılınç (2003) reported 81 phytoplankton species from the lake, of which diatoms were the dominant group. Comparing the ratio of different groups, high levels of eutrophication were observed in the lake because the majority of taxa include cosmopolitan species (also see discussion). Such results were also supported by two other current studies done in the lake (Saygı-Başbuğ & Demirkalp, 2004a, b).

Among the invertebrates, ostracods were the main subjects of this study because they are common microcrustaceans found in almost all different kinds of aquatic environments, from thermal springs (Külköylüoğlu et al., 2003) to very cold waters (Delorme, 1991). Whereas some species are cosmopolitan, some are endemic with specific range preferences. Individual species require certain types of ecological conditions in waters; therefore, ostracods can be used as indicator species of water quality (Benson, 1990; Knox & Gordon, 1999; Külköylüoğlu, 2003). Using indicator species can be cheaper, more reliable and time saver than doing long-term chemical analyses. This concept also includes to show possible relationships between ecological requirements of species and the levels of their response to the changes in aquatic habitats. Once such levels are known, estimating the past, prediction of future, and implication of the present conditions of waters can be made. Thus, although studies attempting to understand ecological requirements of ostracods are scarce, such information can be a great tool for conservation purposes.

During the present study we sought to: (1) study the ostracod community composition in a highly polluted lake, Lake Yeniçağa; (2) discuss ecological requirements of ostracods, focusing on their value as indicator species; and (3) increase our knowledge of the lake's conservation status.

Materials and methods

Site description

The Quaternary landscape of Turkey is marked by significant tectonic activity that continues to the present time. One feature of this activity is the development of tectonically formed lake basins, such as Lake Yeniçağa, a lake basin dating to the early Quaternary, and located on the Northern Anatolian Fault Zone. Historically, Lake Yeniçağa occurs in a depression of 7 × 2.5 km formed after such tectonic activities during the early Quaternary (Lahn, 1948; Erinç et al., 1961; Pekcan, 1996). Since its formation, it appears that the lake surface area has been reduced about threefold to its present size (ca. 5 km^2). These tectonic activities probably caused the depth of the lake's water to fall to ~5 m on average. In the present, Lake Yeniçağa (40°46′410″ N 32°01′788″ E, ~1,010 m a.s.l.) is a shallow eutrophic lake with a maximum depth of 5 m (Fig. 1). Although we did not observe tremendous changes, previous studies reported that the lake surface area fell down up to 3.5 km^2 (Tanoğlu & İnandık, 1952–53) and even to 2.5 km^2 (Kavuşan & Karaseyfioğlu, 1997) during summer season. On the other hand, reduction in both depth and surface area has an historical background that it seemed to be accelerated with the human settlements around the lake 4000 years ago (Bottema et al., 1993). Since then, not only tectonic activities but also human settlement and climatic conditions have been a major impact on the lake water balance and quality.

Data collection

Seven environmental variables [pH, redox potential (Eh), dissolved oxygen (DO), percent oxygen saturation (%Sat), electrical conductivity (EC),

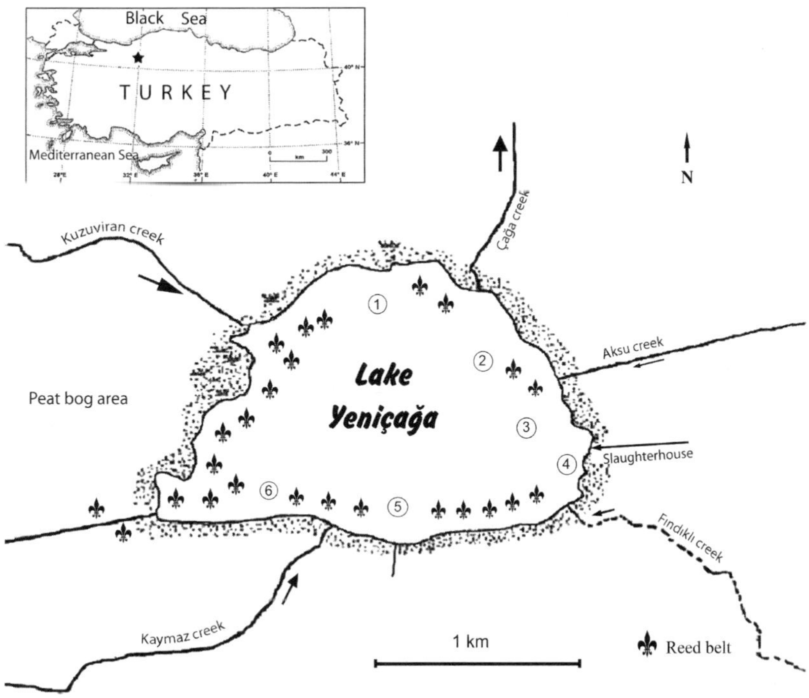

Fig. 1 Study sites around Lake Yeniçağa

Table 1 The occurrence of 13 ostracods and one hemipteran species are shown with the mean and standard deviation (StDev) of eight selected environmental variables.

Date	pH	Eh (mV)	DO (mg/l)	T(w) (°C)	Sat (%S)	Sal (ppm)	EC (μS/l)	TDS (mg/l)	Species
10.14.01	9.04	−111.5	12.32	18.9	131.6	0.2	407.6	273.1	DS, CN, PK
11.11.01	8.39	−83.44	16.478	11.8	157.8	0.26	525.8	352.3	DS, IB, CC, PC, CY, EV, MP
12.23.01	7.68	−42.02	11.368	1.56	85.3	0.2	465.6	312	CC, PF
04.28.02	8.09	−67.6	8.998	16.4	102.5	0.24	500.7	335.5	PK, PC, HE
05.31.02	8.07	−58.34	7.344	22.2	82.3	0.24	512.4	343.3	DS, CN, IB, PK, PC, MP
06.28.02	8.67	−90.66	12.33	28.2	155.2	0.2	500.4	335.3	CN, PK, CY, MP
07.30.02	8.66	−104.2	14.808	26.5	193.8	0.18	424.2	284.2	PK, PC, MP
07.30.03	8.45	−77.46	4.542	22.9	53.86	0.2	414.4	277.7	CN, PZ, CO, IB, CY, MP
08.29.03	8.33	−75.21	11.376	26.4	132.5	0.2	408.8	273.9	DS, CN, CO, MP
10.31.03	8.15	−68.02	6.466	12.2	60.54	0.26	540.8	362.3	DS, CN, CO, IB
11.30.03	8.73	−98.28	8.888	6.42	73.34	0.18	398	266.7	DS, CN, MP
12.27.03	8.61	−90.68	14.266	2.58	104.7	0.2	439.5	294.5	CN, IB
03.31.04	8.59	−90.14	13.034	9.82	115.2	0.22	466.9	312.8	MP
04.30.04	8.08	−60.66	13.14	19	142.3	0.26	487.9	326.9	CO, FC, MP
05.29.04	7.34	−52.57	8.5825	23.1	100.4	0.23	509.6	341.4	DS, CN, CO, EV, MP
06.24.04	7.79	−42.98	9.384	21.5	105.5	0.22	488.8	327.5	EV
07.28.04	7.7	−45.12	12.428	27.4	156.7	0.2	368.9	247.2	DS, CN, MP
08.28.04	8.22	−68.98	10.008	22.3	114.2	0.2	346	231.8	DS, CN
09.25.04	7.95	−53.74	8.658	19.7	94.4	0.2	352.7	236.3	DS, CN, PK, MP
10.29.04	7.96	−54.24	12.238	15.7	123.1	0.2	364.9	244.5	CN, MP
Mean	8.23	−71.79	10.833	17.7	114.3	0.21	446.2	299	
StDev	0.43	20.938	2.9887	8.03	35.66	0.03	61.58	41.26	

Abbreviations and variables (with respective units) include: pH, (Eh) redox potential, (DO) dissolved oxygen, (T(W)) water temperature, (Sat) percent oxygen saturation, (Sal) salinity, (EC) electrical conductivity, (TDS) total dissolved solids, (CN) *Candona neglecta*, (CC) *C. candida*, (IB) *Ilyocypris bradyi*, (DS) *Darwinula stevensoni*, (CY) *Cypridopsis vidua*, (PK) *Physocypria kraepelini*, (CO) *Cypria ophtalmica*, (PC) *Pseudocandona compressa*, (EV) *Eucypris virens*, (FC) *Fabaeformiscandona fabaeformis*, (PF) *Potamocypris fulva*, (HE) *Herpetocypris reptans*, (PZ) *Prionocypris zenkeri*, and (MP) *Micronecta pusilla*

water temperature (Tw), and salinity (Sal); Table 1] were measured in situ. The first two variables were measured with a pH-ORP meter (Hanna model HI-98150), while an oxygen-meter (model YSI-85) was used for the others. Total dissolved solids were measured in the laboratory. We used a global positioning system unit (GPS 45) to record elevation, latitude and longitude information. To reduce the potentially biasing effects of our measurements (or the effects of pseudoreplication) (Hurlbert, 1984), measurements were done before sampling at each station. Water samples for microbiological tests were analyzed with standard methods (APHA, 1989).

Samples taken from six randomly selected sites were collected from the shallow littoral zone (<1 m) between October 2001 and December 2004 (comprising about 20 months of samples) with a hand plankton net (250 μm mesh size). All samples were kept in glass jars (250 ml) in 70% ethanol and were brought to the university laboratory. After filtering the material through four standardized sieves (2.0, 1.5, 0.5, 0.25 mm mesh size, respectively), the samples were fixed in 70% ethanol. Species were identified using the taxonomic keys of Meisch (2000). All specimens were retained in the Department of Biology, Abant İzzet Baysal University, Bolu.

Data analyses

We used Canonical Correspondence Analysis (CCA) within the CANOCO program (Ter Braak, 1989) to show relationships between 10 species (9 ostracods and 1 hemipteran) and five selected environmental variables (EC, DO, Temp, pH, and Eh) when seasons represent supplementary variables. We maintained a larger number of sites (10 species and 4 seasons) than number of predictor (five independent) variables, to reduce arch effect caused by multicollinearity (Ter Braak, 1986, 1987). All variables had skewed distributions and were log transformed [$\ln(x + 1)$] prior to statistical analyses, after which a Monte Carlo test (with 499 permutations) with forward selection was used to provide significance ($\alpha = 0.05$) levels for each variable. Furthermore, estimated species optima (u_k) and tolerance (t_k) levels were calculated with weighted averaging values using CALIBRATE, a program for species-environmental calibration by weighted averaging partial least squares regression (Juggins & Ter Braak, 1992) after Ter Braak & Barendregt (1986) and Birks et al. (1990). One-way ANOVA with unequal variance was used to test the differences among the study sites and numbers of species, and we used Pearson correlation analyses to outline correlations among variables and species. To provide congruity with CCA analyses, we also preferred to use the same environmental variables in Pearson correlation analysis. During the analyses, only live specimens assumed to be alive at the time of collection, were counted, and species that occurred within a single sample were excluded from the multivariate analyses.

Results

Species assemblages

We detected a total of 13 ostracod species (*Candona neglecta*, *C. candida*, *Ilyocypris bradyi*, *Darwinula stevensoni*, *Cypridopsis vidua*, *Physocypria kraepelini*, *Cypria ophtalmica*, *Herpetocypris reptans*, *Eucypris virens*, *Prionocypris zenkeri*, *Pseudocandona compressa*, *Fabaeformiscandona fabaeformis*, *Potamocypris* cf. *fulva*) and one hemipteran species (*Micronecta pusilla*) during this study. The first nine species are known to exhibit broad geographical ranges. *Potamocypris fulva* is a new record for the ostracod fauna of Turkey. Among ostracod species, *C. neglecta* and *D. stevensoni* occurred most frequently. According to the CCA diagram (Fig. 2), water temperature and electrical conductivity were the two most important abiotic predictors of community composition, followed by redox potential. Dissolved oxygen appeared to exhibit the least correlation with community composition. Pearson correlation analyses outlined that occurrences of several pairs of species (e.g. *D. stevensoni* and *C. neglecta*, *I. bradyi* and *Cypridopsis vidua*, *P. kraepelini* and *Ps. compressa*) across sampling locations and over time were significantly correlated.

Species tolerance and environment correlation

No significant differences were observed in the mean numbers of species found among the stations ($P > 0.05$). Estimated species optima (u_k) and tolerance levels (t_k) based on the weighted averaging values showed that *C. neglecta* and *D. stevensoni* were the two species with higher mean effective number of occurrences (N2) as 19.23 and 11.31, respectively (Table 2). Overall, 70.7% of the correlation between species and environmental variables is explained by the first axis of CCA diagram ($F = 3.60$; $P < 0.05$; Eigenvalues = 3.82; Total inertia = 0.558) with a relatively low (7.7%) cumulative variance of species (Table 3).

Pearson correlation analyses illustrated a significant correlation between water temperature and four species ($P < 0.01$ for *P. kraepelini*, *Cypridopsis vidua*, and *M. pusilla*; $P < 0.05$ for *Ps. compressa*). Besides, a significant negative correlation ($r = -0.520$; $P < 0.05$) was found between *M. pusilla* and redox potential, when correlation was also significant between pH and *E. virens* ($r = -0.483$; $P < 0.05$). One ostracod species, *C. candida*, also showed a significant correlation to air temperature ($r = -0.508$; $P < 0.05$).

Water quality and eutrophication

The average water temperature of Lake Yeniçağa was relatively high (17.7°C), and fluctuated from 1.6°C to 28.2°C. Levels of dissolved oxygen averaged 10.83 mg/l across all samples, and ranged from 4.54–14.27 mg/l during the study period (Table 1). The highest (540.76 μs/cm) and the lowest (346.00 μs/cm) electrical conductivity

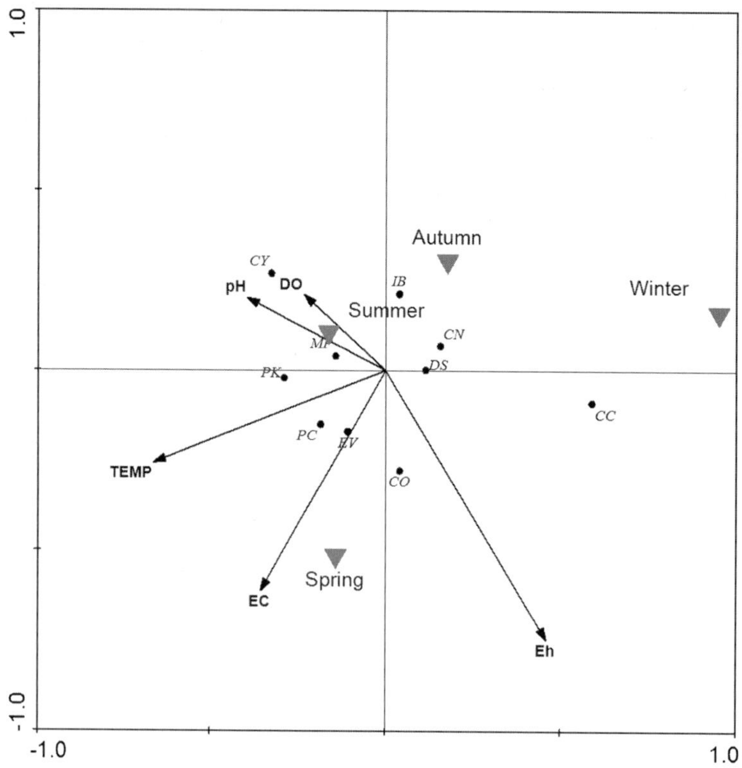

Fig. 2 Diagram of Canonical Correspondence Analyses (CCA) shows relationships between 10 species (9 ostracods and one hemipteran) (points) and five selected environmental (physical) variables (EC, DO, Temp, pH, and Eh) (*arrows*) when seasons (*triangular*) represent supplementary variables. Abbreviations are same as in Table 1

Table 2 Estimated species optima (u_k) and tolerance (t_k) levels for the seven variables and eight most abundant ostracods

Species	Count	Max	N2	pH		Eh		DO		Temp		Salinity		EC		TDS	
				u_k	t_k	u_k	t_k	u_k	t_k	u_k	t_k	u_k	t_k	u_k	t_k	u_k	t_k
CN	32	17	19.23	7.76	0.81	–55.28	28.44	7.23	3.85	15.46	6.50	0.22	0.08	468,53	124,99	313,91	83,74
DS	14	3	11.31	8.10	0.39	–63.18	22.42	9.01	2.14	17.44	6.84	0.20	0.07	436,60	112,05	292,53	75,07
PK	9	8	5.88	8.42	0.48	–80.73	28.35	10.93	4.33	24.53	3.62	0.20	0.06	485,36	83,55	325,19	55,98
IB	7	3	5.40	8.25	1.18	–89.00	34.30	12.73	5.88	15.27	5.09	0.20	0.05	420,83	110,27	281,96	73,88
PC	5	2	4.50	7.94	0.57	–56.43	36.63	9.74	5.36	19.57	6.45	0.25	0.06	522,07	97,03	349,78	65,01
CO	9	21	3.96	6.93	1.38	–59.34	16.75	9.16	3.30	20.92	2.03	0.22	0.05	490.10	59,14	328,37	39,62
CY	3	1	3.00	8.88	0.59	–106.57	29.01	14.13	1.84	21.07	7.70	0.20	0.07	437,73	63,62	293,28	42,63
CC	2	4	1.47	7.73	0.51	–45.00	30.76	9.92	5.29	3.38	8.06	0.20	0.07	432,70	4,95	289,91	3,32
MP	19	146	4.53	8.42	0.51	–81.92	33.71	7.87	5.72	20.48	9.25	0.23	0.05	487,15	90,29	326,39	60,50

Abbreviations are given in Table 1. (Max) maximum numbers of individuals per species, and (N2) effective number of occurrences. Count shows the numbers of individuals found among all stations

values were found in November 2003 and August 2004, respectively (Fig. 3). The lake's mean pH value was found in the alkaline range as 8.22, while redox potential value varied between –42 and –111.47 mV with the average of –71.61 mV during this study. Hardness of lake water ranged from 138.5 to 265 mg/l, implying relatively hard water conditions. The lowest (560 μg/l) and the highest (2030 μg/l) levels of ammonia (NH_3) were found during winter of 2001. The levels of total

Table 3 The main results of Canonical Correspondence Analyses

Axes	1	2	3	4	Total inertia
Eigenvalues	0.295	0.142	0.083	0.027	3.822
Species–environment correlations	0.707	0.618	0.388	0.212	
Cumulative percentage variance of species data	7.7	11.5	13.6	14.3	
Cumulative percentage variance of species–environment relation	52.9	78.4	93.4	98.2	
Sum of all eigenvalues					3.822
Sum of all canonical eigenvalues					0.558

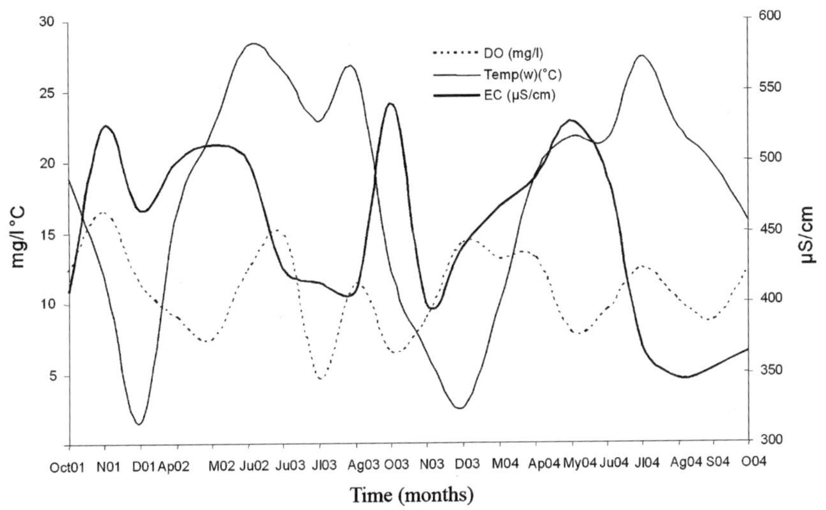

Fig. 3 The relationship among dissolved oxygen (DO), electrical conductivity (EC), and water temperature Temp (w) (°C) during 20 months of sampling. Note the missing dates

coliform bacteria reached up to 10×10^7 cfu/ml while maximum levels of *E. coli* bacteria was calculated as 10×10^3 cfu/ml, respectively.

Discussion and conclusion

Species environment correlation

The numbers of ostracod species (13) found in Lake Yeniçağa corresponds with the estimated average number of species (13.2) reported from other lakes in Turkey (Külköylüoğlu, 2005). This seems to be high species richness but it can also be interpreted as a signal of low water quality or called 'pseudorichness' when quantity rather than quality of species is being considered (Külköylüoğlu, 2004). This can especially be true when cosmopolitan species dominate the habitat (Roca et al., 2000; Külköylüoğlu, 2003; Yılmaz & Külköylüoğlu, 2006). This is because cosmopolitan species generally have wide ranges of tolerances to the changes in aquatic conditions, so that they can increase their survival rate when conditions deteriorate. In the large scale, Meffe & Carroll (1994) pointed out that global increase in the numbers of cosmopolitan species coincides with increasing pollution that causes a global reduction in species richness. This may require a long-time period, but in a small scale (i.e. in a lake ecosystem), sudden changes can occur within a very short period of time due to anthropogenic activities, which cause an increase in densities of cosmopolitans (Külköylüoğlu, 1999; Roca et al., 2000). Therefore, increasing numbers of cosmopolitans caused by such activities may support the idea of 'pseudorichness' when noncosmopolitans are not present or rare, as is the case in Lake

Yeniçağa. Indeed, the general distribution of the nine most common species (*C. neglecta*, *C. candida*, *I. bradyi*, *D. stevensoni*, *Cypridopsis vidua*, *P. kraepelini*, *Cypria ophtalmica*, *E. virens*, *Ps. compressa*) used in CCA analyses (Fig. 2) coincides with their cosmopolitan (or closer to cosmopolitan) characteristics in different habitats. Having a wide geographical distribution, these species are also known to tolerate broad ranges of changes in the values of different environmental variables. For example, according to correlation analyses, there was significant correlation between three species (*P. kraepelini* ($r = 0.77$), *Cypridopsis vidua* (($r = 0.59$), *Ps. compressa* ($r = 0.55$)) and water temperature. Among the species, *Cypridopsis vidua*, a well known cosmopolitan species, has been described from a variety of aquatic ecosystems with relatively large tolerance levels to pH (5.20–12.00), temperature (5.0–36°C), dissolved oxygen (1.6–19.00 mg/l), and conductivities (117–7410 μs/cm) (Külköylüoğlu, 2000, 2004). The species can tolerate high levels of pesticides (e.g. carbofuran and endosulfan) from ricefields (Lim & Wong, 1986). Recently, Kiss (2001) reported the species from polluted shallow waters of Lake Kőhegyi in the Biosphere Reserve (Hungary) where pH level was increased up to 9.09. This corresponds to the optimum and estimated tolerance ranges as given in Table 2. Additionally, earlier experimental studies (e.g. Mbahinzireki et al., 1991) showed that the presence of *Cypridopsis vidua* was closely associated to the presence of aquatic plant *Chara* species because *Cypridopsis vidua* prefers to use the branches of *Chara* as a shelter and food sources. This implies that *Cypridopsis vidua* can have a certain behavioral and ecological divergence (Roca et al., 1993) when suitable conditions are available. Thus, such behavioral displacement along with wide ecological tolerances can explain how *Cypridopsis vidua* can survive in variety of habitats from oligotrophic to eutrophic waters. Although these kinds of behavioral and ecological characteristics are known for *Cypridopsis vidua*, such knowledge is scarce for other freshwater ostracods.

Two other ostracods (*P. kraepelini* and *Ps. compressa*) reported herein are also common in lakes, ponds, and spring waters. A Holarctic species, *P. kraepelini*, which was rarely reported in North America under the name of *P. fadeewi* (Tressler, 1959), was not found over 210 springs of Nevada (Külköylüoğlu, 1999), but it is commonly reported from European shallow waters, reservoirs (Kiss, 2002), canals, diches, ponds, streams, littoral zones of lakes (Meisch, 2000). In Turkey, this species was first reported from the springs of Sakarya river (Gülen, 1977), but since then it has been one of the most common forms encountered almost all year around in the standing habitats of the lakes (Külköylüoğlu, 1998). Shornikov & Trebukhova (2001) listed this species as surviving in significant pollution. Meisch (2000) characterized the species as meso- to polytitanophilic (generally occurring at > 72 mg Ca/l), thermoeuryplastic (referring to a wide range of temperature tolerance), and oligohalophilic (preferring freshwaters). The compounds of calcium and magnesium play important role on water hardness. During the present study, minimum and maximum values of hardness in Lake Yeniçağa were measured as 138.5 and 265 mg/l, respectively. This suggests high levels of input of calcium, magnesium, and a variety of other metals into the lake. Between the years 1997 and 1999, Saygı-Başbuğ & Demirkalp (2004b) were able to measure calcium (25.2–77.2 mg/l) and magnesium (0.9–39.9 mg/l) levels in the lake. These values support the classification of the species as meso to polytitanophilic by Meisch (2000). As a result, such implication for the lake water can be interpreted as high to very high hardness, where *P. kraepelini* can survive. Besides, a recent study undertaken in a small man-made lake in Turkey showed that *P. kraepelini* usually occurs in less alkaline waters with moderate to high temperature (Yılmaz & Külköylüoğlu, 2006). Although its presence is not clear in Australia (Karanovic, pers. commun.) and other countries in the southern hemisphere, we believe that this is because of confusion on its taxonomy and lack of studies in those areas. Consequently, the distribution of the species in lake waters is possibly closer to cosmopolitan, and future studies will probably clarify this assumption.

Pseudocandona compressa is so far uncommon in Turkey and some other European countries. For example, Rossetti et al. (2004) newly reported the species from eutrophic wetlands of northern Italy where a group of species, including *Ps. compressa*, was primarily associated to elevated ionic content with higher trophic conditions. In contrast, the species is widely distributed in other European aquatic habitats, where it prefers both permanent and temporary water bodies, especially shallow littoral zones of lakes (Meisch, 2000). In the southern interior of British Columbia, Green (1994) reported the species mainly from lakes, creeks and a pond, where water temperature varied from 8 to 23°C, pH from 6.0 to 7.0 and conductivity from 150 to 460 μS/cm. Comparing these values with the values measured from the present study, it seems that the tolerance ranges of *Ps. compressa* to at least some of the major variables is higher than previously known. For example, during the present study, the species was only found in the alkaline (pH 8.07–8.66) and relatively warm waters (from 11.8 to 26.5°C) of summer months (Table 1) with relatively high conductivities (424.2–525.8 μS/cm). This does suggest that the presence of *Ps. compressa* in alkaline waters of Lake Yeniçağa can be associated with water temperature and electrical conductivity (Fig. 2). Indeed, tolerance level (t_k = 36.63) of *Ps. compressa* to redox potential was the highest among the other species (Table 2). This may coincide with the ability of species to avoid oxidation and reduction processes in the lake waters. Overall, not much is known about the ecological preferences of this species at the moment, but it appears that ecological tolerance ranges of *Ps. compressa* seems to be much higher than estimated.

On the other hand, the most common three species (*C. neglecta*, *D. stevensoni*, and *I. bradyi*) did not show any significant relationship with those variables used ($P > 0.05$). This is probably because of the high tolerance levels of these three species to different variables (Table 2). For example, in general, the benthic ostracod *D. stevensoni* prefers highly oxygenated waters (Ranta, 1979) to aerate its eggs in brood chamber. However, current studies indicated that *D. stevensoni* can even survive in habitats where dissolved oxygen drops down to 1.0 mg/l (Külköylüoğlu, 1999) in waters from acidic (pH 6.0) to highly alkaline (pH 9.27) conditions (De Deckker, 1981; Külköylüoğlu, 1999). In a recent study undertaken in 16 wetlands in Italy, the water pH level even exceeded the previously known range, reaching up to 9.67 (the highest pH level for the species), where *D. stevensoni* was reported from eutrophic conditions with 20 μM of average annual ammonium concentration (Rossetti et al., 2004). In a comparative study between natural lakes and manmade reservoirs, Külköylüoğlu (2003) did not encounter *D. stevensoni* from reservoirs, while the species was present in one of the natural alkaline lake water (pH 7.90). Based on the data available, *D. stevensoni* can be characterized as cosmopolitan species with high tolerances to survive in different conditions.

Similarly, two other species, *C. neglecta* and *I. bradyi*, also showed relatively high tolerance and optimum values in Lake Yeniçağa. In the CCA diagram (Fig. 2), these two species, *C. neglecta* and *I. bradyi* (and *D. stevensoni*) were located in the opposite site of the diagram, but closer to the center. This implies that five variables used in the analysis did not have any significant influence on the species occurrence during this study. Of the species, *C. neglecta* was the most common species in the lake, but it can also be found in the littoral to deeper parts of lakes, springs, brooks, and slow flowing aquatic habitats. In Lake Yeniçağa, adults of *C. neglecta* were encountered in wide ranges of temperature (2.58–28.2°C), pH (7.34–9.04), dissolved oxygen (4.5–14.3 mg/l), and redox potential from –111.5 to –45.12 mV. Besides, the species displaced the highest tolerance (ca. 125 μs/cm) to electrical conductivity that varied from 352.7 to 540.8 μs/cm (Tables 1, 2). Based on the results, *C. neglecta* can be characterized as oligothermophilic, but it can resist high levels of salinities (mesohalophilic). Frenzel & Oertel (2002) collected the species from 10 of 18 stations in the brackish waters of Strelasund (southern Baltic Sea) where the mean values of calcium (0.39–1.11% in dry weight) and magnesium (0.43–0.83% in dry weight) levels ranged in a wide spectrum of highly saline waters (7.43–8.16 ppt). This coincides with the suggestion of Meisch (2000) that the species can even tolerate very hard saline waters; therefore, it can also be characterized as polyhalinic.

Unlike *C. neglecta*, another bisexual candonid species, *C. candida* is of wider distributional range. In the present study, it showed highest tolerance ($t_k = 8.06$) to water temperature (Table 2) when it was the only species displaying a significant negative association with air temperature ($r = -0.508$, $P < 0.05$). In shallow lakes, such as this one, air temperature may have direct effect on water temperature changes but small ostracods may reduce such effect by means of different behavioral responses as already discussed for *Cypridopsis vidua* above. Therefore, such correlation is not surprising since *C. candida* is reported from variety of habitats including diches, rivers, springs, wells, trough, and lakes. In CCA diagram (Fig. 2), *C. candida* is separated as an outlier. This may be because of high tolerance ranges of the species where the species was not affected by any of those environmental variables used in this analysis. Meisch (2000) underlined that the species is euryplastic for pH, oligothermophilic for temperature, and possibly mesohalophilic for salinity changes in these types of aquatic habitats. During the present study, we encountered this species in November and December of 2001 with a lowest mean temperature value as 1.56°C. On the other hand, Delorme (1991) argued that the species could tolerate very low (5.4) to high (13.0) pH values in warm (up to 30°C) and well oxygenated waters (15 mg/l) in Canada. Supporting this argument, Henderson (1990) reported the species from nutrient poor acid waters of peat bogs where pH was about 4.6. Milhau et al. (1997) referred to *C. candida* as to an indicator of pure waters, but the species has also been reported from relatively polluted waters in Tumen River (Peter the Great Bay, Russia) (Shornikov & Trebukhova, 2001) and in eutrophic maar lakes in Germany (Scharf, 1993). In the springs of Finland, Särkkä et al. (1997) reported *C. candida* being one of the most abundant species in springs under the effect of human inhabitation. Overall, these results indicate that *C. candida* is a highly tolerant species to different values of environmental variables, suggesting a possible cosmopolitan species (Külköylüoğlu, 2004).

Ilyocypris bradyi is usually found in springs, brooks, slow flowing waters, and stagnant littoral zones of the lakes. This species is one of the most common among its conspecific forms in Turkey. According to Meisch (2000) the species is of Holarctic distribution, but the species is also known from India (Victor & Fernando, 1979) and from Chile and Trinidad (Martens & Behen, 1994). Consequently, geographical distribution of this species seems to be much wider and most likely it has a cosmopolitan distribution. Anadón and his colleagues (1986) reported that the species tolerated to nutrient (or bicarbonate) enriched lake deposits of the NE Baza Basin (Spain). Mezquita et al. (2001) reported the species from alkaline (pH 7.3–8.6), highly saline (referring to conductivity from 648 to 5063 μs/cm) eutrophic Iberian Mediterranean brooks, in which both water temperature (3.7–25°C) and dissolved oxygen values (3.2–13 mg/l) showed a wide fluctuation. However, its oxygen requirements can be higher as reported by Roca & Baltanás (1993) in cold to warm (5.8–18.4°C) helocrene springs of Iberian Peninsula, where oxygen level was measured as 5.5 to 14.1 mg/l. Külköylüoğlu (1999) reported the species from 61 of 209 springs in Nevada, of which the species was most common in rheocrene (as well as helocrene) springs within the pH range of 6.8–8.0. In the present study, we also found the species in alkaline waters (pH 8.07–8.61), where salinity was the maximum range of oligohaline (414.4 μs/cm) to mesohaline range (540.8 μs/cm). This implies a high salinity tolerance of *I. bradyi* (euryhaline) in a wide temperature range (polythermophilic) (Meisch, 2000). Indeed, in this study, the species was only present in samples when electrical conductivity was always above 400 μs/cm (Table 1). Besides, comparing with other species reported herein, *I. bradyi* showed the highest tolerance level to dissolved oxygen ($t_k = 5.88$ mg/l) with relatively high electrical conductivity (110.27 μs/cm) (Table 2). Based on its common presence in springs or waters related to springs, ecologically, the species can be called crenophilous (Bronshtein, 1947). Until recently, not much is known about the ecological requirements of this species, and still detailed studies are needed to provide species-specific characteristics.

Although the occurrence of *E. virens* may be associated to the seasonal changes (Külköylüoğlu,

1998, 2005), this cosmopolitan species shows wide tolerance ranges to the pH levels from 7.60 up to 9.27 (Mezquita et al., 1999a, b). The species was insensitive, showing resistance to fipronil, a phenyl-pyrazole compound of insecticide (Chaton et al., 2002). Previous studies (e.g. De Deckker, 1981; Neale, 1988; Mezquita et al., 1999a) also reported the species from fresh to almost saline oligo-mesohaline alkaline waters where temperature does not exceed 24°C (Külköylüoğlu, 2000). In the present study, the species was also found between these ranges (Table 1) and displayed a negative relationship to pH and a significant positive one to electrical conductivity (Fig. 2). Roca et al. (2000) reported this species from nine of 43 aquatic bodies in Spain. They indicated that finding *E. virens* along with two other accompanying species (*H. incongruens* and *Sarscypridopsis aculeata*) from such habitats could reflect strong eutrophic characteristics of species that are cosmopolitan or have a broad distribution in the Holarctic region.

Cypria ophtalmica, which has almost cosmopolitan distribution (Meisch, 2000; Külköylüoğlu, 2003), and was found from cool to relatively warm lake water (12.2–26.4°C) with low to high oxygenated (4.5–13.14°C) alkaline (pH 7.34–8.45) waters of Lake Yeniçağa (Table 1). The species did not show any significant correlation to any variables measured during this study and also it did not have significant optimum and tolerance estimates (Table 2). This is probably due to the species' remarkable tolerance to environmental factors (Meisch, 2000). Indeed, previous studies (e.g. Külköylüoğlu, 2000, 2004) reported the species in different levels of pH (5.0–13.0), temperature (3.5–33.0°C), dissolved oxygen (0.00–20.00 mg/l), and conductivities (87–5260 μS/cm). These are so far the most extreme values for a species among ostracods which at the same time may explain the wide distribution of the species.

One of the most abundant hemipteran species encountered in the lake was *Micronecta pusilla*. The species observed in a small man-made lake in Turkey by Kulköylüoğlu & Dügel (2004) was described as a stenochronal organism based on its occurrence in certain months over 2 years. The authors showed that there was a linear relationship between the occurrence of *M. pusilla* and water temperature and pH, but failed to find any adult specimens below 12°C during winter months. Whereas the species showed highest temperature tolerance (Table 2) from 6.42 to 28.2°C, the frequent occurrence of adults was higher in waters above 12°C. These results support earlier findings, but should not be generalized due to lack of knowledge.

Ostracoda and conservation status of Lake Yeniçağa

Aquatic bodies in Turkey are under the protection of either national or international laws. However, enforcement of such laws occurs only sporadically for most lakes. This is especially true where increasing pollution caused by the direct effect of anthropogenic factors is one of the main problems, and accelerates the negative effects of eutrophication. For example, the reported amount of *E. coli* bacteria over the critical level (10×10^3 cfu/ml) is a direct indication of human interaction on Lake Yeniçağa. Possible consequences of such a problem were outlined in previous studies (e.g. Julià et al., 1998; Tan & Beklioğlu, 2005), and included decline in water quality and reduction in species diversity. For example, Bootsma et al. (1999) indicated that species biodiversity in the shallow lakes of Naardermeer nature reserve in the Netherlands was reduced because of cultural eutrophication caused by high nutrient input into the lake.

Eutrophication, one of the major problems globally for lentic systems and especially so for shallow lakes (Roca et al., 2000), has long occurred in lake Yeniçağa, where human settlement began about 4,000 years ago (Beug, 1967; Bottema et al., 1993). Since then, the lake has experienced numerous anthropogenic activities along with the changes in climatic conditions. However, during the last 30 years the adverse effect of these activities has increased in the lake area, resulting in the chemical composition of the lake's water from unpolluted to polluted (nutrient rich) conditions. This has caused a severe reduction in the species diversity and/or a decline in the quality of species, and an elevated level of 'pseudorichness' in the lake (Külköylüoğlu,

2004). This was the case not only for ostracods, but was also described for other taxonomic groups in Lake Yeniçağa. For example, Kılınç (2003) described 81 phytoplankton taxa from the lake, where predominance of one or two species reduced the species richness in the lake as was the case in June when net plankton was especially dominated by *Anabaena circinalis* and *Aphanocapsa grevillei*.

Correspondingly, Saygı-Başbuğ & Demirkalp (2004a) reported that some species of blue-green algae, such as *Aphanizomenon flos-aquae* and *Oscillatoria* spp., were dominant, and possessed highly elevated mean chlorophyll *a* content from 2.26 to 124.49 mg/m^3. Such increase in algal content is directly related to the nutrient input and eventually causes to rise daily primary production where it was estimated between 120 and 4011 mg C/m^2 per day (Saygı-Başbuğ & Demirkalp, 2004b). These values are above the critical levels and show highly eutrophic conditions in the lake. Indeed, all the cosmopolitan algae groups mentioned above can increase their dominancy in eutrophic conditions. One of the main results of such impact in this polymictic lake is eventually reflected as a decline in the species composition. In such a case, functional usage of such shallow lakes for society cannot be efficiently made if eutrophication continues (Bootsma et al., 1999).

On the other hand, such polluted habitats can provide better places for some cosmopolitan invaders with wide tolerance ranges, because changing ecological conditions will eliminate the effect of competition with the native species. Since many cosmopolitan species such as the nine ostracods discussed above show high levels of tolerances to the abiotic factors (Külköylüoğlu, 2000), increasing levels of pollution (at least up to a point) may be relatively advantageous for them. Consequently, proportional relationship may be expected between the numbers of cosmopolitans and increasing levels of pollution in eutrophicated lakes. This does not mean to ignore the presence of cosmopolitans, nor assert that all invaders are cosmopolitans. Rather, they have value as species that may indicate aquatic conditions and can be considered as a useful tool by conservationists.

Acknowledgements The authors would like to thank the following persons and institutions for their comments, help and support during this study: Dr. Erik Beever (Corvallis University) for his comments and suggestions on the earlier draft of this paper; Dr. Erdoğan Usta for his help during fieldwork; Dr. Gültekin Kavuşan (Ankara University, Turkey) for sharing reprints regarding the lake peat deposit, Dr. Ivana Karanovic (Western Australian Museum) and Dr. Peter Frenzel (Rostock University, Germany) for the personal information about their studies. This project was supported by the State Planning Organization of Turkey (Project No: DPT-97K120330).

References

Anadón, P., P. De Deckker & R. Julià, 1986. The Pleistocene lake deposits of the NE Baza Basin (Spain): salinity variations and ostracods succession. Hydrobiologia 143: 199–208.

APHA., 1989. Standard Methods for the Examination of Water and Waste Water. American Water Works Association and Water Pollution Control Federation. American Public Health Association, Washington D.C.

Balık, S., 1995. Freshwater fish in Anatolia, Turkey. Biological Conservation 2: 213–223.

Beklioğlu, M., O. İnce & İ. Tüzün, 2003. Restoration of eutrophic Lake Eymir, Turkey, by external nutrient loading control and biomanipulation. I. First case study on a warm temperate lake. Hydrobiologia 489: 93–105.

Benson, R. H., 1990. Ostracoda and the discovery of global Cainozoic palaeoceanographical events. In Whatley, R. & C. Maybury (eds), Ostracoda and Global Events. Chapman and Hall University Press, Cambridge, London, 41–59.

Beug, H.-J., 1967. Contributions to the postglacial vegetational history of northern Turkey. In Cushing, E. J. & H. E. Wright (eds), Quaternary Paleoecology. Yale University Press, New Haven, 349–356.

Birks, H. J. B., J. M. Line, S. Juggins, A. C. Stevenson & C. J. F. Ter Braak, 1990. Diatoms and pH reconstruction. Philosophical Transactions of the Royal Society of London 327: 263–278.

Bootsma, M. C., A. Barendregt & J. C. A. van Alphen, 1999. Effectiveness of reducing external nutrient load entering an eutrophicated shallow lake ecosystem in the Naardermeer nature reserve, The Netherlands. Biological Conservation 90: 193–201.

Bottema, S., H. Woldring & B. Aytuğ, 1993. Late Quaternary vegetation history and climate in northern Turkey. Palaeohistoria 35/36: 13–72.

Bronshtein, Z. S., 1947. Fresh-water Ostracoda. Fauna of the USSR, Crustaceans, Vol. 2, No. 1. Cladocopa and Platycopa. Russian Translation Series, 64. Academy of Sciences of the USSR Publishers, Moscow. 1988.

Chan, T. U., B. J. Robson & D. P. Hamilton, 2003. Modelling phytoplankton succession and biomass in seasonal West Australian estuary. Verhandlungen der Internationalen Vereinigung der Limnologie 28: 1086–1088.

Chaton, P. F., P. Ravanel, M. Tissut & J. C. Meyran, 2002. Toxicity and bioaccumulation of Fipronil in the nontarget arthropodan fauna associated with Subalpine mosquito breeding sites. Ecotoxicology and Environmental Safety 52, 8–12.

De Deckker, P., 1981. Ostracods of Athalassic saline lakes. Hydrobiologia 81: 131–144.

Delorme, L. D., 1991. Ostracoda. In Thorpe, J. H. & A. P. Covich (eds), Ecology and Classification of North American Invertebrates. Academic Press, San Diego, London, 691–722.

Erinç, S., T. Bilgin & M. Bener, 1961. Çağa depresyonu ve boğazı. İstanbul Üniversitesi Coğrafya Enstitü Dergisi 6: 170–173.

Ertan, A., A. Kılıç & M. Kasparek, 1989. Türkiye'nin önemli kus alanları [Important Bird Areas of Turkey]. Istanbul, 160 + xvi pp.

Frenzel, P. & P. Oertel, 2002. Die rezenten Ostrakoden und Foraminiferen des Strelasundes (südliche Ostsee). (The recent ostracods and foraminifera of the Strelasund (southern Baltic Sea). Rostocker Meeresbiologische Beiträge 11: 23–37.

Green, G. D., 1994. Freshwater Ostracoda (Crustacea) from the Southern Interior of British Columbia. Royal British Columbia Museum.

Griffiths, H. I., J. M. Reed, M. J. Leng, S. Ryan & S. Petkovski, 2002. The recent palaeoecology and conservation status of Balkan Lake Dojran. Biological Conservation 104: 35–49.

Guo, H. Y, J. G. Zhu, X. R. Wang, Z. R. Wu & Z. Zhang, 2004. Case Study on Nitrogen and Phosphorus Emissions from Paddy Field in Taihu Region. Environmental Geochemistry and Health 26: 209–219.

Gülen, D., 1977. Contribution to the Knowledge of the Freshwater Ostracoda Fauna of Turkey. Istanbul Universitesi Fen Fakültesi Mecmuası, Seri B, 42: 101–106.

Henderson, P. A., 1990. Freshwater Ostracoda: Keys and Notes for the Identification of the Species. Synopses of the British Fauna (New Series). The Linnean Society of London and The Estuarine and Coastal Sciences Association, London.

Hurlbert, S. J., 1984. Pseudoreplication and the design of ecological field experiments. Ecological Monographs 54: 187–211.

Juggins, S. & C. J. F. Ter Braak, 1992. CALIBRATE – A Program for Species – Environmental Calibration by [Weighted Averaging] Partial Least Squares Regression. Environmental Change Research Center, University College, London.

Julià, R., F. Burjachs, M. J. Dasí, F. Mezquita, M. R. Miracle, J. R. Roca, G. Seret & E. Vicente, 1998. Meromixis origin and recent trophic evolution in the Spanish mountain lake La Cruz. Aquatic Sciences 60: 279–299.

Kavuşan, G. & T. Karaseyfioğlu, 1997. Geological setting of Yeniçağa peat deposit, Bolu-Turkey. 3rd European Coal Conference. European Coal Geology 293–299.

Kavuşan, G. & A. Serdaroğlu, 1997. The physical and chemical properties of Yeniçağa-peat deposit, Bolu-Turkey. 3rd European Coal Conference. European Coal Geology 301–310.

Kılınç, S., 2003. The phytoplankton community of Yeniçağa Lake (Bolu, Turkey). Nova Hedwigia 76: 429–442.

Kiss, A., 2001. Limnological investigations of small water bodies in the Pilis Biosphere Reserve, Hungary. Part II. Kőhegyi-tó and Unkás-tócsa. Opuscula Zoologica Budapest 33: 67–74.

Kiss, A., 2002. The cladocera, ostracoda and copepoda fauna of the Fehér-tó (Fertő-Hanság National Park). In Mahunka, S. (ed.), The Fauna of the Fertő-Hanság National Park. Hungarian Natural History Museum, Budapest, 245–248.

Knox, L. W. & E. A. Gordon, 1999. Ostracodes as indicators of brackish water environments in the Catskill Magnafacies (Devonian) of New York State, Palaeogeography, Palaeoclimatology, Palaeoecology 148: 9–22.

Külköylüoğlu, O., 1998. Freshwater Ostracoda (Crustacea) and their quarterly occurrence in Şamlar Lake (Istanbul, Turkey). Limnologica 28: 229–235.

Külköylüoğlu, O., 1999. Taxonomy, ecology and biogeographic distribution of spring water Ostracoda (Crustacea) in Nevada. Ph.D. dissertation. Program in Ecology, Evolution and Conservation Biology. University of Nevada, Reno.

Külköylüoğlu, O., 2000. The importance of cosmopolitan and indicator species of Ostracoda (Crustacea) in Turkey based on some water parameters. Sinop Water Product Conference 2000: 421–437. (Abstract in English).

Külköylüoğlu, O., 2003. Ecology of freshwater Ostracoda (Crustacea) from lakes and reservoirs in Bolu, Turkey. Journal of Freshwater Ecology 18: 343–347.

Külköylüoğlu, O., 2004. On the usage of ostracods (Crustacea) as bioindicator species in different aquatic habitats in the Bolu region, Turkey. Ecological Indicators 4: 139–147.

Külköylüoğlu, O., 2005. Factors affecting the occurrence of Ostracoda (Crustacea) in the Yumrukaya Reedbeds (Bolu, Turkey). Wetlands 25: 224–227.

Külköylüoğlu, O. & M. Dügel, 2004. Ecology and spatiotemporal patterns of Ostracoda (Crustacea) from Lake Gölcük (Bolu, Turkey). Archiv für Hydrobiologie 160: 67–83.

Külköylüoğlu, O., C. Meisch & W. R. Rust, 2003. A new genus (*Thermopsis thermophila* n. gen.) of Ostracoda (Crustacea) from hot springs of western North America. Hydrobiologia 499: 113–123.

Lahn, E., 1948. Türkiye göllerinin jeolojisi ve jeomorfolojisi hakkında bir etüd. Maden Tetkik Arama Enstitüsü Yayınları. Seri B. No. 12. Ankara. p. 19. (in Turkish).

Lim, R. P. & M. C. Wong, 1986. The effects of pesticides on the population dynamics and production of *Stenocypris major* Baird (Ostracoda) in ricefields. Archiv für Hydrobiologie 106: 421–427.

Martens, K. & F. Behen, 1994. A checklist of the recent non-marine ostracods (Crustacea, Ostracoda) from the inland waters of South America and adjacent islands. Travaux Scientifiques du Musée National d'Historie Naturelle de Luxembourg 22: 1–81.

Mbahinzireki, G., F. Uiblein & H. Winkler, 1991. Microhabitat selection of ostracods in relation to predation and food. Hydrobiologia 222: 115–119.

McFadden, M. A., H. T. Mullins, W. P. Patterson & W. T. Anderson, 2004. Paleoproductivity of eastern Lake Ontario over the past 10,000 years. Limnology and Oceanography 49: 1570–1581.

Meffee, G. K. & C. R. Carroll, 1994. Principles of Conservation Biology. Sinauer Press, Sunderland, U.S.A.

Meisch, C., 2000. Freshwater Ostracoda of western and central Europe. In Schwoerbel, J. & P. Zwick (eds), Süßwasserfauna von Mitteleuropa 8/3. Spektrum Akademischer Verlag, Heidelberg.

Mezquita, F., G. Tapia & J. R. Roca, 1999a. Ostracoda from springs on the eastern Iberian Peninsula: ecology, biogeography and palaeolimnological implications. Palaeogeography Palaeoclimatology Palaeoecology 148: 65–85.

Mezquita, F., R. Hernandez & J. Rueda, 1999b. Ecology and distribution of ostracods in a polluted Mediterranean river. Palaeogeography Palaeoclimatology Palaeoecology 148: 87–103.

Mezquita, F., H. I. Griffiths, M. I. Dominguez & M. A. Lozana-Quilis, 2001. Ostracoda (Crustacea) as ecological indicators: a case study from Iberian Mediterranean brooks. Archiv für Hydrobiologie 150: 545–560.

Milhau, B., N. Dekens & K. Wouters, 1997. Evaluation de l'utilisation des ostracodes comme bioindicateurs potentiels de pollution: application aux eaux de la Slack (Boulonnais, France). Ecologie 28: 3–12.

Neale, J. W., 1988. Ostracods and Paleosalinity Reconstruction. In De Deckker, P., J. P. Colin & J. P. Peypouquet (eds), Ostracoda in the Earth Sciences. Elsevier Publications, The Netherlands.

Pekcan, N., 1996. Yeniçağa depresyonunun (Bolu) jeomorfolojik oluşum ve gelişimi. İstanbul Üniversitesi Coğrafya Dergisi. (Ayrı basım), 113–122 (In Turkish).

Ranta, E., 1979. Population biology of *Darwinula stevensoni* (Crustacea, Ostracoda) in an oligotrophic lake. Annales Zoologici Fennici 16: 28–35.

Roca, J. R. & A. Baltanás, 1993. Ecology and distribution of Ostracoda in Pyrenean springs. Journal of Crustacean Biology 13: 165–174.

Roca, J. R., A. Baltanás & F. Uiblein, 1993. Adaptive responses in *Cypridopsis vidua* (Crustacea: Ostracoda) to food and shelter offered by a macrophyte (*Chara fragilis*). Hydrobiologia 262: 127–131.

Roca, J. R., F. Mezquita, J. Rueda, A. Camacho & M. R. Miracle, 2000. Endorheic versus karstic lake: patterns of ostracod distributions and lake typology in a Mediterranean landscape (Castilla - La Mancha, Spain). Marine and Freshwater Resources 51: 311–319.

Rossetti, G., M. Bartoli & K. Martens, 2004. Limnological characteristics and recent ostracods (Crustacea, Ostracoda) of freshwater wetlands in the Parco Oglio Sud (Northern Italy). International Journal of Limnology 40: 329–341.

Särkkä, J., L. Levonen & J. Mäkelä, 1997. Meiofauna of springs in Finland in relation to environmental factors. Hydrobiologia 347: 139–150.

Saygı-Başbuğ, Y. & F. Y. Demirkalp, 2004a. Primary production in shallow eutrophic Yeniçağa Lake (Bolu, Turkey). Fresenius Environmental Bulletin 13: 98–104.

Saygı-Başbuğ, Y. & F. Y. Demirkalp, 2004b. Trophic status of shallow Yeniçağa Lake (Bolu, Turkey) in relation to physical and chemical environment. Fresenius Environmental Bulletin 13: 358–393.

Scharf, B. W., 1993. Ostracoda (Crustacea) from eutrophic and oligotrophic maar lakes of the Eifel (Germany) in the Late, Post Glacial. In McKenzie, K. G. & P. J. Jones (eds), Ostracoda in the earth and life sciences. Balkema, Australia, 453–463.

Shornikov, E. I. & Y. A. Trebukhova, 2001. Ostracods of brackish and fresh waters of Southwestern coast of Peter the Great Bay. In Kasyanov, V. L., M. A. Vaschenko & D. L. Pitruk (eds), The State of Environment and Biota of the Southwestern Part of Peter the Great Bay and the Tumen River Mouth, 2002, Vol. 3. Dalnauka, Vladivostok, 56–84.

Sümer, N., 2002. Flora of Lake Yeniçağa. MS Thesis. Abant İzzet Baysal University.

Tan, C. O. & M. Beklioğlu, 2005. Catastrophic-like shifts in shallow Turkish lakes: a modeling approach. Ecological Modeling 183: 425–434.

Tanoğlu, A. & H. İnandık, 1952–1953. Bolu-Gerede Arasındaki Çağa Depresyonu Hakkında Küçük Bir Not. İstanbul Üniversitesi Coğrafya Enstitüsü Dergisi 3(4): 198–199.

Ter Braak, C. J. F., 1986. Canonical correspondance analysis: a new eigenvector technique for multivariate direct gradient analysis. Ecology 67: 1167–1179.

Ter Braak, C. J. F., 1987. The analysis of vegetation–environment relationships by canonical correspondence analysis. Vegetatio 69: 69–77.

Ter Braak, C. J. F., 1989. CANOCO – an extension of DECORANA to analyze species environment relationships. Hydrobiologia 184: 169–170.

Ter Braak, C. J. F. & L. G. Barendregt, 1986. Weighted averaging of species indicator values: its efficiency in environmental calibration. Mathematical Biosciences 78: 57–72.

Tressler, W. L., 1959. Ostracoda. In Edmondson, W. T. (ed.), H. B. Ward & G. C. Whipple's Freshwater Biology, 2nd edn. John Wiley & Sons, New York, 657–734.

Victor, R. & C. H. Fernando, 1979. The freshwater ostracods (Ostracoda: Crustacea) of India. Records of the Zoological Survey of India 74: 147–242.

Yılmaz, F. & O. Külköylüoğlu, 2006. Ecological requirements of freshwater Ostracoda (Crustacea) in Lake Aladağ (Bolu, Turkey). Ecological Research 21: 165–173.

Hydrobiologia (2007) 585:135–140
DOI 10.1007/s10750-007-0634-5

OSTRACODA (ISO15)

Food selection in *Eucypris virens* (Crustacea: Ostracoda) under experimental conditions

O. Schmit · G. Rossetti · J. Vandekerkhove · F. Mezquita

Abstract Ostracods have long been studied by scientists because their fossil remnants provide a valuable tool for the reconstruction of past environmental changes, including climate change and anthropogenic eutrophication. Relatively little is known about the physiology, behaviour and reproductive ecology of recent forms. We argue that filling this gap in knowledge requires stable cultures that can be used in laboratory studies. Here we provide quantitative information on the food preference of the common non-marine ostracod *Eucypris virens*. Using an experimental device allowing a free choice of eight food items, including both auto- and heterotrophic organisms, observations were carried out on groups of animals from different populations. Our results indicate that *E. virens* highly prefers spinach and the cyanobacterium *Tolypothrix tenuis* to other food items. The latter also plays an important role in maintaining the quality of the culture medium and provides a convenient substrate for moulting and egg-laying. As such, we recommend Cyanobacteria like *T. tenuis* as a food source for long-standing cultures of *E. virens*, and other non-marine ostracod species.

Keywords *Eucypris virens* · Freshwater ostracods · Culturing · Food selection · Laboratory experiments

Guest editors: R. Matzke-Karasz, K. Martens & M. Schudack
Ostracodology – Linking Bio- and Geosciences

O. Schmit (✉) · J. Vandekerkhove · F. Mezquita
Department of Microbiology and Ecology, University of València, C/ Doctor Moliner 50, Burjassot 46100 Valencia, Spain
e-mail: osch@alumni.uv.es

O. Schmit · G. Rossetti · J. Vandekerkhove · F. Mezquita
Department of Environmental Sciences, University of Parma, Viale G.P. Usberti, 33/A, Parma 43100, Italy

Introduction

Until now, our knowledge of the feeding preference of non-marine ostracods was limited to merely anecdotal observations. In natural environments, freshwater ostracods are described as generalists (Liperovskaya, 1948; Meisch, 2000), feeding on algae, decaying organic material, plants and animals (both alive and dead). More specific information comes from the few laboratory studies that cultured individual ostracod species on well-defined food items (review in Danielopol et al., 2002) and from analyses of stomach contents (Liperovskaya, 1948; Margalef, 1953). Although data on mortality rates and fitness are lacking, it seems that providing algae, either living or dry (Otero et al., 1998; Mezquita et al., 1999; Yin et al., 1999; Baltanás et al., 2000; Chial & Persoone, 2002) and commercial fish food (Cywinska & Hebert, 2002), allows the survival and growth of various freshwater ostracod species. Laboratory observations also suggest that some ostracods can feed on periphyton (Roca et al., 1993; Uiblein et al., 1994, 1996). Roca et al. (1993) even report that *Cypridopsis vidua* actively looks for *Chara* beds to feed on the periphyton and to hide from predators. *Notodromas monacha* is primarily a neuston feeder (Kiss, 2004). The biofilm could be of high importance in the diet as a protein food source (Lawrence et al., 2002). *Eucypris virens* have been observed devouring recently dead or moribund crustaceans such as cladocerans, anostracans, and other ostracods (O. Schmit, pers. obs.).

Due to the particularly good fossilisation of their calcareous valves, ostracods are commonly used to reconstruct changes in past environments (Griffiths & Holmes, 2000). Today, their availability as living material allows the application of genetic, cytological and physiological tools, and the implementation of laboratory experiments under strictly controlled conditions. In field conditions, freshwater ostracods usually show a high reproductive potential, a relatively short generation time and the possibility to grow large populations. Despite these characteristics, they have only rarely been used in laboratory experiments (but see Gandolfi et al., 2001; Van Doninck et al., 2003). We argue that this is partly due to the limited knowledge of their ecological requirements, which makes it hard to obtain stable and extensive cultures.

The few attempts to keep ostracods under laboratory conditions made use of a broad range of temperatures, ranging between 4 and 30°C (Baltanás et al., 2000, Van Doninck et al., 2003). Otero et al. (1998) cultured *E. virens* at 16°C and showed that it is essential to mimic a winter photoperiod to allow the generations to succeed continuously. Indeed, short day length induces production of rapidly hatching eggs in *E. virens*. In contrast, long days result in the production of predominantly resting eggs, which need desiccation before hatching can occur. As culture medium, mineral water can be used (Otero et al., 1998; Gandolfi et al., 2001) or filtered water from the original habitat (Xia et al., 1997; Yin et al., 1999).

This study aims at expanding the poor knowledge of ecological prerequisites of non-marine ostracods by obtaining, for the first time, quantitative data on the food preference of a common freshwater ostracod, *Eucypris virens* (Jurine, 1820). This species has a large geographic distribution that covers Europe, North Africa, Middle East, Central Asia, China, Australia, New-Zealand and North America (Meisch, 2000). *Eucypris virens* is typically found in temporary pools. Due to the (co-)existence of sexual and parthenogenetic populations (Schön et al., 2000), it is an excellent model species to study fundamental questions related to the evolution of reproductive modes. Males are only recorded from some populations around the Mediterranean (see the NODE database, Horne et al., 1998). Elsewhere, only parthenogenetic females are found. The mechanisms determining this pattern of geographical parthenogenesis are still poorly understood, in part because of the scarcity of model organisms that are appropriate to conduct laboratory experiments to test for differences in life history and fitness components between sexual and asexual populations.

Materials and methods

Adult females were obtained from four Italian populations, distributed along a latitudinal gradient of more than six degrees to increase the generality of the findings. The populations inhabited temporary pools, in meadows and ditches.

Two experiments were performed, both on groups of 20 animals. In a first experiment, seven food types were provided: frozen *Artemia* sp., *Mysis* sp., *Chironomus* sp., *Daphnia* sp., spinach and Brussels sprouts (thawed before use), and dried "Spirulina" (a commercial food for fish containing 7% of *Spirulina* sp.). Three gravels of different colours (white, black, grey) were added as blank items. The second experiment only differed from the first in that *Chironomus* sp., an item that was almost completely neglected by the ostracods in the first experiment, was replaced by living bunches of the freshwater cyanobacterium *Tolypothrix tenuis* cultured in the laboratory (Strain: CCAP 1482/3A). Each experiment was replicated with animals from two source locations.

For both experiments, food selection was observed in a plastic Petri dish (Ø 9 cm) with 10 open sectors radiating from the centre (see Fig. 1), allowing the animals free access to the different items offered. Animals were starved for 3–5 hours before the experiments. The experimental device was filled with mineral water (1 cm deep, conductivity: 435 $\mu s\ cm^{-1}$, pH: 7.21) and stocked with the different items one hour before the introduction of test animals and the start of the observations. Food items were provided in similar amounts (as were the gravels) and were not limiting throughout the observational period. Moreover, the quantity provided allowed all animals to feed simultaneously on the same item. The experiments were performed at room temperature (20 ± 2°C) under artificial light. The observations started with the transfer of the *E. virens* specimens to the centre of the Petri dish. Observations continued for three consecutive hours at 5′ intervals (hour 0–1) or 30′ intervals (hours 1–3). For further analyses, the first hour observations were averaged over 30′ intervals. At each check time, we assigned each individual to one of three action categories: swimming, resting and feeding. For those animals that were feeding, we took note of the food type that was being consumed.

The data were analysed statistically using two-way ANOVA (Statistica 6.0; StatSoft 2003), with the number of individuals as the dependent variable and either the actions, time or food items as independent factors. The data were square root transformed prior to analyses to obtain normally distributed data, as confirmed by Kolmogorov-Smirnov tests. Bartlett's tests (Sokal & Rohlf, 2003) were applied to check if variances were homogeneous. Only very few animals were located on the blank items (<1% in all experiments). Therefore, these observations were not considered in the analyses of food selection.

Fig. 1 Modified Petri dish used in the experiments. The food items were placed between the digits and the external wall of the Petri dish

Results

Actions

In both experiments, most of the individuals were found on the food items (Table 1: action effect in ANOVA. On average Exp. 1: 65 ± 6%, Fig. 2a; Exp. 2: 76 ± 39%, Fig. 2b). Relatively few individuals were either swimming or resting. This pattern remained unaltered throughout the entire observational period (Table 1: no interaction Time X Action in ANOVA).

Food selection

Both experiments showed that *E. virens* fed very selectively on particular food sources (Table 1: food selection effect in ANOVA). On average, most of the individuals were observed on either spinach (Exp. 1, Fig. 3a: 60 ± 4%) or *Tolypothrix tenuis* (Exp. 2, Fig. 3b: 39 ± 0.2%), whereas for example only 4 ± 3% (Exp. 1) to 8 ± 4% (Exp. 2) of the individuals were found on any of the four

Table 1 Summary of results of two-way ANOVA analyses on data on actions and food selection of groups of 20 *Eucypris virens* individuals

	Df	*F*-values (Exp. 1)	*F*-values (Exp. 2)
Actions	2	44.2***	26.0***
Time	3	0.2 ns	0.2 ns
Action × Time	6	2.1 ns	0.5 ns
Food selection	6	34.2***	23.4***
Time	3	0.2 ns	0.8 ns
Food selection × Time	18	1.5 ns	1.8 ns

*** $P < 0.001$; ns, not significant

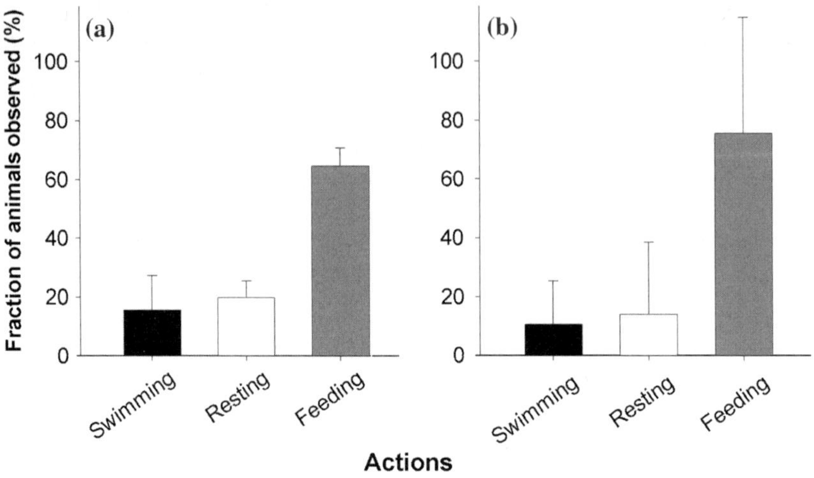

Fig. 2 The prevalence of observed actions expressed as the percentage of the 20 individuals observed. To reduce the complexity of the graphs and because of the absence of a time × action effect (see Table 1), bars show averages over the different time intervals. Error bars show 95% CI using populations as replicates

animal foods that were offered. In none of the experiments were all 20 animals observed together on the same item at the same time, although the quantity of food allowed it.

Discussion and conclusions

Actions

In both group experiments, most of the animals were found on the food items, whereas only a few were resting or swimming. The fact that swimming behaviour was relatively rare, and when observed, of short duration, could indicate that swimming is either risky (e.g. detection by predators) or energetically costly, or a combination of both. The results of the group experiments contrast with those of our preliminary observations on single individuals, in that single individuals were more often observed resting (on average 52 ± 18% of the observational period). It is unlikely that this is related with the presence or absence of competitors, as in both experiments

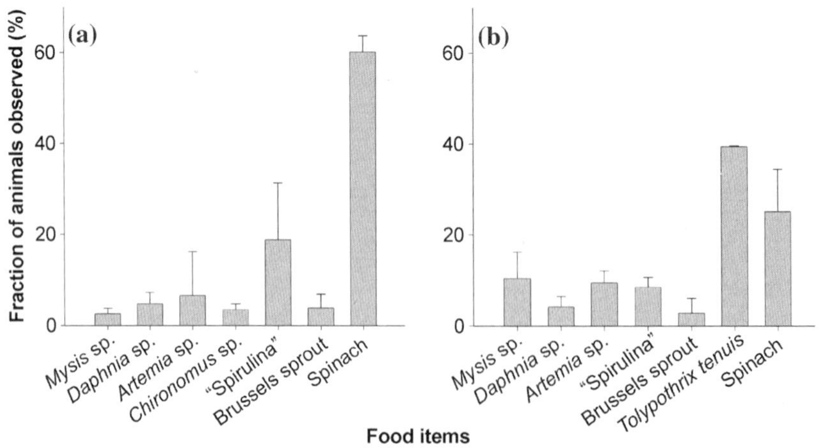

Fig. 3 The average number of *E. virens* individuals observed on each of the food items offered. To reduce the complexity of the graphs, and because of the absence of a time x food selection effect (see Table 1), bars show averages over the different time intervals. Error bars show 95% CI using different populations as replicates

food items were provided *ad libitum*. Possibly, the high number of resting animals in the preliminary experiments was an artefact related to the short observation period (10′ vs. 3 h in the group experiments). Indeed, in all experiments, complete immobility was the first behaviour observed. Animals were lying on the bottom, "playing dead" for several minutes probably as a consequence of handling stress.

Food selection

The observed preference of *E. virens* for spinach and *T. tenuis* can be related to the specific environments colonised by this species. *Eucypris virens* typically inhabits temporary ponds in meadows, the bottom of which is often covered by plants.

Our results are in accordance with Liperovskaya (1948) who found much more frequently algae than animal remains in the gut of several species of generalist freshwater ostracods. Filamentous and mat-forming cyanobacteria (and other algae) are common inhabitants of temporary waters (Margalef, 1983) where *E. virens* is found. This could be an important food source in these stressed temporary habitats. On the other hand, the preference for *Tolypothrix* may be somewhat surprising, as many cyanobacteria are known to be toxic. For example, Graber and Gerwick (1998) identified a toxin from a marine assemblage of *Lyngbya majuscula* and *Tolypothrix* sp. As far as we know, no toxicity is recorded from the freshwater species *Tolypothrix tenuis*. In addition, tolerance to toxic cyanobacteria is known from some freshwater crustaceans: when ingested, the toxic *Oscillatoria sancta* is not harmful for the crayfish *Pacifastacus leniusculus* (Lirås et al., 1998). Possibly, a similar immunity response is exhibited by ostracods. It must be noticed that a record of "on a food item" could, in our experiments, not necessarily mean "feeding on that item"; indeed, the animal can be located on it but resting, not eating. However, the day after the experiments it was common to observe green faeces on the bottom of the Petri dish, as evidence that the spinach and the algae had been consumed. Also, at each control time there were always a few animals feeding on other food types (dead animals, "Spirulina", Brussels sprouts). This suggests that exploratory behaviour remained active even if one or two food types were clearly preferred over others.

As a food source for cultures of *E. virens*, and probably many other freshwater ostracods, we recommend the cyanobacterium *T. tenuis*. This alga is not only a highly preferred food item, it may also be important for maintaining an acceptable water quality through release of oxygen and uptake of degradation products. In one of the experiments carried out without *T. tenuis*, three animals died during the following night. Water was turbid and odorous as a consequence of the high quantity of decaying organic matter. In the experiments, which included living algae, this never occurred. Furthermore, algae also serve as physical support for laying eggs (O. Schmit, pers. obs.), and as anchorage for moulting (D. J. Horne, pers. comm.). For long-standing cultures it may be beneficial occasionally to supplement the algal diet with other food items to ensure fully balanced nutrition.

Acknowledgments OS particularly acknowledges Valentina Pieri for her general help in ostracodology and her continuous collaboration. David J. Horne provided useful comments and the English translation of the paper by Liperovskaya. Roger K. Butlin is acknowledged for his comments and revision of the English text. Gessica Gorbi is thanked for her help with the culturing of algae and Andrea Piotti for his assistance in statistical analyses. Giorgio Benassi brought specimens of *Eucypris virens* from central Italy. Two anonymous referees contributed to the improvement of this paper. This work was funded by the EU Marie Curie Research Training Network SexAsex (*From Sex to Asex: a case study on interactions between sexual and asexual reproduction*, contract MRTN-CT-2004-512492). We are very grateful to all members of the network for their practical contributions to the work presented here and for many stimulating discussions.

References

Baltanás, A., M. Otero, L. Arqueros, G. Rossetti & V. Rossi, 2000. Ontogenetic changes in the carapace shape of the non-marine ostracod *Eucypris virens* (Jurine). Hydrobiologia 419: 65–72.

Chial, B. & G. Persoone, 2002. Cyst based toxicity tests XII - Development of a short chronic sediment toxicity test with the ostracod crustacean *Heterocypris incongruens*: Selection of test parameters. Environmental Toxicology 17: 520–527.

Cywinska, A. & P. D. N. Hebert, 2002. Origins of clonal diversity in the hypervariable asexual ostracode *Cypridopsis vidua*. Journal of Evolutionary Biology 15: 134–145.

Danielopol, D. L., E. Ito, G. Wansard, T. Kamiya, T. M. Cronin & A. Baltanás, 2002. Techniques for collection and study of Ostracoda. Geophysical Monographs 131: 65–97.

Gandolfi, A., E. B. A. Todeschi, V. Rossi & P. Menozzi, 2001. Life history traits in *Darwinula stevensoni* (Crustacea: Ostracoda) from Southern European populations under controlled conditions and their relationship with genetic features. Journal of Limnology 60: 1–10.

Graber, M. A. & W. H. Gerwick, 1998. Kalkipyrone, a toxic γ-pyrone from an assemblage of the marine cyanobacteria *Lyngbya majuscula* and *Tolypothrix* sp. Journal of Natural Products 61: 677–680.

Griffiths, H. I., J. A. Holmes, 2000. Non-marine ostracods and quaternary palaeoenvironments. Quaternary Research Association, Technical Guide No. 8, London, UK.

Horne, D. J., A. Baltanás & G. Paris, 1998. Geographical distribution of reproductive modes in living non-marine ostracods. In Martens, K. (ed.), Sex and parthenogenesis: evolutionary ecology of reproductive modes in non-marine ostracods. Backhuys, Leiden, the Netherlands, 77–99.

Kiss, A., 2004. Field and laboratory observations on the microhabitat and food selection as well as predator avoidance of *Notodromas monacha* (Crustacea: Ostracoda). Revista Española de Micropaleontología 36: 147–156.

Lawrence, J. R., B. Scharf, G. Packroff & T. R. Neu, 2002. Microscale evaluation of the effects of grazing by invertebrates with contrasting feeding modes on river biofilm architecture and composition. Microbial Ecology 44: 199–207.

Liperovskaya, E. S., 1948. On the feeding of freshwater ostracods. Zoologicheskie Zhurnal 27: 125–136 [In Russian].

Lirås, V., M. Lindberg, P. Nystrom, H. Annadotter, L. A. Lawton & B. Graf, 1998. Can ingested cyanobacteria be harmful to the signal crayfish (*Pacifastacus leniusculus*)? Freshwater Biology 39: 233–242.

Margalef, R., 1953. Los crustáceos de las aguas continentales ibéricas. Instituto Forestal de Investigaciones y Experiencias, Madrid.

Margalef, R., 1983. Limnología. Omega, Barcelona, Spain.

Meisch, C., 2000. Freshwater Ostracoda of Western and Central Europe. Spektrum Akademischer Verlag, Gustav Fischer, Heidelberg, Germany.

Mezquita, F., J. R. Roca & G. Wansard, 1999. Moulting, survival and calcification: the effects of temperature and water chemistry on an ostracod crustacean (*Herpetocypris intermedia*) under experimental conditions. Archiv für Hydrobiologie 146: 219–238.

Otero, M., V. Rossi, A. Baltanás & P. Menozzi, 1998. Effect of genotype and photoperiod on diapause strategies in *Eucypris virens* (Jurine, 1820) (Crustacea: Ostracoda). Archiv für Hydrobiologie 52: 229–236.

Roca, J. R., A. Baltanás & F. Uiblein, 1993. Adaptive responses in *Cypridopsis vidua* (Crustacea, Ostracoda) to food and shelter offered by a macrophyte (*Chara fragilis*). Hydrobiologia 262: 127–131.

Schön, I., A. Gandolfi, E. Di Masso, V. Rossi, H. I. Griffiths, K. Martens & R. K. Butlin, 2000. Persistence of asexuality through mixed reproduction in *Eucypris virens* (Crustacea, Ostracoda). Heredity 84: 161–169.

Sokal, R. R. & F. J. Rohlf, 2003. Biometry: the principles and practice of statistics in biological research, 3rd edn. Freeman and Co., New York, NY, USA.

StatSoft Inc., 2003. Electronic statistics textbook. Tulsa, OK: Statsoft. WEB: http://www.statsoft.com/textbook/stathome.html.

Uiblein, F., J. R. Roca & D. L. Danielopol, 1994. Experimental observations on the behaviour of the ostracode *Cypridopsis vidua*. Verhandlungen der Internationalen Vereinigung für Limnologie 25: 2418–2420.

Uiblein, F., J. R. Roca, A. Baltanás & D. L. Danielopol, 1996. Tradeoff between foraging and antipredator behaviour in a macrophyte dwelling ostracod. Archiv für Hydrobiologie 137: 119–133.

Van Doninck, K., I. Schön, F. Maes, L. De Bruyn & K. Martens, 2003. Ecological strategies in the ancient asexual animal group Darwinulidae (Crustacea, Ostracoda). Freshwater Biology 48: 1285–1294.

Xia, J., E. Ito & D. R. Engstrom, 1997. Geochemistry of ostracode calcite: Part 1. An experimental determination of oxygen isotope fractionation. Geochimica et Cosmochimica Acta 61: 377–382.

Yin, Y., W. Geiger & K. Martens, 1999. Effects of genotype and environment on phenotypic variability in *Limnocythere inopinata* (Crustacea: Ostracoda). Hydrobiologia 400: 85–114.

Hydrobiologia (2007) 585:141–157
DOI 10.1007/s10750-007-0635-4

Extra-lobal and complex dimorphic features in Middle Devonian palaeocopine ostracods

Gerhard Becker · Willi K. Braun

Abstract With respect to Middle Devonian ostracods from western Canada, special cases of palaeocopine extradomiciliar dimorphism are discussed, as in the opinion of many workers are unusual for younger hollinoideans and, in particular, primitiopsoideans. Although the general view that velar dimorphism is the main characteristic of Mid Palaeozoic Palaeocopina is basically not wrong, dimorphic histial structures, even though less conspicuous in aspect, have been recognised in some Mid-Devonian hollinoideans, specifically hollinids and ctenoloculinids. This is now also proven to be true of distinct hollinellids and, by the way, discrete primitiopsids, while distinguishing between narrow-based projections (ridges, flanges, spines) and broad-based protrusions (e.g. lobes). Whereas the former may occur in velar as well as histial situations, lobal features are especially found in the histial position. Special attention is paid to false primitiopsids, being (with some caution) grouped in the Hollinoidea (Family unknown). Essentially, three genera are new or redefined (former left under open nomenclature): Gen. aff. *Abditoloculina*: multilobate in aspect, lobal (histial) dimorphic; Gen. aff. *Gravia*: coalesced velar dimorphic; *Bisphenella*: histial dimorphic. With respect to the hold discussion about ostracod phylogeny, the updated standard classification scheme is given, as applied to Palaeozoic, non-myodocopine ostracods. Reduction of dimorphism in palaeocopines is considered to be an apomorphic character, lack of dimorphism assumed to indicate asexual reproduction. At the same time, a plea is made in favour of reviving Treatise revision.

Keywords Morphology · Dimorphism · Evolution · Asexual reproduction · Ostracoda

Guest editors: R. Matzke-Karasz, K. Martens & M. Schudack
Ostracodology – Linking Bio- and Geosciences

G. Becker (✉)
Forschungs-Institut Senckenberg, Senckenberg-Anlage 25, 60325 Frankfurt am Main, Germany
e-mail: Gerhard.Becker@Senckenberg.de

W. K. Braun
Department of Geological Sciences, University of Saskatchewan, Saskatoon S7N 5E2, Canada

Introduction

The model of adventral structures, as proposed by Jaanusson (1957) and emended by Martinsson (1962), is proven to be a powerful tool in the hands of taxonomic workers to classify palaeocopine ostracods. Morphologically speaking, the term "adventral structures" refers to linear ornamental extensions, as located upon the ventral part of the valve in definite positions, accordingly expressed by distinctive terms (margina, velum, histium). Being dimorphic in appearance, these features are considered to be applicable for the classification of the Palaeocopina Henningsmoen, 1953 into (assumed) natural family-groups. Following a proposal of Becker (1990), which was adopted by Whatley et al. (1993), only a few taxa of superfamily rank are currently in use (see below, approach to systematics).

Although the concept of adventral structures is, basically, easy to understand, there are, with natural variations, numerous deviations from the standard configurations. Extradomiciliar dimorphic features may become complex, or may appear unclear in development and origin. Being unique for palaeocopines, adventral structures and extradomiciliar dimorphism may not really reasonably be explained in terms of functional morphology of living ostracods (Becker, 2005), notwithstanding contrary, though highly speculative attempts (Kesling, 1969). Moreover, they must be considered, in terms of phylogeny, a discontinued concept (Becker, 1998). While identifying margina or velum generally appears unproblematic, there is sometimes serious disagreement as to whether an adventral structure is velar or histial in origin. The relevant literature is legion, where such an indication is sometimes highly controversially debated, in particular, with hollinoideans (e.g. Kesling, 1951 versus Jaanusson, 1957). "It is possible that the truth lies somewhat in between the two views" (Henningsmoen, 1965: 351).

All material is part of a (as a whole) very large and diverse ostracod fauna, which has been collected by W.K.B. – as a result of many years' fieldwork – from Yukon and Northwest Territories, Canada. The faunal assemblages, as stored with the University of Saskatoon micropalaeontological collections, were repeatedly evaluated under stratigraphically emphasised aspects (Braun, 1977, 1978, 1993; Braun & Mathison, 1986) and are about to be treated systematically (Becker & Braun, unpublished). The following account, considering the prejudice that extradomiciliar dimorphism in mid-Palaeozoic palaeocopines should be solely velar in origin, is part of this effort (see also Becker & Braun, 2005).

Extradomiciliar dimorphism

Relevant authors, following in the footsteps of Jaanusson (1957), like to describe extradomiciliar features in terms of linear ornaments (like ridges, flanges, frills, rows of tubercles). These features are subsequently classified, with respect to their relative position on the valve surface from distal to proximal (Fig. 1a), as margina (occurring distally to the velum, immediately adjacent to the free edge), velum (occurring distally to the histium, at lateroventral bend) or histium (occurring above the velum, forming a direct ventral/lateroventral continuation of the connecting lobe). These features, if dimorphic in appearance, refer to several types of extradomiciliar dimorphism (Fig. 1b), called marginal, velar or histial, being simple or locular, or appearing uncomplicated to complex (combined, coalesced). Later, thinking in terms of functional morphology, Jaanusson (1966) discarded terms like histial and velar (marginal was excluded), laying emphasis on the involved domiciliar space, with reference to the antrum of Henningsmoen (1965). The purpose of the extradomiciliar dimorphism was obviously to form certain smooth, concave channel-like areas upon the shell and along a portion of the free margin. In thick-shelled ostracods, these areas are simple impressions on the surface of the valve, which remain necessarily shallow and exposed. Bordering the antrum with a flange (border projection, fence, as shaped by any adventral structure above) can attain an increase of the depth. Such spaces (Fig. 1c) appear simple antral or locular, being labelled infravelar (antrum occurs between the velum and the free edge), supravelar (antrum occurs between the velum and the histium) or biantral (supravelar and infravelar antra occur in the same valve).

As far as the basic design is concerned, corresponding point projections (spines) are neglected. Therefore, Henningsmoen (1965: 335) holds it more practical to distinguish between narrow-based projections (ridges, flanges, spines) and broad-based protrusions (e.g. lobes). Such structures may occur in velar as well as histial positions. In connection with lobal features, as occurring especially in the histial position, special attention should also be

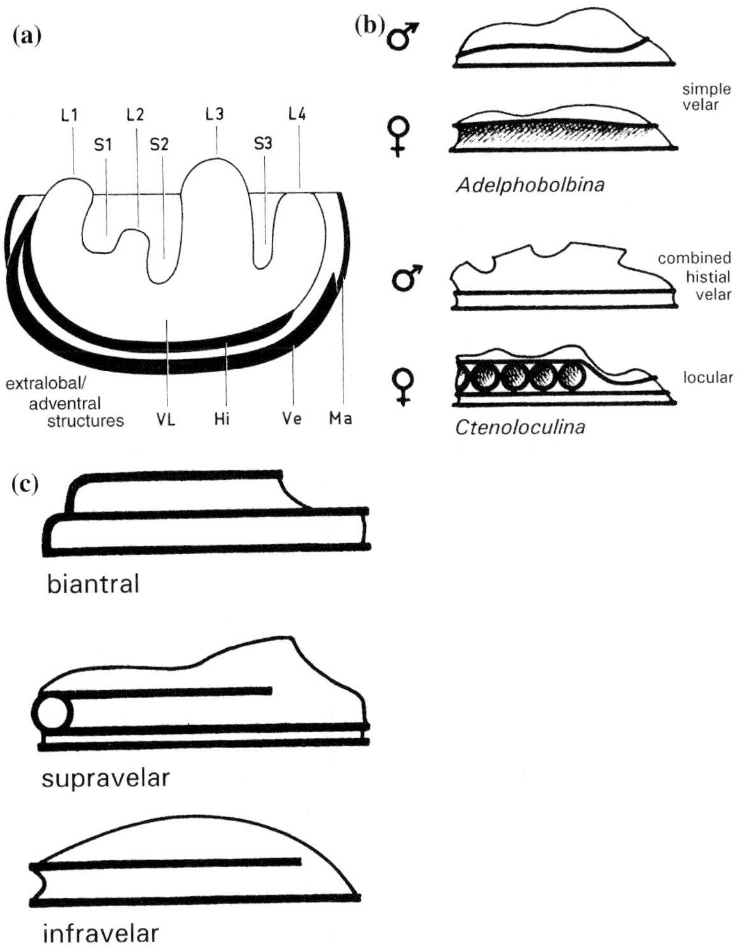

Fig. 1 (**a**) Palaeocopine carapace architecture. As explained on a (hypothetical) quadrilobate valve. Lobate area (located above hid-height) shows sulci (S1–S3; S2 containing adductor muscle attachment point) and lobes (L1–L4; L3 most prominent). Connecting lobe, as adjoining ventral ends of lobes, includes ventral lobe (VL). The latter is located below mid-height and belongs basically to the extra-lobate area, which demonstrates adventral structures, i.e. linear extensions paralleling free margin from proximal to distal. While occupying certain positions upon the ventral surface in relation to VL, they are (following Jaanusson, 1957 and Martinsson, 1962) named margina (Ma, as occurring immediately adjacent to the free edge), velum (Ve, as occurring distally to the histium, or in the equivalent position if occurring without histium), and histium (Hi, as forming a direct continuation of the connecting lobe). Schematically, after Becker (1968). (**b**) Extradomiciliar dimorphism, in relation to adventral structures. (Above) velar, simple in appearance, as demonstrated in the hollinellid genus *Adelphobolbina* Stover, 1956. (Below) coalesced velar and outer lobal (histial), locular in appearance, as demonstrated in the hollinid genus *Ctenoloculina* Bassler, 1941. Schematically, after Henningsmoen (1965). (**c**) Extradomiciliar spaces, in relation to the antrum. (Bottom) infravelar (situated between velum and the free edge). (Middle) supravelar (situated between velum and histium, provided with one loculus). (Top) biantral (supravelar antrum and infravelar antrum occur in the same valve). Schematically, after Becker (1968)

paid to the (so-called) connecting lobe, as defined by Jaanusson (1957: 190) as lobe connecting the ventral ends of some or all lobes in quadrilobate valves. The importance of the connecting lobe (in practice often expressed as ventral lobe (VL)) finds an expression in Henningsmoen's (1965: 353) "main lobal structure". Dimorphism, as involved in hollinoideans with the connecting lobe, is called "lobal" (histial), applied terminology: "lobal dimorphism" Henningsmoen (1965); the term "lobar dimorphism" Becker et al. (1989) concerning dimorphic L3, describes a secondary hollinoidean (and ? kirkbyoidean) character.

Hollinid dimorphism

From analogy to the established term "kloedenellid dimorphism" (Adamczak, 1991), the term "hollinid dimorphism" was proposed following a recommendation of that author to avoid "illogical names" (Adamczak pers. comm., 12.10.1992). This was agreed by the majority of the then Treatise authors (see below). Like other types, it is used for the whole superfamily and refers to the for hollinoidean heteromorphs typical, ventrally located dolon being simple antral or locular in aspect. In the simplest case it appears infravelar, what is true of the bulk of mid-Palaeozoic representatives, in particular, the Hollinellidae – but is, on no account, applied to all, especially older forms. For the Ordovician Tetradellidae, Schallreuter (1966) diagnoses the presence of velar and histial structures in both sexes, the latter of which is non-dimorphic, while Jaanusson (1966) describes a supravelar antrum in tetradellids. Following from that author (1966: 4), one can say, "among genera with a histial dolon the heteromorphic subhistial features are more complicated then previously thought".

The opinion that infravelar dimorphism generally or solely occurs in hollinids (including ctenoloculinids), as was stated by Kesling (1951) and Henningsmoen (1953), but debated by Jaanusson & Martinsson (1956) and Jaanusson (1957), and definitively refuted by Henningsmoen (1965: 352), while laying emphasis on ventral surfaces. "It seems probable that whereas hollinids (with narrow venter) have a velar dimorphism (with the velar structure as outer [proximal] antral fence), [while] others (with wide venter at least in heteromorphs) have a histio-velar dimorphism (i.e. both the velar and lobal [histial] structures take part in the formation of the outer antral fence)". The latter applies to *Hollina* as well as *Ctenoloculina* (Fig. 2), "and in these forms, it seems as if the outer antral fence is formed by both (more-or-less) coalesced velar and outer lobal (histial) structure, so that the lateroventral flange may be said to display a combined velar structure along the antrum". Likewise, "the relations of the antrum to the velar and histial structures are genetically fixed and apparently phylogentically controlled" (Henningsmoen, 1965: 345).

The utility of those characters for classification at the family level, however, is debated. Sometimes subfamilial separation is preferred, while classifying forms with simple antra in the Hollininae and such with locular dimorphism in the Ctenoloculininae. *Hollina* and *Ctenoloculina*, however, show combined dimorphism, as is also the case in *Flaccivelum* (assumed to be a Hollinellidae), as was demonstrated by Becker (in Bless & Jordan, 1972: pls. 34–35).

Unfamiliar histio-velar configurations

In the material under consideration, special cases of dimorphism have been recognised, actually being histio-velar in origin but hard to explain in detail. Sometimes the profound thoughts of relevant workers, in particular, Henningsmoen (1965) and Jaanusson (1966), appear helpful.

Extra-lobate hollinids

Extradomiciliar dimorphism occurs principally on the extra-lobate surface, i.e. the distal part of the palaeocopine valve, basically being (with respect to the lobate area) quadrilobate in appearance. It refers to linear extensions or extra-domiciliar spaces placed between border projections. The lobate area in the upper (proximal) part of the valve contains the lobes (L1–L4 to a maximum), from which the most prominent lobe (always L3) can show a secondary type of domiciliar dimorphism, namely, lobar dimorphism sensu Becker (in Becker et al., 1989). The connecting lobe (represented in less lobate valves by the anteroventral or posteroventral lobe) is considered to be part of the extra-lobate area, as occupying the valve surface below mid-height and between the free margin. Besides the fact that the histial structure (if present) becomes attached to the VL, the latter is not concerned with dimorphism.

In the material from the Middle Devonian of western Canada, exotically ornamented palaeocopines occur which are multi-lobate in aspect (Fig. 3). Apparently several species are distinguishable showing hollinid characteristics, as being preplete in lateral outline; complex quadrilobate in appearance, with small L1/2,

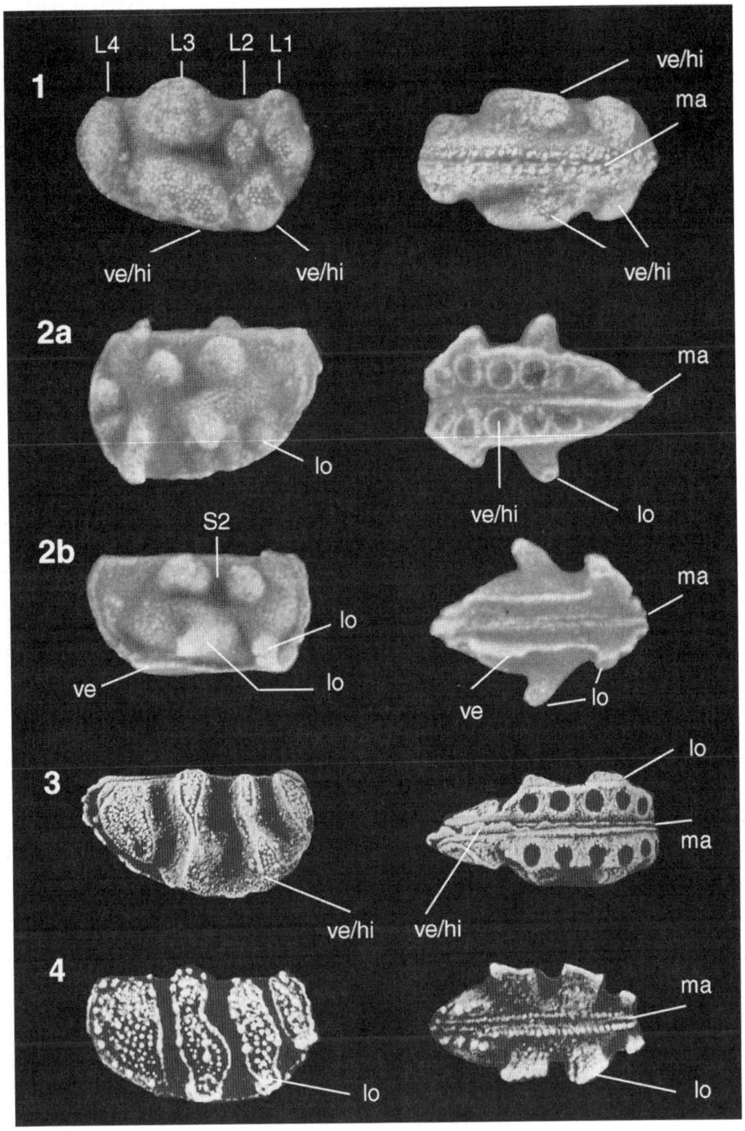

Fig. 2 Histio-velar dimorphism in hollinids (magnification approx. × 36–40). As demonstrated in species showing exotic ("baroque" sensu Bless & Jordan, 1971) lobation patterns and combined velar and histial dimorphism [ve/hi], margina [ma] distinct. (**1**) *Hollina pyxidada* Kesling, 1952 from Ferron Point Formation of Michigan; tecnomorphic carapace. (**2**) *Abditoloculina pulchra* Kesling, 1955 from Centerfield Formation of western New York; a = heteromorphic carapace [read ve/hi instead of lo], b = tecnomorphic carapace. Histial dimorphism lobal [lo]; spurs present in both sexes, but not combined with the velum [ve]. (**3**) *Ctenoloculina thliberilobata* Kesling, 1953 from Arkona shale of Ontario; heteromorphic carapace. Specimen explained in Henningsmoen (1965). (**4**) *Ctenoloculina acanthina* Kesling, 1953 from Arkona shale of Ontario; tecnomorphic carapace. For abbreviations see above. After Kesling & Chilman (1987); Kesling (1953)

prominent L3 and subdued L4; adventral structures provided as delicate (ad)marginal ridge, shortened velar flange appearing typically narrow; valve surface smooth. However, the most intriguing features are bulbous or spur-like protuberances, clearly in the course of the connecting lobe in anteroventral, mid-ventral and posteroventral positions. Although rare, the anteroventral and mid-ventral protrusions are found to be dimorphic. While the anterior one, as lying in the (vertical) continuation of L1 appears to be a velar feature, the mid-ventral one is of clear histial

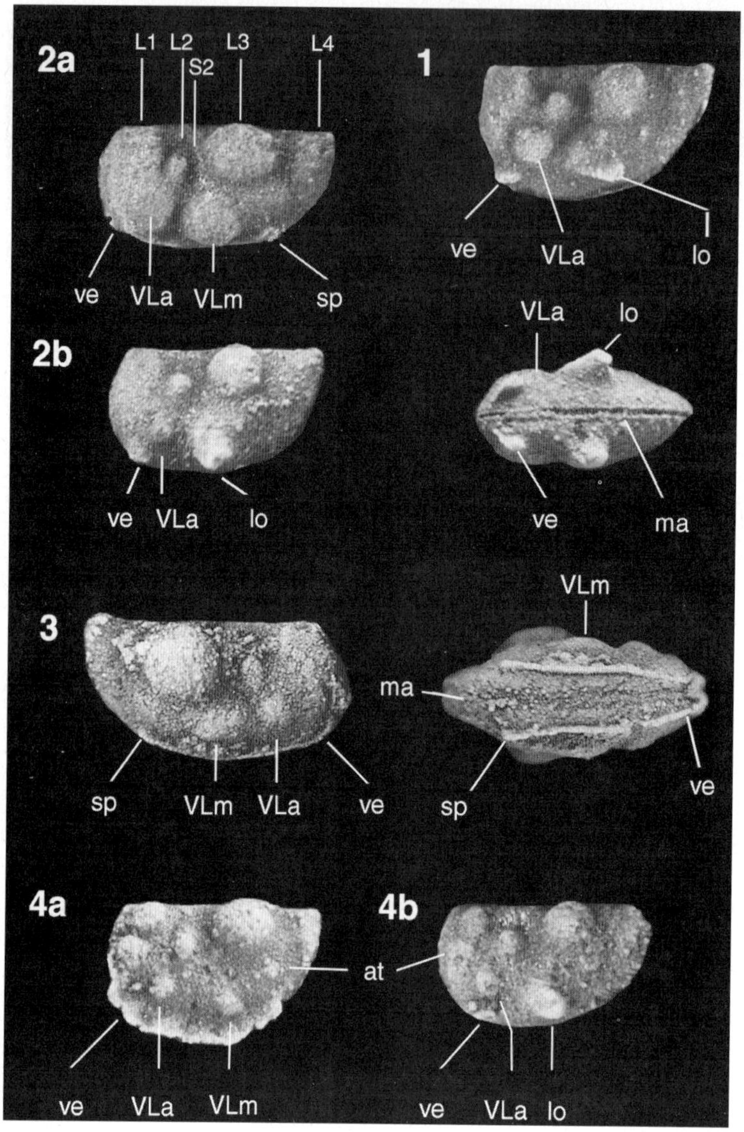

Fig. 3 Lobal (histial) dimorphism in hollinids (magnification approx. × 36–40). Multi-lobate genus, Gen. aff. *Abditoloculina* gen. nov. from the Middle Devonian of western Canada. (**1**) Gen. aff. *A.* sp. 441 from the DM3 zone of eastern Yukon; tecnomorphic carapace; lobes isolated from each other; distinct velar spur [ve] and lobal (histial) protuberance [lo] developed; anteroventral lobe [VLa] an isolated bulb. (**2**) Gen. aff. *A.* sp. 302 from the DM5 zone of the Northwest Territories; a = heteromorphic carapace; b = tecnomorphic carapace; L1 and L2 close together, fused with anteroventral lobe [VLa]; mid-ventral lobe prominent, histial dimorphic in appearance; heteromorphs provided with delicate ventral ridge [ve] terminating in spur [sp]; tecnomorphs showing velar spur [ve] and histial lobal feature [lo]; margina [ma] narrow ridge; note lobar dimorphic L3. (**3**) Gen. aff. *A.* sp. 347 from the DM5 zone of the Northwest Territories; heteromorphic carapace. Anterior [VLa] and mid-ventral [VLm] lobes comparatively small; heteromorphs provided with narrow ventral flange [ve] terminating in spur [sp]; margina [ma] indistinct; L3 apparently lobar dimorphic. (**4**) Gen. aff. *A.* sp. 516 from the DM7 zone of the Northwest Territories; a = heteromorphic carapace; b = tecnomorphic carapace. Since connecting lobe extremely subdivided and besides VLa and VLm additional swellings [at] are developed, the actual species is of particular baroque appearance; the distinct velar antrum [ve] of a single heteromorph appears scalloped; tecnomorphs with lobal (histial) [lo] feature. Photographer W.K.B.

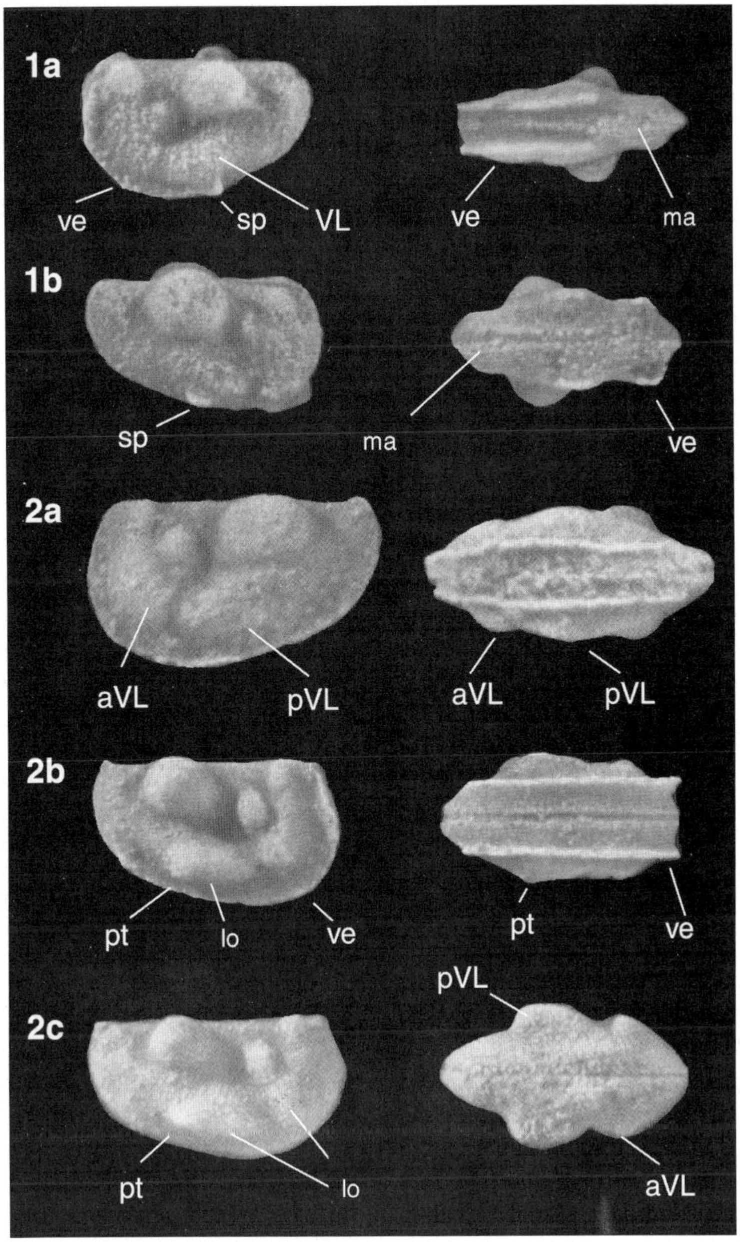

Fig. 4 Unusual hollinoidean species. Considered to show for their genera strange characteristics (magnification approx. × 36–40). (**1**) *Falsipollex minimus/bulbosus* Kesling & Tabor, 1953 from the Genshaw Formation of Michigan; a = heteromorphic carapace; b = tecnomorphic carapace. Although provided with velar features only being an untypical *Falsipollex*, because multi-lobate in appearance and showing the tecnomorphic spurs [sp] (instead subconic) unusually wide and flat and the heteromorphic frill [ve] (instead strongly incurved) narrow. (**2**) *Hollinella* [*Bisphenella*] *rana* Weiss (in Kesling & Chilman, 1987) from the Gravel Point Formation of Michigan; a = heteromorphic carapace; b = tecnomorphic carapace, c = juvenile carapace. Although provided with trimorphic velar features [ve], the species must be considered to be an untypical hollinellid because being multi-lobate in appearance (ventral lobe separated into two parts [aVL, pVL]), and showing lobal (histial) dimorphism [lo] (pVL tapering to posterior protuberant tip [pt]). After Kesling & Chilman (1987)

origin. Since it shows combined velar and lobal (histial) dimorphism, the new hollinid genus, provisionally called Gen. aff. *Abditoloculina* gen. nov., resembles somewhat in lobation pattern *Hollina* Ulrich & Bassler, 1908, as well as *Abditoloculina* Kesling, 1952 in having spurs, but are not combined with the (velar) frill in the heteromorphs (Fig. 2/2). While the latter genus is provided with loculate antra in heteromorphs (as known in Ctenoloculininae), the actual species of Gen. aff. *Abditoloculina* gen. nov. from eastern Yukon and Northwest Territories (see explanation of Fig. 3) show clear histio-velar dimorphism with simple velar antrum and histial spur/lobe. In the most evolved species, Gen. aff. *Abditoloculina* sp. 516 from the Eifelian of Northwest Territories, however, the distinct antrum of a single heteromorph is assumed to appear scalloped – resembling a ctenoloculinid species.

Untypical *Falsipollex* (Hollinidae; Fig. 4/1), as represented by *F. minimus/bulbosus* Kesling & Tabor, 1953 (Middle Devonian, Genshaw Formation of Michigan), may (at first view) appear similar to Gen. aff. *Abditoloculina* gen. nov. Nevertheless, these (for *Falsipollex* strange) forms show tecnomorphic spurs (instead of being subconic) unusually wide and flat, and the heteromorphic frill (instead of usually strongly incurved) narrow; the L3 appears distinctly lobar dimorphic. But both critical spurs are clearly velar structures. Histial features are not developed in both sexes of that species, as was demonstrated (by means of excellent microphotographs) by Kesling & Chilman (1987). In contrast, some multi-lobate, Middle Devonian *Hollinella* (classified as *Bisphenella* Kesling & Chilman, 1987; Fig. 4/2), displays – like *Flaccivelum* (above) – lobal (histial) dimorphism.

Uncertain hollinoideans

In the material under consideration, a small number of palaeocopine specimens occur showing features of primitiopsoidean as well as hollinoidean aspect. With respect to the lateral outline, they appear like an illusion, quite similar to *Gravia* Polenova, 1953 [non *Gravia* Polenova, 1952, = nom. nud.]. But all are provided with rather complex (histio-velar) adventral structures, somewhat resembling (Ordovician) Tetradellidae. The right over left overlap of valves seems in agreement with that of primitiopsoideans. The lobation patterns appear simple, as in smooth primitiopsids, or (more-or-less) complex bilobate, as defined in hollinoideans by Bless & Jordan (1971). At least in adults, the adductorial sulcus (S2) is always a conspicuous feature (unlike primitiopsids, but similar to hollinoideans). The distinctive linear extensions are rows of narrowly spaced tubercles or denticles, unlike known palaeocopines running (more-or-less) close together upon a small venter. The marginal structure is always delicate. The velar structure seems dimorphic, apparently forming a (approximately

Fig. 5 False primitiopsids, Gen. aff. *Gravia*, as demonstrated by endemic Canadian material (magnification approx. × 36–40). (**1**) Gen. aff. *G.* sp. 554 Braun, 1978 from the DM3 zone of eastern Yukon; (assumed) tecnomorphs; valve with obscured lobation; spines postulated to be of histial [?hi] origin; margina [ma] delicately denticulate. (**2**) Gen. aff. *G.* sp. 377 Braun, 1978 from the Upper DM5 sub-zone of the Northwest Territories; a = (assumed) adult tecnomorphic carapace; b = juvenile carapace; adults weakly lobate; adventral structures of unusual appearance, velar structure [ve] clearly tuberculate and marginal structure [ma] denticulate, practically attached to each other; small, but distinct spur midventrally developed, apparently of histial origin [hi]. (**3**) Gen. aff. *G.* sp. 376 Braun, 1978 from the Upper DM6 sub-zone of the lower Mackenzie Valley; a = heteromorphic carapace; b = tecnomorphic carapace; adductorial sulcus [S2] prominent; velar and marginal structures formed by rows of tubercles, but former appearing more distinct; both structures [ve/ma] fused, consequently subvelar field comparatively very small; single heteromorph shows (at least) posteroventrally located, tubulous dolon (do); histial structure [hi] wing-like. (**4**) Gen. aff. *G.* sp. 468 from equivalents of upper Keg River (reef) unit of Rainbow South Oilfield; tecnomorphic carapace; bilobate in appearance; adductorial sulcus prominent, extending to narrow ventral lobe; adventral structures [ve/ma] of similar appearance as shown in sp. 367; histial structure [hi] expressed by oblique ridge upon ventral lobe. (**5**) Gen. aff. *G.* sp. 368 Braun, 1978 from Upper DM6 sub-zone to Lower DM7 sub-zone of the lower Mackenzie River region; tecnomorphic carapace; complex bilobate in aspect, looking rather like an (unknown) hollinoidean; histial structure [hi] similar in appearance as in sp. 468, velar structure [ve] distinct row of tubercles, clearly separated from denticulate margina [ma], consequently subvelar field comparatively broad. Probably the two latter species should be assigned to a separate genus (in its own right). Photographer W.K.B.

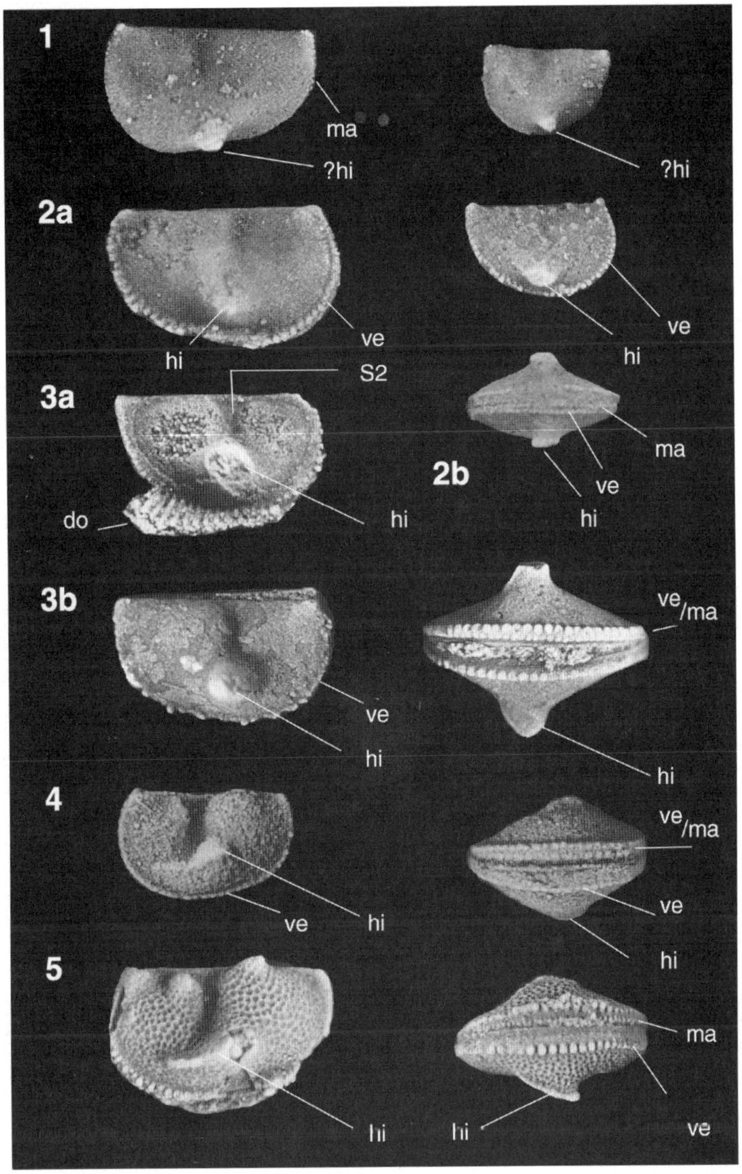

posteroventrally located) flange-like frill suggesting an infravelar dolon. In addition, a mid-ventrally located, apparently non-dimorphic spine or spur is developed, occupying a certain histial position. While primitiopsid extradomiciliar dimorphism is regarded to be typically velar in origin (but see hybrid primitiopsids below), histial features are to a certain extent known in hollinoidean families. Diagnostically important perimarginal structures (tubercles, ridges) are not observed in the Canadian material. Thus, the material under consideration follows, with respect to the complex adventral structures, an unknown concept in hollinoideans, rather than in primitiopsids.

Consequently, the proposed new genera are assumed to represent hollinoideans, however, remaining in a, in some respects, doubtful systematic position (Family unknown). Regarding overall carapace morphology, the smooth to unisulcate species of Gen. aff. *Gravia* gen. nov. (Fig. 5/1–4; Eifelian), appears (more-or-less) *Gravia*-like, while the more trilobate Gen. aff. *Gravia* sp. 368 (Fig. 5/5; Givetian) follows an unknown hollinoidean bauplan.

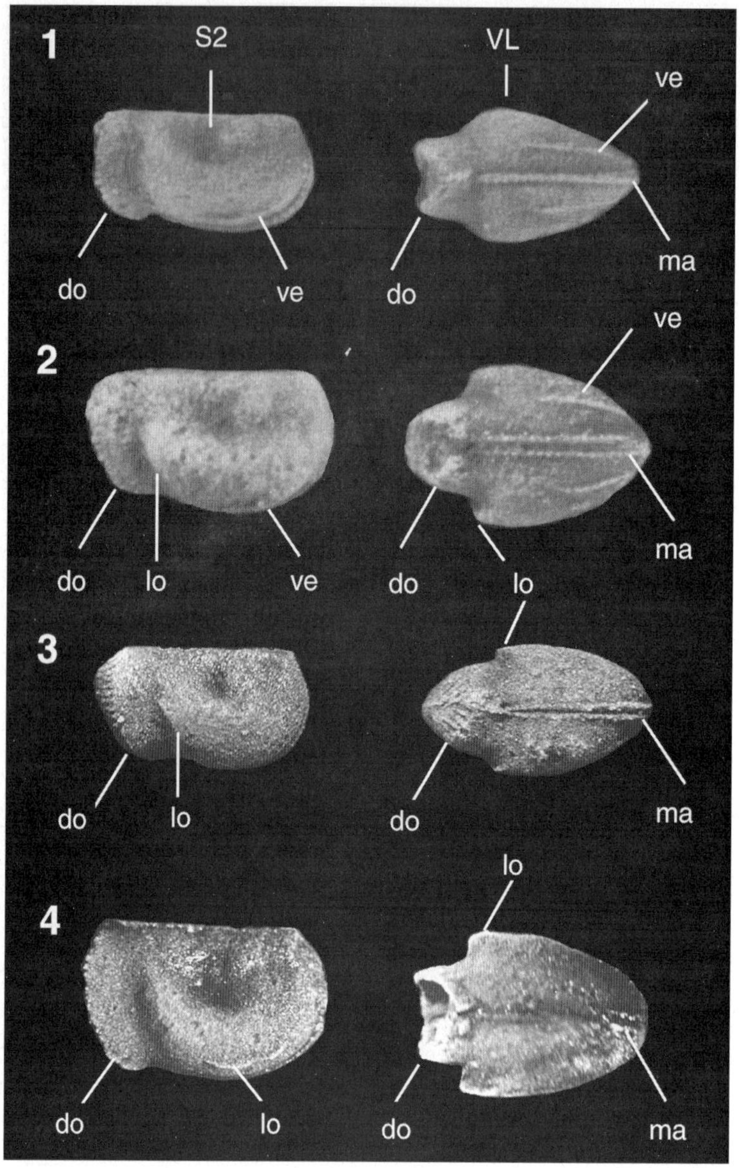

Fig. 6 Hybrid *Sulcicuneus*. Demonstrated in species, as occurring in the Middle Devonian of the Great Lakes Region and western Canada (magnification approx. × 36–40). (**1**) *S. porrectinatium* Kesling, 1951, type species, from the Bell Shale of Michigan; heteromorphic carapace; median sulcus [S2] well defined; ventral lobe [VL] gently rounded; velar ridge [ve] developed; heteromorphs provided with prominent, almost closed dolon [do], velar in origin; margina [ma] row of indistinct tubercles. (**2**) *S. latus* Kesling & Chilman, 1987 from the Gravel Point Formation of Michigan; heteromorphic carapace; tumid in appearance; ventral lobe provided with abrupt descent behind S2, probably resembling lobal (histial) feature [lo]; velar ridge [ve] comparatively short; heteromorphic dolon [do] almost closed; margina [ma] distinct row of tubercles. (**3**) *S.* sp. 308 Braun, 1978 from the Lower–Upper DM5 sub-zone of the Northwest Territories; heteromorphic carapace; complex bilobate in appearance; ventral lobe provided with a posterior edge, bordered by narrow ridge-like extension of lobal (histial) aspect [lo]; delicately striate "false" brood pouch present, velar in origin. (**4**) *S.* sp. 299 Braun, 1978 from the Upper DM5 sequence of the Northwest Territories; heteromorphic carapace; S2 most prominent; edge of pronounced ventral lobe provided with narrow (somewhat interrupted) ridge [lo] being clearly of histial origin; velar dolonate pouch [do] open towards its back, demonstrating perimarginal ridges; margina [ma] apparently delicate ridge. After [1–2] Kesling & Chilman (1987); [3–4] photographer W.K.B.

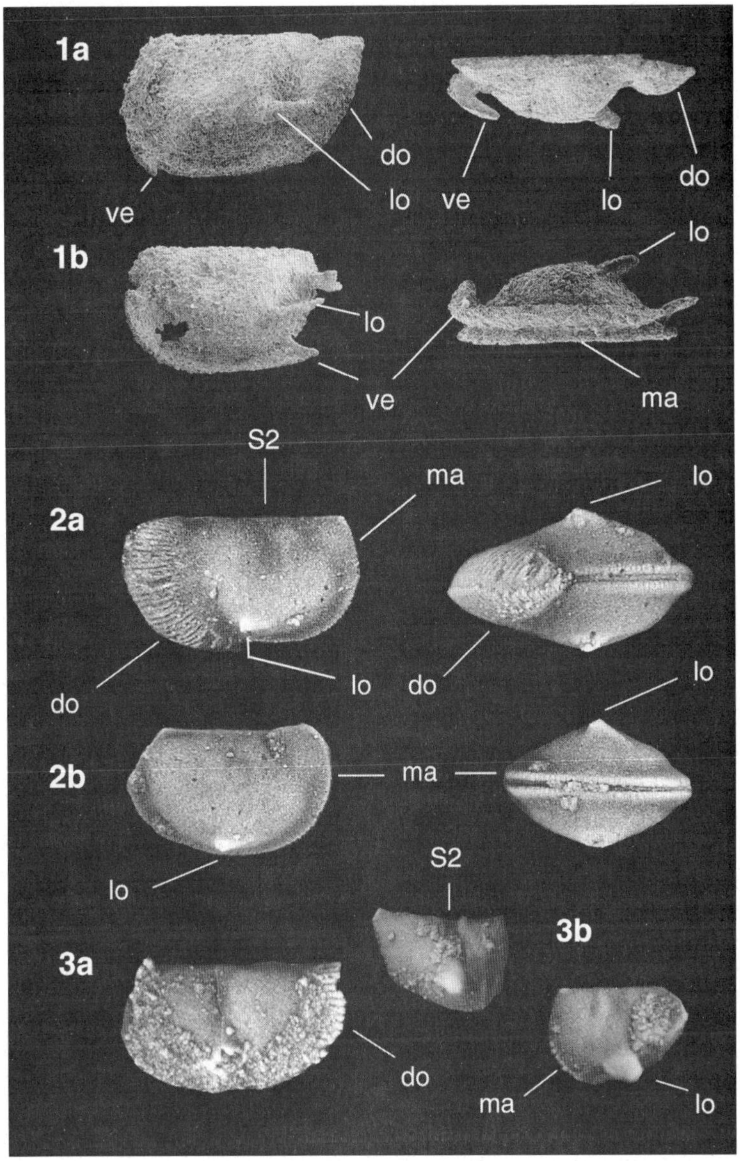

Fig. 7 Enigmatic *Parasulcicuneus*. As demonstrated in the Chinese type species and Canadian representatives (magnification approx. × 36–40). (**1**) *P. spiniatus* Wang & Jones, 1993 from the Early Devonian of South China; a = heteromorphic carapace; b = tecnomorphic carapace; lobation rather inconspicuous; both sexes provided with posteroventral spur [lo], histial in origin as located upon the ventral lobe; velar structure prominent flange, terminating in antero- and posteroventral spines [ve]; margina [ma] ridge-like; heteromorphic dolon [do] prominent, velar in origin. (**2**) *P.* sp. 489 from Lower–Upper DM5 boundary units of eastern Yukon; a = heteromorphic carapace; b = tecnomorphic carapace; characteristically appearing winged in ventral view; adductorial depression [S2] shallow, heteromorphic brood pouch [do] delicately, but clearly striate; small lobal (histial) protuberance [lo] present; margina [ma] distinct ridge. (**3**) *P.* sp. 812 from Upper DM6 – Lower DM7 sub-zones of western Manitoba; a = heteromorphic carapace; b = tecnomorphic carapace; apparently revealing strong proportional dimorphism; (assumed) tecnomorphs with almost asymmetrical-triangular (strongly preplete) valves appearing complex bilobate; adult heteromorphs with subamplete valve seemingly simple bilobate, and provided with (at least) posteriorly located, striate dolon [do] (apparently being in a velar position); mid-ventral lobal (histial) protrusion [lo] present, in juveniles wing-like; margina [ma] appearing delicately denticulate. Photographers [1] G.B., [2–3] W.K.B.

Table 1 Revised classification of Palaeozoic Ostracoda

Class Ostracoda Latreille, 1802
Superorder Palaeocopa Henningsmoen, 1953 (Horne et al., 2002)
Order Palaeocopida Henningsmoen, 1953
Suborder Palaeocopina Henningsmoen, 1953 (including Binodicopina Schallreuter, 1972)
Superfamily Eurychilinoidea Ulrich & Bassler, 1923: among others, family Eurychilinidae
Superfamily Hollinoidea Swartz, 1936: families Tetradellidae, Hollinidae, Hollinidae, Ctenloculinidae
Superfamily Primitiopsoidea Swartz, 1936: family Primtiopsidae
Superfamily Beyrichioidea Matthew, 1886 (including "Aparchitoidea"): several families
Superfamily Drepanelloidea Ulrich & Bassler, 1923: families Drepanellidae, Bolliidae, Aechminidae, ? Punciidae
Superfamily Kirkbyoidea Ulrich & Bassler, 1906: families Arcyzonidae, Amphissitidae, Kirkbyidae, Kellettinidae, Herrigokellidae
Superfamily Nodelloidea Becker, 1968: family Nodellidae
Suborder unknown: families paraparchitids, kirkbyellids and allies, "non-klodenelloideans"
Suborder Platycopina Sars, 1866
Superfamily Kloedenelloidea Ulrich & Bassler, 1908: families Kloedenellidae, Cavellinidae, Cytherellidae, Barychilinidae, ? Monotiopleuridae
Superorder Podocopa Sars, 1926 (Horne et al., 2002)
Order Podocopida Sars, 1866
Suborder Metacopina Sylvester-Bradley, 1961
Superfamily Healdioidea Harlton, 1933: family Healdiidae
Superfamily Thlipsuroidea Ulrich, 1894: families Thlipsuridae, Quasillitidae, Bufinidae, Ropolonellidae
Suborder Podocopina Sars, 1866 (Kornicker & Sohn, 1976)
Superfamily Bairdiocypridoidea Shaver, 1961: families Pachydomellidae, Bairdiocyprididae
Superfamily Bairdioidea Sars, 1888: families Bairdiidae, Beecherellidae
Superfamily Cypridoidea Baird, 1845
Superfamily Cythereoidea Baird, 1850
Order Eridostraca Adamczak, 1961
Superorder Myodocopa Sars, 1866 (Kornicker & Sohn, 1976)
Order Myodocopida Sars, 1866
Suborder Myodocopina Sars, 1866 [syn. Paleomyodocopina Kornicker & Sohn, 2000, Palaeomyodocopina nom. correct., Becker, 2004]
Superfamily Cypridinelliformoidea Kornicker & Sohn, 2000
Superfamily Cytheroidea Baird, 1850
Suborder Halocyprinida Skogsberg, 1920
Suborder Halocyprinina Skogsberg, 1920
Suborder Cladocopina Sars, 1866
Order Leperditicopida Scott, 1961 (with some caution)

Scheme, based on the proposals of Whatley et al., 1993, Becker, 1990, 2002 and taking recent relevant literature into account (in particular, Horne et al., 2002; Becker, 2005; Liebau, 2005); (listed families are main examples only). For authorship taxa on the familiar level see the Kempf (1988) index. Depending on the preferred view of the revising author, the Ostracoda Latreille, 1802 are mutually classified in the relevant literature either as Class or Subclass. Higher taxa names (concerning suprafamilial levels) are not protected by the code. Therefore, the naming of such taxa, e.g. suborders, may be different from author to author, not supporting each other. Superorders, as recently introduced particularly by neontologists (e.g. Kornicker & Sohn, 2000), will not generally be adopted by Palaeozoic workers. With respect to the obligatory regulations of the current code (ICZN, 1999: Art. 29.2), superfamilies are labelled with the suffix "-oidea" (instead of "-acea")

Hybrid primitiopsids

Since Adamczak's revision of mid-Palaeozoic Primitiopsidae (Adamczak, 1968, 2005a, b), *Sulcicuneus* Kesling, 1951 was considered without doubt to be a primitiopsid genus. Appearing almost unisulcate, it is distinguished by distinct velar dimorphic features, closely matching the primitiopsid type of extradomiciliar dimorphism. The exclusive appearance of velar features (in addition to the always-present margina) was beyond doubt, so that it seemed admissible for Adamczak (1968: 11), while referring to adventral structures, to avoid expressions like velar or histial structure. *Sulcicuneus*, however, apparently offends against the primacy of right over left overlap.

Current studies on material from western Canada show that the group around the type species *S. porrectinatium* Kesling, 1951 (Fig. 6/1) seems pretty complex, as shown by species from North America (Great Lake region, western Canada). In any case *Sulcicuneus*, as presented by the type species and its allies, is not a "pure-bred" primitiopsid. Occurrences of a further, somewhat enigmatic genus, *Parasulcicuneus* Wang & Jones, 1993 from South China and western Canada (see Fig. 7), increase the opinion of "hybrid" primitiopsids combining different elements. For further, more detailed information see Becker & Braun (unpublished, in press).

Genus *Sulcicuneus* Kesling, 1951 (type species: *S. porrectinatium* Kesling, 1951 from the Middle Devonian of Michigan, U.S.A.) was described by Kesling (1951: 222), as showing the females [heteromorphs] with a large incurved frill on each valve, and males [tecnomorphs] with no posterior frill; median sulcus pronounced, terminating in the ventral region of the carapace; carapace slightly inequivalved, left valve reported larger.

Genus *Parasulcicuneus* Wang & Jones, 1993 (type species: *P. spiniatus* Wang & Jones, 1993; Fig. 7/1) shows (fide Adamczak, 2005a) rather inconspicuous lobation; adductorial sulcus (S2) short and shallow; lobes (L2 and L3) confluent with prominent VL, provided with posteroventral spur in both sexes as well as in juveniles, histial is aspect; velar structure prominent flange running from anterior to posteroventral margin, but is not present in juveniles; heteromorphs show closed dolon, marginal flange distinct. *Parasulcicuneus* (Fig. 7) is provided with histial structures, a feature now also recognised in most *Sulcicuneus* species (Fig. 6).

Approach to systematics

Recently, while following a "call upon Palaeozoic ostracod workers to provide evidence for or against the inclusion of ancient groups within the Ostracoda" (K. Martens, electr. comm., 19.5.1998), two alternative schemes were presented independently of one another by Becker (2005: 23) and Liebau (2005: 115). "Neither scheme is likely to be widely accepted without modification; nevertheless these two papers provide valuable overviews of current thinking and should serve to ferment further discussion" (Ikeya et al., 2005: ix). Since "the long-awaited Treatise revision remains 'work in progress' and there is little sign that, when eventually completed, it will offer universally acceptable classification" (Ikeya et al., 2005: ix), every serious effort in classifying ostracods should be understood as a plea in favour of reviving Treatise revision – a circumstance stressed at the occasion of ISO [International Symposium on Ostracoda] 2005 (Becker, pers. comm.).

Becker's (2005) scheme, first introduced by Becker (1990) and later adapted for the POC [Treatise pre-printing] series (Becker, 1997, and later) is actually based on Whatley et al. (1993), summarising proposals of a team of specialists, at that time entrusted with Treatise revision by the then editors. Some "thoughts and comments", as provided by Dr. Peter. J. Jones (pers. comm., August 1992), may document this way of shaping scientific opinion: "As a general principle, I strongly support Gerhard's (Becker, 1990) plea to keep the classification as simple as possible, and to throw out as many of the higher taxonomic groupings as we can. We do not need superorders. The highest rank below the class Ostracoda (Phylum Crustacea) should be that of order, and subdivision of this category into suborders (and in turn into superfamilies) should be kept to a minimum. The overall classification should be based firstly on relationships between species. Starting at this level, it should be agglomerative, working upwards to genera in the search of natural suprageneric groupings (families, superfamilies)".

Unfortunately, the scheme given by Whatley et al. (1993), being the result of comprehensive pre-Treatise work of a team of specialists, seems to be forgotten in recent (neontological) studies. Since it is with ostracod workers (not key-note writers) to adopt a "universally acceptable classification" (Ikeya et al., 2005: ix), Whatley et al.'s scheme should be considered to represent a "benchmark classification scheme" (Ikeya et al., 2005: viii). The more so, since having only higher categories in mind and not being based on any material, Liebau's (2005) classification scheme appears rather artificial – in contrast to the new Treatise concept (above), which concerns genera as represented by their type species. Moreover, a lot of relevant recent publications (see lists of references) could not be taken into account in Liebau's (2005) classification scheme, thought to replace the old-fashioned Treatise concept, which has "been rendered woefully inadequate by the succeeding four decades of discovery, description and revision" (Ikeya et al., 2005: viii). Liebau's (2005) scheme needs to be updated.

Following Whatley et al. (1993) and taking (together with Becker, 1990 and especially Becker, 2002 recent relevant literature into account, in particular, Horne et al., 2002, Becker, 2005, Liebau, 2005), the systematics of Palaeozoic Ostracoda, as chosen as standard for the POC series and used by present authors, is updated as shown in Table 1.

While the ostracod nature of the Palaeocopina was never really debated, it appears with respect to the admittedly aberrant Leperditicopida, that there is an increasing trend towards removing them from the Ostracoda. The also enigmatic Eridostraca, however, now have a safe place within the (sub)class.

The Palaeocopina are the most intriguing group, characterised by adventral structures appearing dimorphic or not, which are considered to be a most innovative conception, proved to be enormously constant in ostracod history. The conspicuously complex-dimorphic Hollinoidea (see this paper) are known up from the Ordovician. Among typical velar dimorphic groups, the (basically) Silurian Beyrichioidea are regarded a special case (Schallreuter, 1995), providing true brood pouches in heteromorphs (e.g. Becker, 1968, 2005). While discussing Horne's (pers. comm.; Horne et al., 2002) "phylognetical classification scheme", Becker (2005: 47) hesitantly considered to accept the idea of dividing the conventional Palaeocopina into two separate suborders, (dimorphic) Palaeocopina, including all groups showing dimorphic adventral structures, and (non-dimorphic) Drepanellocopina [not "Kirkbyocopina" which is a discredited name] for Drepanelloidea, Kirkbyoidea and Puncioidea.

On the other hand, prominent extradomiciliar dimorphism is considered to be an archaic (plesiomorphic) character (Becker, 1998) becoming superfluous; advanced (derived) palaeocopines have non-dimorphic adventral features. The (with respect to adventral structures) non-dimorphic Kirkbyoidea, believed to have drepanelloidean ancestors, are considered to be the most advanced group, which shows the adductorial pattern in an extremely distally located position (key character; Becker & Swanson, 1992). The extant Punciidae appear also non-dimorphic, but have the adductorial pit above mid-height of the valve, obviously a plesiomorphic character. Apparently known from the Late Carboniferous, this family appears (in an exciting manner) to be the only living Palaeocopina (see "drepanellid archetype" Becker, 2005). With respect to the morphologically non-dimorphic forms, basically dominating younger palaeocopine communities, recent studies on "asexual reproduction" are of interest.

Parthenogenesis: a cautious explanation of non-dimorphic palaeocopine ostracods

At the occasion of the 15th International Symposium on Ostracoda [ISO 2005] in Berlin, an intriguing discussion on the meaning of asexual reproduction in the evolution of Ostracoda was started. Since being the most advanced palaeocopines, the question arose whether or not their non-dimorphic offshoots may be subject of parthenogenesis. To reveal this (not only for Palaeozoic workers) puzzling matter, a careful approach to produce an explanation was stimulated.

"The paradox of sex remains the queen of problems in evolutionary biology" (Martens, 2005: 71). As a rule, sexual reproduction is widespread throughout the animal and plant kingdoms, but under certain conditions parthenogenesis occurs, namely a form of reproduction in which the ovum develops into a new individual without fertilization. First Charles Bonnet (1720–1793) drew attention to this phenomenon, but his theories were highly unrealistic and unscientific.

In certain crustaceans, parthenogenesis is common and in some species males are uncommon or unknown. In younger (post-Palaeozoic to present-day) ostracods, parthenogenesis is well-known and especially for non-marine species, the subject of detailed phylogenetical studies (a.o. Martens et al., 2005: 72). In contrast, Palaeozoic workers, because they are dealing mainly with old, marine representatives, are not very familiar with this phenomenon. Soft parts are, as a rule, not preserved in ancient, fossil groups (Becker, 2005). Speaking in terms of morphology, the intriguing Palaeocopina Henningsmoen, 1953 are characterised by varied, distinct adventral structures causing conspicuous extradomiciliar dimorphism. A variety of adventral structures appear dimorphic and are considered to be a most innovative phenomenon, believed to be constant in ostracod history. While successively reduced in younger forms (Becker in Becker et al., 1974), prominent extradomiciliar dimorphism is, however, considered to be an archaic (for ostracods plesiomorphic) character (Becker, 1998), apparently becoming superfluous from the Devonian upwards. Therefore, it seems consequent that advanced (derived) palaeocopines have non-dimorphic adventral features. This is true of mid-Palaeozoic Drepanelloidea Ulrich & Bassler, 1923 and the Kirkbyoidea Ulrich & Bassler, 1906 as a whole. The extant Puncioidea Hornibrook, 1949, considered as being the only living palaeocopines, appear also non-dimorphic. In addition, an increasing number of mid Devonian palaeocopines, being basically dimorphic in appearance, are found (this paper) to lack (despite substantial material) any dimorphic counterpart. Having recent studies above in mind, the question concerning asexual reproduction rises in all cases where sexes cannot be distinguished.

Non-marine ostracods flourish in habitats stressed by varied environmental factors. Connections between parthenogenesis and such circumstances of life might be expected. On the other hand, the palaeocopine groups under consideration are believed to have lived in stable biotopes, clearly open marine and often situated below the wave base. Lagoonal or sheltered niches, as explained by Becker (1982), are not involved. And the missing morphological counterpart in hollinids and occasionally primitiopsid (see above) is the (presumed) heteromorph. Moreover, kirkbyoideans are not strictly non-dimorphic (see lobar dimorphism above).

In summary, the question of whether or not palaeocopines show parthenogenesis needs further investigation. In any case, parthenogenesis would mean competitive advantage, since "asexuals can reproduce twice as fast as sexuals, because all, rather than half, of their offspring are themselves capable of reproduction" (Martens, 2005: 71). Such populations, however, are made up of genetic replicas of the parent, and should some adversity occur in the environment, the entire population or species risks extinction. In fact, extradomiciliar dimorphic palaeocopines have (cum grano salis) died out at the occasion of the P/T (Permian/Triassic) major extinction event – assumed offshoots are clearly non-dimorphic in aspect (Becker, 2005: Fig. 8). For similar (but not homologous) features resembling hollinoidean locular dimorphism, as displayed in an enigmatic Neogene *Loculicytheretta* species, see Becker (2005: Fig. 6/14).

Conclusions

As a result of work on Middle Devonian ostracod assemblages from western Canada, (at first view) doubtful or problematic material is presented showing special, unusual or puzzling cases of extradomiciliar dimorphism. The development of adventral structures, as seen in palaeocopines, was one of the most innovative inventions in ostracod history; only the podocopine duplicature seems of comparably importance. The palaeocopine velum, which "may have been an early evolutionary innovation in the palaeocopids" (Briggs, 1997: 114), was accompanied by further domiciliar features, like margina and histium, having a climax in Late Ordovician times. Subsequent successively obscured adventral structures must be considered as being a discontinued model. It was pointed out by Becker (1998), while referring to Becker et al. (1974) who described obscuring features in Late Devonian beyrichoideans, that the palaeocopine extradomiciliar dimorphism became plesiomorphic in nature. However, along the evolutionary pathway(s), new forms have appeared, and apomorphic features due to innovation (events) replaced plesiomorphic characters. The idea of distinction and initiation/radiation interplay runs like a thread through geological times describing the motor of ostracod evolution from Palaeozoic to Recent, only to be stopped by man-made events (Kempf in Becker, 2005). Consequently, sometimes an idea under consideration appears a "new-fashioned" concept. The idea of innovative replacement arises, but, for objective reasons, interpretations given herein may be provisional in parts. Nevertheless, an inspiration of distinction and initiation/radiation interplay appears to run like a thread through geological times describing the motor of ostracod evolution.

Acknowledgement The author thanks Prof. Alan Lord (FIS, Frankfurt am Main) for fruitful discussions and improving the English text, as well as anonymous referees for useful comments. Sincere thanks are also expressed to colleagues, considering a poster dealing with the matter under consideration, as was demonstrated at the occasion of the 15th International Symposium on Ostracoda (ISO 2005), Berlin, 12–15.9.2005.

References

Adamczak, F. J., 1968. Palaeocopina and Platycopina (Ostracoda) from Middle Devonian rocks in the Holy Cross Mountains, Poland. Stockholm Contribution Geology 17: 127–150.

Adamczak, F. J., 1991. Kloedenellids: morphology and relation to non-myodocopide ostracodes. Journal of Paleontology 65: 255–267.

Adamczak, F. J., 2005a. [Edited by G. Becker]: Contributions to Palaeozoic Ostracod Classification [POC], No. 37. The Superfamily Primitiopsacea Swartz, 1936. Family Primitiopsidae Swartz, 1936. Mid-Palaeozoic genera. Neues Jahrbuch für Geologie und Paläontologie, Monatshefte 2005: 17–45.

Adamczak, F. J., 2005b [Edited by G. Becker]: Contributions to Palaeozoic Ostracod Classification [POC], No. 38. The Superfamily Primitiopsacea Swartz, 1936. Family Primitiopsidae Swartz, 1936. Mid-Palaeozoic genera, cont. Neues Jahrbuch für Geologie und Paläontologie, Monatshefte 2005: 151–174.

Becker, G., 1968. Geschlechtsdimorphismus bei Ostracoden. Natur und Museum 98: 47–56, 119–172.

Becker, G., 1982. Ostracoden-Entwicklung im Kantabrischen Variszikum (Nordspanien). Neues Jahrbuch für Geologie und Paläontologie, Abhandlungen 163: 153–163.

Becker, G., 1990. Zur Morphologie und Taxonomie paläozoischer Ostracoda. Mit kritischen Bemerkungen zur Wertigkeit von Carapax-Merkmalen. Senckenbergiana lethaea 70: 147–169.

Becker, G., 1997. Contributions to Palaeozoic Ostracod Classification [POC], No. 1. "Non-kirkbycean" ostracods: Families Ordovizonidae n. fam., Kirkbyellidae Sohn, Scrobiculidae Posner, and Cardiniferellidae Sohn. With description of the new family Ordovizonidae n. fam. and a new kirkbyellid genus: *Eukempfella* n. g. Sonderveröffentlichungen des Geologischen Instituts der Universität zu Köln 114 [Festschrift Eugen Kempf]: 29–57.

Becker, G., 1998. Contributions to Palaeozoic Ostracod Classification [POC], No. 8. The Superfamily Kirkbyacea Ulrich & Bassler, 1906. 7. The drepanellid roots. Neues Jahrbuch für Geologie und Paläontologie, Abhandlungen 209: 381–400.

Becker, G., 2002. Contributions to Palaeozoic Ostracod Classification [POC], No. 24. Palaeozoic Ostracoda: The standard classification scheme. Neues Jahrbuch für Geologie und Paläontologie, Abhandlungen 226: 165–228.

Becker, G., 2004. Contributions to Palaeozoic ostracod classification [POC], Suppl., No. 4. The Genus *Sulcuna* Jones & Kirkby, 1874 (Palaeomyodocopina). *S. (Devonosulcuna)* n. subgen. Neues Jahrbuch für Geologie und Paläontologie, Monatshefte 2004: 289–307.

Becker, G., 2005. Functional morphology of Palaeozoic ostracods: phylogenetic implications. Hydrobiologia 538: 23–53.

Becker, G., M. J. M. Bless, M. Streel & J. Thorez, 1974. Palynology and ostracode distribution in the Upper Devonian and basal Dinantian of Belgium and their dependence on sedimentary facies. Mededelingen van de Rijks Geologische Dienst, Nieuwe Serie 25: 9–100.

Becker, G. & W. K. Braun, 2005. Extra-lobal or complex dimorphic features in Middle Devonian ostracods: incompatible oddities? or, considering the prejudice that extradomiciliar dimorphism in mid-palaeozoic palaeocopines should be solely velar in origin. In Kohring, R. & B. Sames (eds), 15th International Symposium on Ostracoda, September 12–15 2005, Program and Abstracts, Berliner paläobiologische Abhandlungen 6, Berlin, 14–15.

Becker, G. & W. K. Braun, Contributions to Palaeozoic Ostracod Classification [POC], Suppl., No. 13. Superfamily Primitiopsoidea SWARTZ, 1936, cont. Histiovelar features in Mid Palaeozoic representatives. "Hybrid" primitiopsids. Neues Jahrbuch für Geologie und Paläontologie, Monatshefte, (in press).

Becker, G., C.-D. Clausen & K. Leuteritz, 1989. Silikatisierte Ostracoden und Fazies im Grenzbereich do/cd. Beiträge zur elektronenmikroskopischen Direktabbildung von Oberflächen 22: 101–110.

Becker, G. & K. Swanson, 1992. Are the Kirkbyacea extinct palaeocopine ostracods or close relatives of the extant Punciacea? A disputation. Senckenbergiana lethaea 72: 121–128.

Bless, M. J. M., & H. Jordan, 1971. Classification of palaeocopid ostracodes belonging to the Family Hollinellidae. Bulletin du Centre de Recherches de Pau, Societe Nationale des Petroles d'Aquitaine, 5 (supplement): 869–890.

Bless, M. J. M., & H. Jordan, 1972. Ostracodes of the Family Hollinellidae. Mededelingen van de Rijks Geologische Dienst, Serie C, V, 3(1): 8–83.

Braun, W. K., 1977. Usefulness of ostracodes in correlating Middle and Upper Devonian rock sequences in western Canada. In Murphy, J. M., W. B. N. Berry & Ch. A. Sandberg (eds), Western North America: Devonian. University California, Riverside Campus Museum Contribution 4: 65–79.

Braun, W. K., 1978. Devonian ostracodes and biostratigraphy of western Canada. The Geological Association of Canada, Special Paper 18: 259–288.

Braun; W. K., 1993. Evolutionary patterns and mechanisms in Devonian Ostracoda of Western Canada. In McKenzie, K. G. & P. J. Jones (eds), Proceedings 11th international Symposium Ostracoda, Warrnambool, Victoria, Australia, 8–18 July, 1999, Balkema, Rotterdam etc, 53–64.

Braun, W. K. & J. E. Mathison, 1986. Mid-Givetian events in western Canada: the Dawson BayWatt Mountain – "Slave point" interlude. Canadian Society of Petroleum Geologists 34: 426–451.

Briggs, W. M., 1997. *Axelheibergella* (Ostracoda, Crustacea), a new genus of marginally septate cladocopid ostracods from the central Arctic Ocean. Journal of Micropalaeontology 16: 109–120.

Henningsmoen, G., 1953. Classification of Paleozoic straight-hinged ostracods. Norsk geologisk Tidsskrift 31: 185–288.

Henningsmoen, G., 1965. On certain features of palaeocope ostracodes. Geologiska Föreningens i Stockholm förhandlingar 86: 329–394.

Horne, D. J., A. C. Cohen & K. Martens, 2002. Taxonomy, morphology and biology of Quaternary and Living Ostracodes. In Holmes, J. A. & A. R. Chivas (eds), The Ostracoda: Applications in Quaternary Research. AGU Geophysical Monograph Series 131: 5–36.

Ikeya, N., A. Tsukagoshi & D. J. Horne, 2005. Preface: The phylogeny, fossil record and ecological diversity of ostracod crustaceans. Hydrobiologia 538: vii–xiii.

International Code of Zoological Nomenclature (1999). IV–XXIX, 1–306; London (International Trust zoological Nomenclature) [ICZN 1999].

Jaanusson, V., 1957. Middle Ordovician ostracodes of central and southern Sweden. Bulletin of the Geological Institutions of the University of Uppsala 37: 173–442.

Jaanusson, V., 1966. Ordovician ostracodes with supravelar antra. Publications from the Palaeontological Institution of the University of Uppsala 66: 1–30.

Jaanusson, V. & A. Martinsson, 1956. Two hollinid ostracodes from the Silurian Mulde Marl of Gotland. Bulletin of the Geological Institutions of the University of Uppsala 36: 401–410.

Kempf, E. K., 1988. Index and bibliography of marine Ostracoda, 4. Bibliography. Sonderveröffentlichungen des Geologischen Instituts der Universität zu Köln, 53: 1–454.

Kesling, R. V., 1951. A new genus and species of primitiopsid ostracod from the Devonian Traverse Group of Michigan. Contributions from the Museum of Paleontology, University of Michigan 8: 221–230.

Kesling, R. V., 1953. Ostracodes of the family Hollinidae from the Arkona Shale of Ontario. Contributions from the Museum of Paleontology, University of Michigan 10: 203–219.

Kesling, R. V., 1969. Copulatory adaptions in ostracods. Part 3. Adaptions of some extinct ostracods. Contributions from the Museum of Paleontology, University of Michigan 22: 273–312.

Kesling, R. V. & R. B. Chilman, 1987. Dimorphic Middle Devonian paleocopan Ostracoda of the Great Lakes Region. Papers on Paleontology 25: 1–82.

Kesling, R. V. & N. R. Tabor, 1953. Ostracods of the Family Hollinidae from the Genshaw Formation of Michigan. Contributions from the Museum of Paleontology, University of Michigan 10: 83–100.

Kornicker, L. S. & I. G. Sohn, 1976. Phylogeny, ontogeny and morphology of the living and fossil Thaumatocypridacea (Myodocopa: Ostracoda). Smithsonian Contributions to Zoology 219: 1–14.

Kornicker, L. S. & I. G. Sohn, 2000. Myodocopid Ostracods from the Late Permian of Greece and a Basic Classification for Paleozoic and Mesozoic Myodocopida. Smithsonian Contributions to Paleobiology 91: 1–33.

Liebau, A., 2005. A revised classification of the higher taxa of the Ostracoda (Crustacea). Hydrolobiologia 538: 115–137.

Martens, C. K., 2005. From sex to asex: a case study of interactions between sexual and asexual reproduction. In Kohring, R. & B. Sames (eds), 15th International Symposium on Ostracoda, September 12–15 2005, Program and Abstracts, Berliner paläobiologische Abhandlungen 6, Berlin, 71 pp.

Martens, K., I. Schön, & D. J. Horne, 2005. Biogeography and phylogeography of nonmarine Ostracoda (Crustacea). In Kohring, R. & B. Sames (eds), 15th International Symposium on Ostracoda, September 12–15 2005, Program and Abstracts, Berliner paläobiologische Abhandlungen 6, Berlin, 72 pp.

Martinsson, A., 1962. Ostracodes of the Family Beyrichiidae from the Silurian of Gotland. Bulletin of the geological institutions of the University of Uppsala 41: 1–369.

Polenova, E. N., 1952. Ostracoda from the upper part of the Givetian Stage of the Russian Platform. Trudy VNIGRI, Novia Seria 60 (Mikrofauna SSSR, 5): 65–156 (in Russian).

Polenova, E. N., 1953. Ostracoda from the Devonian deposits of the central Devonian field and the middle Volga area. Trudy VNIGRI, Novia Seria 68: 1–157 (in Russian).

Schallreuter, R. E. L., 1966. Zur Taxonomie und Phylogenie der Ostrakodenfamilie Tetradellidae Swartz, 1936 (Palaeocopina, Hollinacea) und eine neue Familie der Hollinacea. Geologie 15: 846–875.

Schallreuter, R .E. L., 1995. Beiträge zur Geschiebekunde Westfalens III. Ostrakoden aus silurischen Geschieben. Geologie und Paläontologie in Westfalen 34: 5–145.

Whatley, R. C., D. J. Siveter & I. D. Boomer, 1993. Arthropoda (Crustacea: Ostracoda). In Benton, M. J. (ed), The Fossil Record 2: Chapman & Hall, London, 343–356.

Hydrobiologia (2007) 585:159–180
DOI 10.1007/s10750-007-0636-3

OSTRACODA (ISO15)

Evolutionary and taxonomic aspects within the species group *Pseudocandona eremita* (Vejdovský) (Ostracoda, Candonidae)

Sanda Iepure · Tadeusz Namiotko · Dan L. Danielopol

Abstract The contribution deals with evolutionary and taxonomic aspects of the stygobitic ostracod lineage *Pseudocandona eremita* (Ostracoda, Candonidae). It is based on a comparative study of the morphology of both the carapace and limbs of individuals belonging to eight populations from Romania (three reproducing bisexually, the others by parthenogenesis). Left valves display various outline shapes, classified in two morphotypes, the "dorsally rounded shape" and the "triangular shape". The intra- and interpopulation shape variability of these valves is described using geometric morphometrics and multivariate statistics. It is shown that the more flat carapace occurs frequently in populations living in gravel-sand habitats. Moreover, variability of 12 morphological limb characters taken on females is quantitatively evaluated by discriminant analysis. The character state of the m process of the male copulatory organ gives the most reliable information for the separation of the three bisexual populations in different species: *P. serbani* Danielopol, *P. danubialis* sp. nov. and *P. transylvanica* sp. nov. Short descriptions of the latter two species are presented. For the five parthenogenetic populations where no particular discriminant trait could be found, a solution of compromise is adopted, i.e. they are temporary assigned to *Pseudocandona eremita* (Vejdovský, 1882, Thierische Organismen der Brunnenwässer von Prag. Prag) *sensu lato*.

Keywords Crustacea · Ostracoda · Candonidae · Comparative morphology · Taxonomy · Evolution

Guest editors: R. Matzke-Karasz, K. Martens & M. Schudack
Ostracodology – Linking Bio- and Geosciences

S. Iepure (✉)
Institute of Speleology "E. Racoviţă", Romanian Academy of Sciences, Str. Clinicilor 5, 400006 Cluj, Romania
e-mail: siepure@hasdeu.ubbcluj.ro

T. Namiotko
Department of Genetics and Cytology, Institute of Biology, University of Gdańsk, Kładki 24, 800-822 Gdańsk, Poland

S. Iepure · D. L. Danielopol
Institute of Limnology, Austrian Academy of Sciences, Mondseestrasse 9, 5310 Mondsee, Austria

Introduction

Subterranean ostracod fauna is a privileged group for evolutionary studies. This is due to the fact that for some of the present-day subterranean ostracod lineages one can recognise fossil species (Danielopol, 1980a), which further allows with some certainty, reconstructing diverse evolutionary pathways. The present contribution deals with evolutionary and taxonomic aspects of an ostracod lineage, the species group *Pseudocandona eremita* (Vejdovský), widely distributed in Europe and which was recently redefined taxonomically by Namiotko & Danielopol (2004).

Here, we will compare the morphology of eight populations belonging to what Danielopol (1982) defined as two species, *Pseudocandona serbani* Danielopol and *P. eremita* (Vejdovský). The ostracods we used for this study originate from porous groundwater habitats in sandy-gravel sediments located in Romania (cf. next section). Three of these populations are represented by both males and females (they have bisexual reproduction) while the other five populations have exclusively females (we assume they reproduce parthenogenetically). The information on the morphological variability of the valves and limbs of these ostracods will be further used for a discussion on the evolutionary aspects related to life in the interstitial environment of sandy-gravel sediments as well as to the improvement of the taxonomy of the *P. eremita* lineage.

Considering the taxonomy of the eight populations here presented we have to mention that we will approach the problem in a different way as compared to the data in Danielopol (1982) and Namiotko & Danielopol (2004). In these papers we considered that *P. eremita* (Vejdovský) and *P. serbani* Danielopol display wide geographical distributions in Europe, a view also shared by other specialists (e.g. Meisch, 2000). Here we will start from the assumption that populations geographically closely located or belonging to the same hydrological unit or riverine network (cf. next section for more precise information) should be morphologically more similar than those separated by large distances or belonging to different hydrological systems and therefore they could be assigned to the same taxonomic unit. In order to test the validity of this assumption we will treat all the populations in the same way and not assign them a priori a taxonomic status.

Materials and methods

Sampling sites

The sampling sites are located in the extra- and intra-Carpathian regions of Romania (Fig. 1,

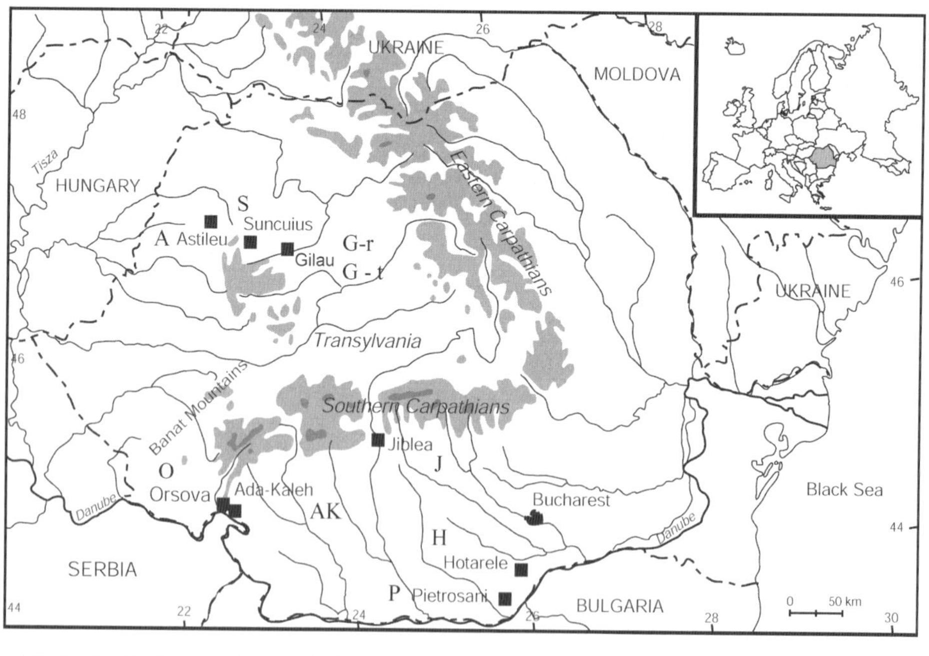

Fig. 1 Geographical distribution of the studied populations belonging to the *Pseudocandona eremita* species-group. Localities from which material was obtained for the present study coded as follows: A—Aştileu, AK—Ada-Kaleh, G-r—Gilău well no. 1, G-t—Gilău well no. 2, H—Hotarele, J—Jiblea, O—Orşova, P—Pietroşani, S—Şuncuiuş. Black filled triangle—Oradea, open triangle—Cluj

Table 1). The ostracods from the extra-Carpathians were caught in wells at Ada-Kaleh (AK) and Orşova (O) accessing to porous aquifers closely located to the Danube River or to its tributaries. Ada-Kaleh was a small island (1.7 by 4–5 km) in the middle of the Danube River, about 3.5 km downstream from Orşova (south-western Romania). The construction of the Iron Gates dam among 1972–1974 transformed the Danube valley into a reservoir, and additionally caused a 35 m raise in the water level of the river near the dam. The Ada-Kaleh island, old Orşova and at least five other villages consequently disappeared. Male and female specimens similar in morphology with those of Ada-Kaleh described here were collected also in a well from Orşova (Strada Pietii). These specimens will not be further here presented. The remaining studied populations from the extra-Carpathians were collected from: a well in the village Hotarele (H), on the Argeş river terrace, about 30 km south-east of Bucharest, and from a well in the village Pietroşani (P), on the upper Danube terrace, 50 km south-west of Bucharest. The sites from the intra-Carpathians are located in the north-western and northern flank of the Apuseni Mountains. These are: a well in the Aştileu village (A), in the valley of the Crişu Repede river, about 40 km from Oradea, a well in the Şuncuiuş village (S) also in the valley of the Crişu Repede river, about 70 km from Oradea, and finally two wells, being 100 m away each from other in Gilău village (G-r, G-t), 15 km south-west from Cluj, on the Someş river terrace. Specimens from the type locality of *Pseudocandona serbani* (Danielopol, 1982) at Jiblea (J), in the Olt valley were included in the present study too.

Abbreviations for ostracod morphology

The acronyms for valves, limbs, segments, claws and setae are mostly those used by Danielopol (1969) and Meisch (2000): a—outer lobe of hemipenis; A2—second antenna; A2EI—first endopodite of A2; A2EII—penultimate endopodite of female A2 (second and third endopodite combined); b—inner lobe of hemipenis; ci (co)—-inner (outer) of two postero-distal setae on A2EI; CR—caudal ramus (previously called furca); CRm—anterior margin of CR; d_2—second ventral seta on protopodite of L7; d_p—posterior seta on protopodite of L7; Exo—largest seta of A2 exopodite; g—distal seta on penultimate endopodite of L7; G_a—anterior claw of CR; G_M (G_m)—major (minor) claw on terminal endopodite of A2; G_p—posterior claw of CR; G_1—intero-distal claw on penultimate endopodite of A2; G_2—antero-distal claw on penultimate endopodite of A2; G_3—smaller intero-distal claw on penultimate endopodite of A2; h—height of the valves; he—medial lobe of hemipenis; h_1—short seta on terminal endopodite of L7; h_2—medium seta on terminal endopodite of L7 and apical claw on terminal endopodite of L6; h_3—large seta on terminal endopodite of L7; l—length; LV—left valve; L6—walking leg (6th limb); L6EII—second endopodite of L6; L7—cleaning leg (7th limb); L7EI—first endopodite of L7; RV—right valve; s_a (s_p)—anterior (posterior) seta of CR; s_p–G_p—dis-

Table 1 Geographical location and codes of the studied populations of the *Pseudocandona eremita* species-group

Site	Code	Sex	Co-ordinates		Altitude (m a.s.l.)
			Latitude (N)	Longitude (E)	
Ada-Kaleh	AK	♀♂	44°71′	22°45′	54
Aştileu	A	♀	47°03′	28°38′	244
Gilău-1	G-r	♀	46°75′	23°38′	397
Gilău-2	G-t	♀	46°75′	23°38′	397
Hotarele	H	♀	44°17′	26°37′	63
Jiblea	J	♀♂	45°22′	24°36′	294
Orşova	O	♀♂	44°43′	22°23′	54
Pietroşani	P	♀	45°10′	24°50′	54
Şuncuiuş	S	♀♂	46°67′	22°25′	247

tance between insertion points of s_p and G_p on CR.

Data acquisition and morphometric analysis of the valves

Outlines of the valves of eight adult female specimens from the eight sites (Table 1) were photographed with a NIKON digital camera and further digitized using the Tps-dig software (Rohlf, 2001). Because the species belonging to the *eremita* group have asymmetric valves, we used only the larger valve, i.e. the left one. The geometric morphometrics analysis of the valve outlines was performed using the Linhart's B-spline algorithm with MORPHOMATICA software versions 1.3–1.5 (Brauneis et al., 2005a). Details on the theoretical and practical aspects of the application of B-splines to the mathematical description of ostracod outlines and further to morphometric comparative studies are presented in Baltanás et al. (2003) and Brauneis et al. (2005b). To compare the outlines for the various valves we used 16 control points which allow constructing a polygon (Fig. 2) from which the outline is reconstructed. The difference between two superimposed outlines may be described by the diference vectors of the corresponding control points (e.g. the vectors in Fig. 5). If $d_0, d_1, \ldots, d_n$ and $d'_0, d'_1, \ldots, d'_n$ are the two sets of control points (the vectors $d_j - d'_j$ are called *delta vectors* also denoted by δ_j). For the pair-wise characterization of outline differences of valves we used the distance called mean δ^2 defined as

$$\sqrt{\frac{1}{n+1} \sum_{j=0}^{n} \left(d_j - d'_j\right)^2}.$$

The approximation of the mean and maximum error of the fitted mathematical curve to the digitised outline for each individual (Fig. 2) allows one to decide about the adequacy of the number of control points necessary for an acurate description. In our case the maximum error was below 1%, what we take as an excellent approximation. To avoid differences due to valve size, the outline approximations were transformed, using the sub-routine menu "normalised for area" in the MORPHOMATICA (Brauneis et al., 2005a, b), such that all

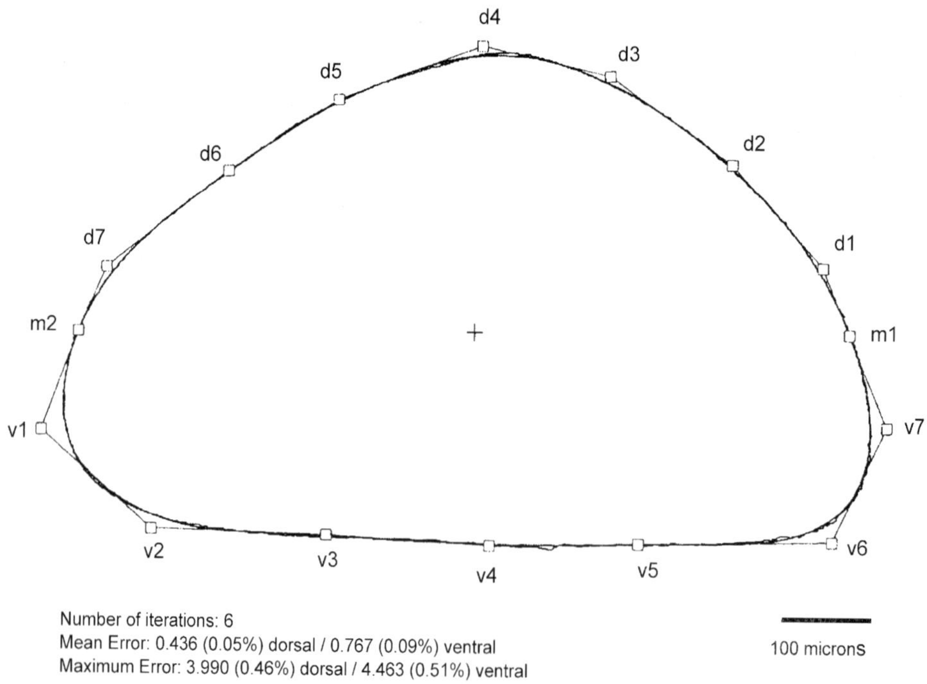

Fig. 2 Reconstruction of the valve outline using the Linhart's B-spline algorithm, with location and symbols of the 16 control points. Cross marks the centroid of the outline

computed outlines had an equal area. A virtual mean outline (so-called consensus shape) was computed from the data of individual outlines for each population (Fig. 8) and further used for comparative purposes with the individual or other consensus outlines.

Maximum length (l) and height (h) of each valve were recorded and the h/l ratio was calculated. For each of these variables the mean, range and standard deviation (SD) were calculated (Table 2). The sexual dimorphism of the valve shapes for the bisexual populations is described using the length and height meristic parameters (Table 2).

Morphology of limbs

Eight adult females from each population, dissected and mounted in glycerine on glass slides, were used for data collection. The morphology of limbs was investigated using a transmitted light microscope at a magnification of 400×. A total of 25 different linear distances (lengths of various setae, claws and segments mentioned in the list of abbreviations presented above) was taken on each of 64 female using an eyepiece micrometer with an accuracy of 2.5 μm. The obtained distance measurements were further converted into 25 ratios (12 traits on A2, one on L6, seven on L7 and five on CR) and expressed as percentages. For each ratio its mean, range and standard deviation (SD) were calculated. The length ratios are listed in Tables 4 and 5.

Morphology of the hemipenis was studied using males from Ada-Kaleh, Orşova, Jiblea and Şuncuiuş.

Statistical analyses

Before we will proceed with the presentation of our data, it is necessary to precise that we will speak about morphological diversity (we call it disparity) of the ostracod carapace (more precisely we compare the shape of the left valves of the females within and between various populations). Disparity of the valves will be examined at two hierarchically different levels: (a) as phenotypic variety within a population, those occurring during one time slice and (b) as phenotypic variety originated through evolutionary change, respectively morphological transformation over longer periods of time, which is visible when one compares populations or species belonging to an evolutionary group (cf. Ciampaglio et al., 2001; Winther, 2001).

Morphological disparity of the valve shape within and among populations was examined using multivariate statistical algorithms. A matrix of pair-wise Euclidean distances obtained in MORPHOMATICA (Brauneis et al., 2005a, b) was used for non-metric multidimensional scaling

Table 2 Carapace length and height in eight populations belonging to the *P. eremita* species-group (N—number of studied specimens, SD—standard deviation, codes for population names as in Table 1)

Site	Sex	N	Length (mm)			Height (mm)			Height/Length		
			Mean	Range	SD	Mean	Range	SD	Mean	Range	SD
Bisexual populations											
AK	♀	9	0.89	0.84–0.92	0.02	0.53	0.50–0.54	0.01	0.60	0.58–0.61	0.01
	♂	4	0.93	0.91–0.98	0.03	0.52	0.51–0.56	0.02	0.56	0.55–0.57	0.01
J	♀	8	0.82	0.77–0.84	0.02	0.46	0.44–0.49	0.02	0.57	0.55–0.59	0.02
	♂	4	0.86	0.81–0.91	0.04	0.45	0.44–0.48	0.02	0.53	0.52–0.54	0.01
S	♀	9	0.84	0.81–0.90	0.03	0.47	0.44–0.50	0.01	0.56	0.53–0.58	0.02
	♂	3	0.93	0.87–1.00	0.06	0.51	0.47–0.57	0.05	0.55	0.54–0.57	0.01
Parthenogenetic populations											
A	♀	8	0.85	0.83–0.87	0.01	0.46	0.44–0.48	0.01	0.54	0.53–0.57	0.02
G-r	♀	8	0.78	0.75–0.82	0.02	0.41	0.38–0.43	0.02	0.53	0.50–0.57	0.03
G-t	♀	8	0.79	0.73–0.89	0.05	0.46	0.43–0.50	0.03	0.57	0.56–0.59	0.01
H	♀	8	0.88	0.86–0.90	0.02	0.47	0.46–0.49	0.01	0.54	0.52–0.55	0.01
P	♀	8	0.81	0.74–0.84	0.03	0.46	0.44–0.48	0.02	0.57	0.54–0.59	0.02

(MDS), UPGMA (unweighted group average method) and One-way ANOSIM (analysis of similarities) in the programme PRIMER v.5 (Clarke & Gorley, 2001). ANOSIM is a non-parametric permutation/randomisation method which evaluates how well separated are the populations chosen a-priori. We adopted the Clarke & Warwick (2001) rule of thumb that groups are well separated when R statistic is larger than 0.75. We ran the test with 5000 actual permutations.

The pattern of valve disparity, i.e. the degree of shape dispersion (or variation) in the morpho-space within populations was estimated using the mean pair-wise distance (MPD). This index is defined as the sum of Euclidean distances in morphospace between all possible pair-wise combinations of the sample divided by the total number of combinations (Ciampaglio et al., 2001). We calculated for each MPD value also the 95% confidence limits by boostraping with replacement and 5000 permutations using the computer programme which accompanies the "Ecological Methodology" text-book of Krebs (1999).

To determine if females could be accurately classified by sampling site based on the limbs and if so, which of the distance measurements were important in distinguishing between populations, a discriminant function analysis (Manly, 1998) was carried out with STATISTICA for Windows 5.1 programme (StatSoft, 1997). Out of all 25 calculated ratios, we used in the discriminant analysis only 12 (listed in Table 6) which complied with the assumptions of this analysis, i.e.: normal distribution (tested by Shapiro–Wilk's W test), homogeneity of variances (tested by Cochran C test), correlations between means and variances (examined by inspecting the descriptive statistics) and redundancy of variables (by computation routinely implemented in the STATISTICA menu).

As in the valve shape analysis, for the limbs we also employed the mean pair-wise distance (MPD) coupled with the 95% confidence limits as an additional measure of the degree of the intra-populational variation. For the computation of the MPD we used the 12 traits applied in the discriminant analysis.

Results

Disparity of female valves

The population of Ada-Kaleh (AK) displays for most of its individuals a well developed umbone, the antero- and postero-dorsal margins form straight lines and valves are clearly triangular (Fig. 3a, b). The size of the valves approaches 0.9 mm length and the ratio height/length (h/l) varies around 0.6 (Table 2). When the outlines are standardised for their surface and further superimposed on their mean outline (the consensus shape), one notices that additional differences exist not only on the dorsal side but also on the anterior, posterior and ventral sides (the different lengths of the vectors, cf. Fig. 5a). The MPD value is high with large confidence intervals due to the variation of the mean δ^2 values expressed as distances between the pair-wise individual outlines (Table 3). The shape of the virtual consensus valve appears clearly triangular too (Fig. 8a). The differences between the outlines of the AK individuals and those of individuals of the other populations when expressed in two dimensional morphological space (Fig. 7) show clearly that triangular shapes are spread on the left side of the plot, clustered in three sub-groups. Note that the Orşova specimens (here not figured) display valve shape very similar to that of the AK specimens.

The population of Jiblea (J) consist of both, individuals with triangular valves and individuals with left valves displaying poorly developed umbones and rounded dorsal appearances (Figs. 3c, d, 6a). The mean h/l ratio is less than that of the AK population and the range between the extreme morphotypes is larger (Tables 2, 3, Figs. 6a, 7).

The Şuncuiuş (S) individuals display a reduced umbone (Figs. 3e, f, 8e) and the mean h/l ratio is less than those of AK and J populations (Table 2). Hence, the valves of most of individuals and the consensus one (Fig. 8e) appear more or less rounded. The spread within the morphological space is wide and the MPD value is high (Fig. 7, Table 3).

The Hotarele (H) population has rounded valves, practically no umbone (Figs. 3g, h), the

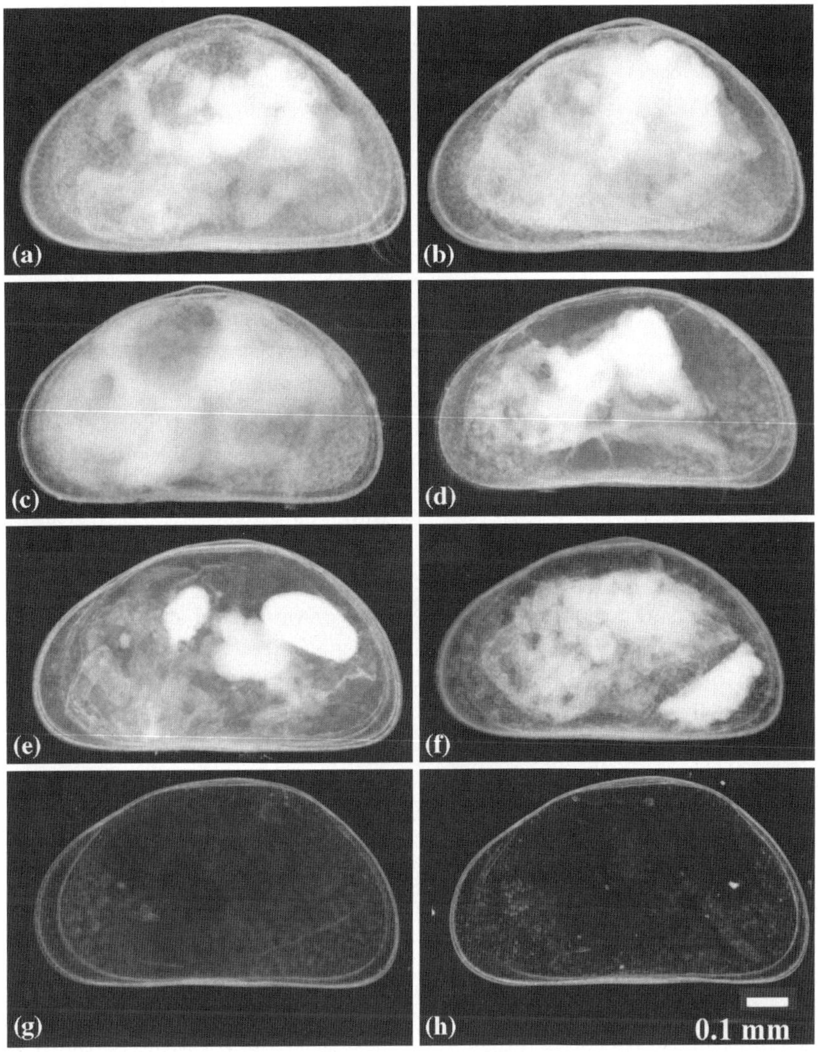

Fig. 3 Lateral view of female carapaces and/or valves of the studied populations belonging to the *Pseudocandona eremita* species-group: (**a**, **b**) Ada-Kaleh, (**c**, **d**) Jiblea, (**e**, **f**) Şuncuiuş, (**g**, **h**) Hotarele (anterior to the left)

mean h/l ratio, less than 0.55 (Table 2) and low intra-population variability (Figs. 5b, 7, Table 3). The virtual mean valve (Fig. 8c) appears clearly rounded and more flatted than those of the previous populations (Figs. 9a).

The Aştileu (A) population shows a similar disparity pattern, the rounded dorsal shape, the low h/l ratio (Figs. 4e, f, 6b, 8f, Table 2) but the higher shape dispersion in the morphospace and correspondingly the higher MPD value (Fig. 7, Table 3) when compared to Hotarele.

Individuals from Pietroşani (P) have a higher h/l ratio than those of the previous two populations (Table 2), the umbone is better developed (Figs. 4g, h) than that of the Aştileu individuals (Figs. 4e, f) and resembling that of the Şuncuiuş individuals (Figs. 3e, f). The dispersion in morphospace and the MPD value appear relatively significant, similar to those of the Aştileu and Gilău-r populations, and the mean shape resembles that of the Şuncuiuş population (Figs. 7, 8d, e, Table 3).

The remaining two populations from the wells No. 1 and No. 2 in Gilău are clearly different from each other, hence we called them G-r and G-t, respectively. The former is dominated by females with round shape valves and the low h/l ratio, whereas the latter—by more triangular

Table 3 Mean pair-wise distance (MPD) values with 95% confidence limits ($CL_{95\%}$) as a measure of intra-populational variation of the valve shape and the limb traits in females from the eight studied populations of the *P. eremita* species-group (sample size in each case—8 individuals, codes for the population names as in Table 1)

	Valves		Limbs	
	MPD	$CL_{95\%}$	MPD	$CL_{95\%}$
Bisexual populations				
AK	17.25	15.52–18.57	0.515	0.465–0.565
J	14.59	12.68–17.02	0.419	0.370–0.468
S	14.46	12.99–15.80	0.344	0.307–0.380
Parthenogenetic populations				
A	11.71	9.35–11.25	0.473	0.420–0.525
P	11.79	10.42–13.15	0.379	0.330–0.427
H	7.96	7.40–8.68	0.406	0.370–0.441
G-r	12.07	11.08–13.72	0.411	0.359–0.462
G-t	8.96	8.20–9.82	0.365	0.331–0.400

phenotypes with the higher h/l ratio (Fig. 4a–d, 8g, h, 9b, Table 2). The two populations which are closely located (cf. previous section) differ also in their degree of variation within the morphospace, the G-r population is more variable, having the higher MPD value than the G-t population (Fig. 7, Table 3).

A comparative examination of the individual (Figs. 5–7) and of the mean (Figs. 8–10) valve shapes among the eight populations allows us to point out the more frequent occurrence of the round-shape valves as compared with the triangular ones. In the morphological space (Fig. 7) the most triangular individuals are located on the left side and therefore there is a clear separation between the AK triangular shape and those of the other populations (cf. also Figs. 8, 10), particularly well noticeable when compared with the shape of the H (cf. the virtual consensus shapes in Fig. 9a) and of the G-r individuals. A classification with UPGMA cluster analysis of the dissimilarities between the mean shapes of the valves for the eight populations (Fig. 10) confirms the previous observation and it points out the closer similarity between the J and the G-t consensus morphotypes as compared to the other more rounded ones. One should note that within the group of the round-shape consensus morphotypes (Fig. 10), that of Şuncuiuş clusters with the Pietroşani, while the Aştileu, being still similar to Şuncuiuş (Fig. 9c) occupies an intermediary position between the H with G-r and the P with the S morphotypes.

It has to be stressed that inspecting the MPD values (Table 3) and the dispersal in the morphological space (Fig. 7) the morphological valve diversities of the females belonging to the three bisexual populations appear higher when compared with those of the five parthenogenetic ones.

The one-way ANOSIM test was applied to the eight populations in order to check how well they are pair-wise separated. The AK population, with its triangular valves is clearly separated from the populations with more round-shape valves, i.e. from the A (R statistic = 0.801), H (R = 0.772) and G-r (R = 0.799). Moreover, the G-t population is also well separated from the A (R = 0.815) and H (R = 0.912). All the other populations do not show clear separation one from the others, the computed R-statistic varies between 0.6 (e.g. R statistic for a pair of the AK and P equals 0.586) and 0.1 (e.g. R = 0.096 for the P and S).

The male carapaces of the three bisexual populations as compared with those of females are longer and have lower h/l ratios (Table 2).

Morphological variation of female limbs

Tables 4 and 5 summarise the measurements of the 25 morphological traits (ratios expressed as percentages). The descriptive statistics show for various traits considerable differences between the arithmetic means of the populations, even if the extreme values overlap. The discriminant function analysis applied on 12 of these variables offers the possibility to separate some of the populations morphometrically. A two-dimensional ordination of the 64 females belonging to the eight populations is presented in Fig. 11. Discrimination between the populations was highly significant: Wilks' lambda = 0.00264, $F(84, 283) = 5.5228$, $P < 0.001$. The standardised coefficients for the first three roots (canonical functions), which accounted for 82.8% of the explained variance are listed in Table 6, whereas the squared Mahalanobis distances (D^2) between the studied populations, the F values and probability levels as well as percent of females that were correctly classified in each population by the classification functions are presented in Table 7.

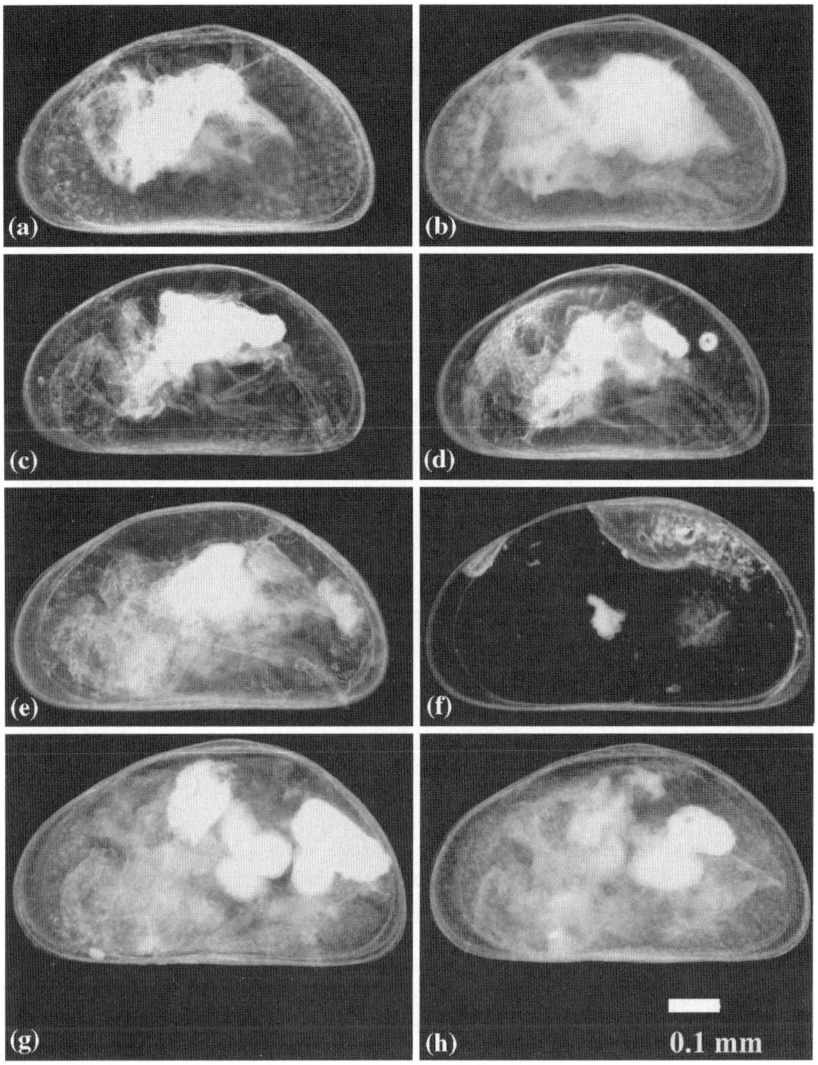

Fig. 4 Lateral view of female carapaces and/or valves of the studied populations belonging to the *Pseudocandona eremita* species-group. (**a**, **b**) Gilău well no. 2 (G-t), (**c**, **d**)—Gilău well no. 1 (G-r), (**e**, **f**) Aştileu, (**g**, **h**) Pietroşani (anterior to the left)

In the scatter-plot (Fig. 11a, b) females of the AK population appeared clearly on the right-hand side of the figure. The first canonical function (Root 1) accounted for 41.2% of the explained variability and discriminates mostly between females of this population and the remaining populations. The second (Fig. 11a) and third (Fig. 11b) roots accounted for 24.5% and 17.1% of the explained variability, respectively and provided some separation of the females of the P population (Root 2) and the A population (Root 3), which both had the most positive values. However, the discrimination in the two latter cases is not as clear as that provided by the Root 1. The first function was marked mainly by: a) positive coefficient (0.89) for the ratio of the lengths of the medium and large setae on the terminal segment of L7 (h_3/h_2), (b) negative coefficient (–0.87) for the relative length of the large seta on the terminal segment of L7 (h_3/L7EI), and c) positive coefficient (0.73) for the relative length of the distal seta on the penultimate segment of L7 (g/L7EI). The second function was most heavily loaded by: (a) negative coefficient (–1.28) for the ratio h_3/L7EI, (b) positive coefficient (0.58) for the relative length

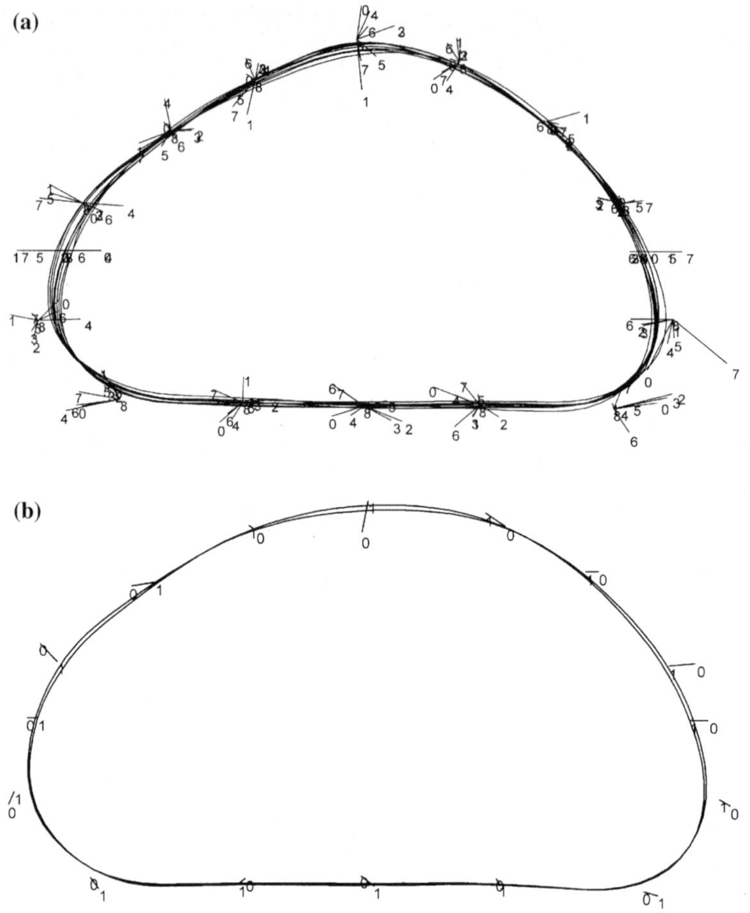

Fig. 5 Variation of the left valve shape standardised for surface, with vector dimensions (enlargement 5×): (**a**) Ada-Kaleh, superimposition of the outlines of eight specimens (numbers 0–7) and of the reference consensus outline (number 8), (**b**) Hotarele, superimposition of the outlines with the minimal (0) and maximal (1) height

of the posterior seta on protopodite of L7 (d_p/L7EI), and c) negative coefficient (–0.49) for the relative length of the largest seta of A2 exopodite (Exo/A2EI), while the third function by: a) positive coefficient (0.97) for the relative length of the major claw on the terminal segment of A2 (G_M/A2EII), b) negative coefficient (–0.87) for the ratio h_3/h_2, and c) positive coefficient (0.63) for the relative length of the outer of two postero-distal setae on EI of A2 (co/A2EI) (Table 6, Fig. 11). It appears that the most significant discrimination is that between the AK individuals and the other populations. This is mainly due to the important relative length of several traits, especially setae h_3/h_2 and g on L7 as well as the Exo seta on A2 and claw G_1 on A2 (Tables 4–6). Considering the remaining populations, the discrimination was not as successful as that of the AK (Fig. 11). The females of the Pietrosani population are separated from those of the other populations by the relatively short setae h_3 and Exo as well as long seta d_p (Tables 4–6). The population of Aştileu could be differentiated mainly by the low ratio h_3/h_2 (Table 5). The females from the remaining populations can hardly be discriminated one from the other (Fig. 11).

Comparing the MPD values based on the limb traits with those calculated on the valve shape characteristics (Table 3) there is no clear congruence visible. However, two of the three bisexual populations display high intra-populational variation. The AK females appear to be the most disparate (same as for the valves) and

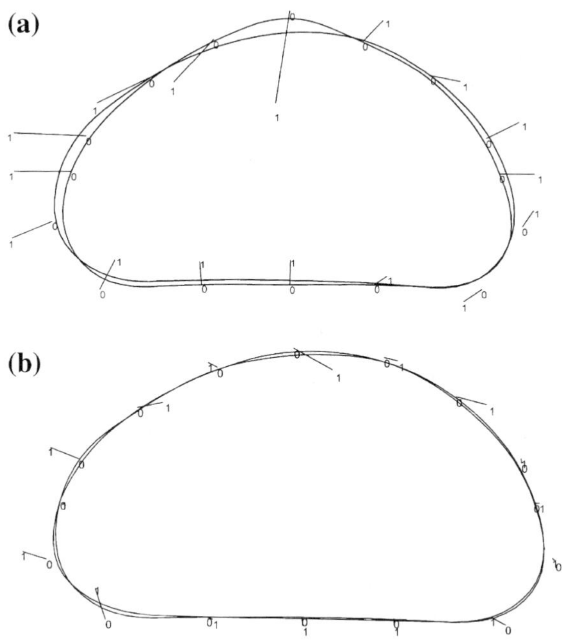

Fig. 6 Variation of the left valve shape standardised for surface, with vector dimensions (enlargement 5×): (**a**) Jiblea, superimposition of the outlines with the minimal (1) and maximal (0) height, (**b**) same for Aştileu

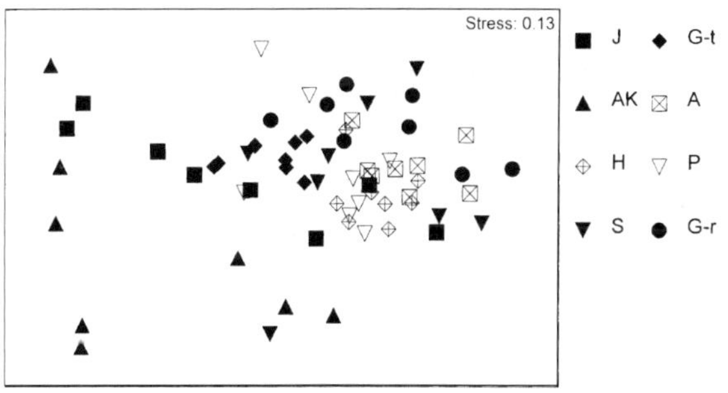

Fig. 7 Ordination of females from the eight studied populations belonging to the *Pseudocandona eremita* species-group by non-metric multi-dimensional scaling (MDS) technique based on pair-wise Euclidean distances (codes for population names as in Table 1)

the J females represent the third most variable population. Conversely, the S females, from the third bisexual population, seem to be the most even.

Morphology of the hemipenis

The males of the AK, J and S sites display clear differences on their hemipenes traits (Fig. 12a–c). Males from the Orsova well (here not shown) display the characteristics of the AK males. The Ada-Kaleh and the Jiblea males have the peniferal medial lobe b shortly extruded from the peniferum on its ventral side. It is more angular in the males of the former population as compared to that in Jiblea which is widely rounded. This expansion is more extruded in the S males. The distal part of the internal m process is well sclerotised and crenulated. In both the AK and J males it is distally rod-shaped, however, in the AK males this distal part of the m process is oriented posteriorly, whereas in the J males it is directed more ventrally. The distal part of the m process in the Şuncuiuş males is S-shaped and displays crenulations (Fig. 12b).

Descriptions of two new species

Family Candonidae Kaufmann, 1900
Subfamily Candoninae Kaufmann, 1900
Genus Pseudocandona Kaufmann, 1900

Pseudocandona danubialis *sp. nov.*
(Figs. 3a, b, 12a, 13a, b)

Type locality: Ada-Kaleh Island, in a well, about 7 m deep acceding to groundwater which flows below the Danube river-bed (Mehedinti County; south-western Romania) (geographical coordinates, Table 1). Leg. D. L. Danielopol and Dan Dancău (2.07.1967).

Type material: *Holotype*: a male, with limbs dissected and mounted in glycerine-gelatine, valves stored in ethanol (NHMW 20705). *Allotype*: a female with limbs dissected and mounted on a permanent slide and the left valve stored in ethanol (NHMW 20706). *Paratypes*: 8 females, 3 males. The holotype and allotype are deposited at the Natural History Museum in Vienna (NHMW), whereas the paratypes are stored at the Institute of Limnology, Austrian Academy of Sciences, Mondsee. Other material: 3 males and 3 females collected on 2.07.1967 by D. L. Danielopol and D. Dancău from a well within the porous aquifer of the Danube River at Orşova, Strada Pietii (Mehedinti County).

Derivation of name: From Danubius, the Latin name of the River Danube.

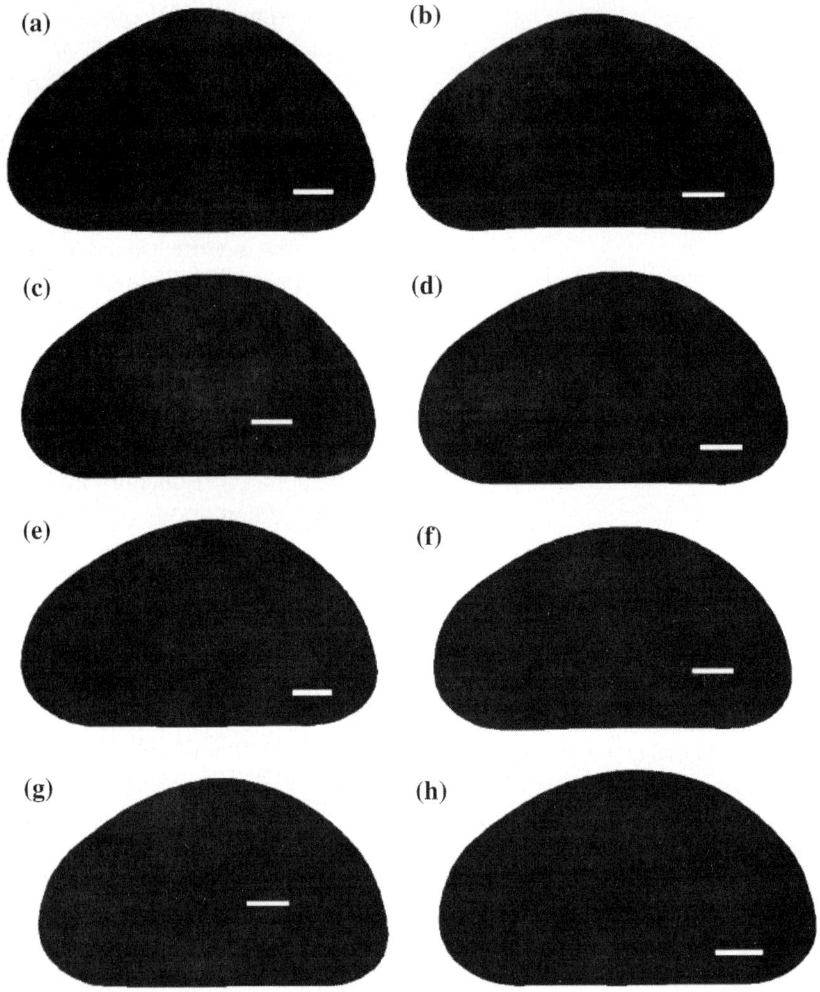

Fig. 8 Consensus oultlines for the left valves: (**a**) Ada-Kaleh, (**b**) Jiblea, (**c**) Hotarele, (**d**) Pietroşani, (**e**) Şuncuiuş, (**f**) Aştileu, (**g**) Gilău—G-t, (**h**) Gilău—G-r (bar length—0.1 mm)

Diagnosis: Relatively large representative of the eremita-species group of the genus Pseudocandona with distinctly triangular valves in lateral view (h/l = 0.55–0.61). LV of the males with greatest h situated at about 57% of valve l (Fig. 13a). Dorsal margin in front of the highest part straight and sloping more steeply than behind the highest part. LV overlapping RV along entire valve margin. RV similar in outline to LV but slightly smaller. Calcified part of inner lamella of both valves amounting to c. 9% of the valve l. Valves with shallow pits and a few setae at both ends. Holotype measurements: LV: l = 0.98 mm; h = 0.56 mm; h/l = 0.57. Left valve of the allotype (female) relatively high with the greatest h situated at the mid-length. Female valves similar in shape to those of males (Fig. 13b). Width of inner lamella of LV represents anteriorly 9% of the valve length. Allotype measurements: LV: l = 0.84 mm; h = 0.50 mm; RV: l = 0.80 mm, h = 0.47 mm; h/l = 0.59. Apical claws on A2 remarkably long, in females G_1 = 1.3–1.7, G_2 = 0.6–0.8, G_3 = 1.2–1.5, G_M = 1.1–1.3, G_m = 0.8–0.9 times the length of A2EI. Terminal claw on L6 female about 3.1–3.5× the length of L6EII. Second and third endopodites of L7 fused, with shortest terminal seta h_1 hook-like. The outer lobe a of hemipenis (Fig. 12a) well developed; inner lobe b short, with pointed ventral margin; medial lobe he widely trapezoidal and

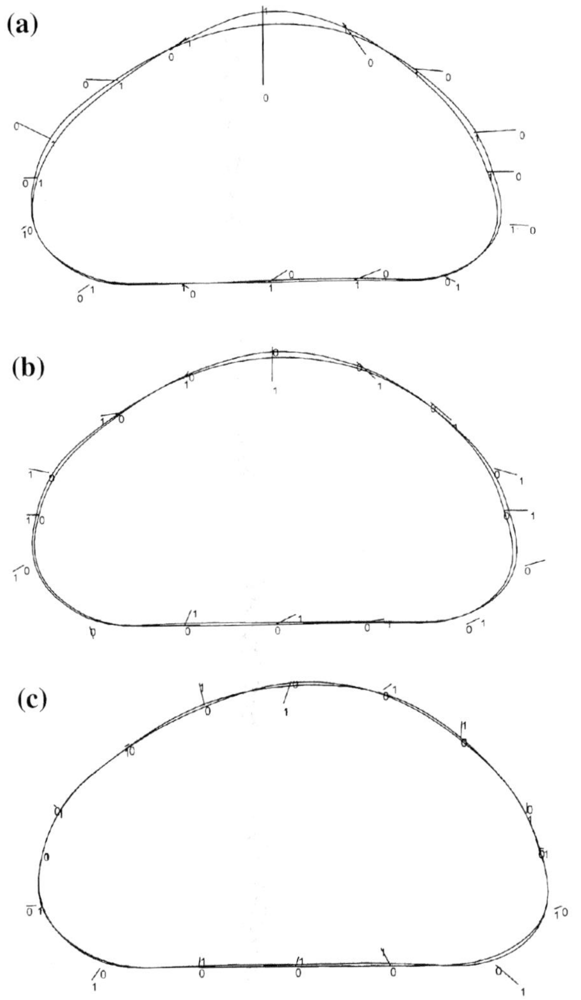

Fig. 9 Differences between the consensus outlines of the left valves standardised for surface with vector dimensions (enlargement 5×): (a) superimposition of the outline of Hotarele (0) on that of Ada-Kaleh (1), (b) superimposition of the outline of Gilău-r (1) on that of Gilău-t (0), (c) superimposition of the outline of Aştileu (1) on that of Şuncuiuş (0)

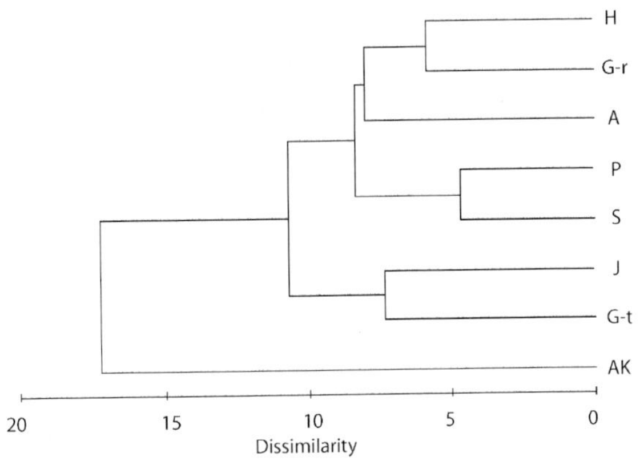

Fig. 10 Dendrogram for hierarchical clustering of the mean (consensus) outlines of the eight populations using group-average linking (UPGMA) algorithm (scale—degree of dissimilarity, codes for population names as in Table 1)

reaching beyond the a and b lobes. The sclerotised and crenulated appendix of the m process, rod-shaped and curved to the posterior end of the animal body. Bursa copulatrix elongated and funnel-shaped.

Pseudocandona transylvanica *sp. nov.*
(Figs. 3e, f, 12b, 13c, d)

Type locality: Şuncuiuş village, well acceding to the porous aquifer of the Crişul Repede River (Bihor County, north-western Romania) (geographical coordinates, Table 1). Leg. D. L. Danielopol and E. Şerban (28.07.1967).

Type material: *Holotype*: a male with limbs dissected and mounted in glycerine-gelatine, valves stored in ethanol (NHMW 20707). *Allotype*: one female with limbs and valves stored as those of the holotype (NHMW 20708). *Paratypes*: 8 females, 2 males. The holotype and the allotype are deposited at the Natural History Museum in Vienna (NHMW) while paratypes are stored at the Institute for Limnology, Austrian Academy of Sciences, Mondsee.

Derivation of name: From Transylvania, the Latin name of the region.

Diagnosis: Carapace approximately triangular (h/l = 0.53–0.58). LV of the males with greatest h situated at about 56% of valve l (Fig. 13c). Dorsal margin in front of the highest part straight with the postero-dorsal margin sloping less steeply than the antero-dorsal one. LV overlapping the RV along the entire valve margin. RV similar in outline to the LV but slightly smaller. Calcified part of the inner lamella of both valves amounting to c. 9% of the length. Holotype measurements: LV: l = 1.00 mm; h = 0.57 mm; h/l = 0.57; RV: l = 0.97 mm; h = 0.54 mm; h/l = 0.56. Female valves similar in shape to those of males (Fig. 13d). Allotype measurements: LV: l = 0.90 mm; h = 0.48 mm; h/l = 0.58; RV: l = 0.88 mm.

Table 4 Descriptive statistics of the morphological traits of the female second antenna and walking leg for the eight studied populations of *P. eremita* species group (sample size in each case—8 individuals, SD—standard deviation, codes for the population names as in Table 1)

Variable	Statistic	AK	J	S	A	G-r	G-t	H	P
		Bisexual populations			Parthenogenetic populations				
Exo/A2EI	Mean	95	88	83	78	86	82	85	83
	Range	82–115	76–96	77–93	63–90	80–94	70–90	75–100	73–89
	SD	10	7	6	9	5	7	10	5
co/A2EI	Mean	81	77	77	78	77	82	75	74
	Range	69–92	68–85	69–80	70–86	70–80	78–90	70–82	69–81
	SD	6	5	4	5	4	4	4	5
ci/A2EI	Mean	48	66	53	55	57	58	51	49
	Range	42–60	51–76	51–56	44–70	50–63	56–64	45–56	45–51
	SD	7	8	2	11	4	3	4	2
G_1/A2EI	Mean	141	127	126	127	130	132	122	122
	Range	131–165	117–140	120–136	119–134	126–135	125–140	114–136	118–128
	SD	11	7	6	5	3	5	8	3
G_1/A2EII	Mean	292	245	258	256	261	261	242	250
	Range	270–316	232–265	248–272	236–272	252–271	244–272	232–254	241–264
	SD	14	11	9	13	6	11	9	9
G_2/A2EI	Mean	63	48	53	56	51	55	48	48
	Range	55–76	42–53	50–59	51–60	46–58	52–56	42–54	42–53
	SD	8	4	3	3	4	1	4	4
G_3/A2EI	Mean	133	123	119	122	124	125	113	117
	Range	121–154	115–130	114–128	111–130	120–131	120–130	105–126	113–121
	SD	10	5	5	7	4	3	7	3
G_3/A2EII	Mean	277	236	244	246	250	248	225	239
	Range	252–300	226–254	222–260	232–260	240–263	233–268	214–235	221–252
	SD	16	10	13	10	7	13	7	10
G_M/A2EI	Mean	118	103	106	105	105	110	100	100
	Range	112–131	96–113	100–116	93–112	100–113	101–116	91–108	96–101
	SD	7	5	6	6	4	5	6	2
G_M/A2EII	Mean	244	198	217	211	212	218	197	205
	Range	225–272	186–215	207–224	189–232	200–225	200–232	179–215	189–224
	SD	14	9	5	15	9	10	11	10
G_m/A2EI	Mean	81	73	73	72	74	75	69	72
	Range	78–94	68–81	67–80	65–76	72–76	70–76	64–76	69–76
	SD	5	4	4	4	2	2	4	3
G_M/G_2	Mean	188	214	199	187	209	198	207	186
	Range	163–213	200–232	183–215	177–200	193–221	189–207	192–233	151–207
	SD	16	10	10	9	10	7	14	17
h_2/L6EII	Mean	318	296	275	272	298	277	285	282
	Range	307–348	259–343	240–308	252–280	272–350	259–300	274–296	272–309
	SD	13	24	19	10	28	14	7	12

Apical claws on A2 moderately elongated, equalling in females $G_1 = 1.2$–1.4, $G_2 = 0.5$–0.6, $G_3 = 1.1$–1.3, $G_M = 1.0$–1.2, $G_m = 0.7$–0.8 times the length of A2EI. Terminal claw on L6 amounting to 2.4–3.1 × the length of L6EII. Second and third endopodites of L7 not differentiated with shortest terminal seta h_1 hook-like. Hemipenis with well developed, ventrally oriented and roundly pointed extension of medial lobe b, and crenulated, robust and S-shaped distal appendix of m process. Additional data can be found in the description of *Pseudocandona eremita* from Şuncuiuş in Danielopol (1982).

Discussion

Evolutionary aspects—origin and development of adaptive traits

It is commonly accepted that the aquatic interstitial habitats with their reduced habitable space

Table 5 Descriptive statistics of the morphological traits of the female cleaning leg and caudal ramus for the eight studied populations of *P. eremita* species group (sample size in each case—8 individuals, SD—standard deviation, codes for the population names as in Table 1)

Variable	Statistic	AK	J	S	A	G-r	G-t	H	P
		Bisexual populations			Parthenogenetic populations				
d_2/L7EI	Mean	72	70	68	73	76	69	70	70
	Range	67–84	63–77	63–74	59–94	70–82	65–75	65–80	63–74
	SD	6	5	4	12	4	4	5	4
d_p/L7EI	Mean	115	118	118	111	116	118	120	129
	Range	102–124	105–140	108–132	94–129	100–129	110–126	108–131	120–145
	SD	7	12	8	12	10	5	8	8
h_1/L7EI	Mean	17	25	22	25	27	22	24	24
	Range	13–24	23–29	18–26	19–29	23–33	18–26	21–29	18–32
	SD	3	1	2	3	3	3	2	2
h_2/L7EI	Mean	105	119	121	115	120	109	121	121
	Range	100–116	100–139	105–147	102–129	110–133	100–125	105–180	100–145
	SD	6	14	13	9	7	8	24	16
h_3/L7EI	Mean	148	145	137	129	161	138	138	118
	Range	135–160	134–160	105–153	102–143	149–200	125–158	125–157	100–137
	SD	8	9	17	13	17	9	9	11
h_3/h_2	Mean	143	125	126	111	133	128	125	130
	Range	133–155	109–135	119–143	100–129	124–158	119–135	111–138	121–135
	SD	9	8	8	9	11	6	9	5
g/L7EI	Mean	95	84	92	89	93	73	91	92
	Range	82–113	63–100	85–97	71–102	77–100	60–84	84–103	88–100
	SD	10	12	4	10	8	7	7	5
s_a/CRm	Mean	14	13	14	13	15	14	14	15
	Range	12–18	11–14	12–16	9–17	14–17	13–17	13–21	13–17
	SD	2	1	1	3	1	1	2	1
s_p/CRm	Mean	42	38	44	45	43	44	42	48
	Range	38–47	41–46	37–48	41–50	38–59	35–50	40–45	42–52
	SD	3	10	4	3	6	4	1	3
G_a/CRm	,Mean	70	66	67	68	68	70	71	74
	Range	67–83	63–69	56–75	65–72	63–78	63–83	67–73	70–82
	SD	5	1	5	2	5	6	2	3
G_p/CRm	Mean	62	57	57	58	58	61	60	65
	Range	52–76	52–60	52–61	55–63	51–67	56–68	58–65	56–73
	SD	6	2	2	2	4	3	2	4
s_p-G_p/CRm	Mean	26	27	26	26	26	27	24	25
	Range	21–32	25–30	24–33	23–28	23–29	27–28	20–28	20–29
	SD	3	2	3	1	2	1	3	3

(as compared to that occurring in cavernous systems) represent a constraining environment to the animals which colonise them (Coineau, 2000; Gibert & Culver, 2005). Stygobitic crustaceans belonging to different phyletic groups evolve convergently a suite of traits, called by Coineau (2000) phreatomorphic characters when animals occupy interstitial habitats or troglomorphic traits for cavernous dwelling animals (Christiansen, 2005). As examples of phreatomorphic traits one can mention the miniaturisation and/or elongation of the body, the elongation of ambulatory limbs related to the reduction of the number of setae or the important development of chemo-sensorial organs (Hartmann, 1973; Danielopol, 1980a).

The most ancient *Pseudocandona* representatives belonging to the *eremita* species group that could be identified with certainty are *P. roaixensis* Carbonnel (Danielopol, 1982) from the Lower Miocene (Tortonian, in Carbonnel, 1969) in southern France and *P. ratisbonensis* (Lutz) from Lower Miocene (Eggenburgian) in Bohemia, Czech Republic (Pokorný, 1986). Both these species were recorded from lacustrine (epigean) palaeohabitats. Originally, *P. ratisbonensis* was sampled

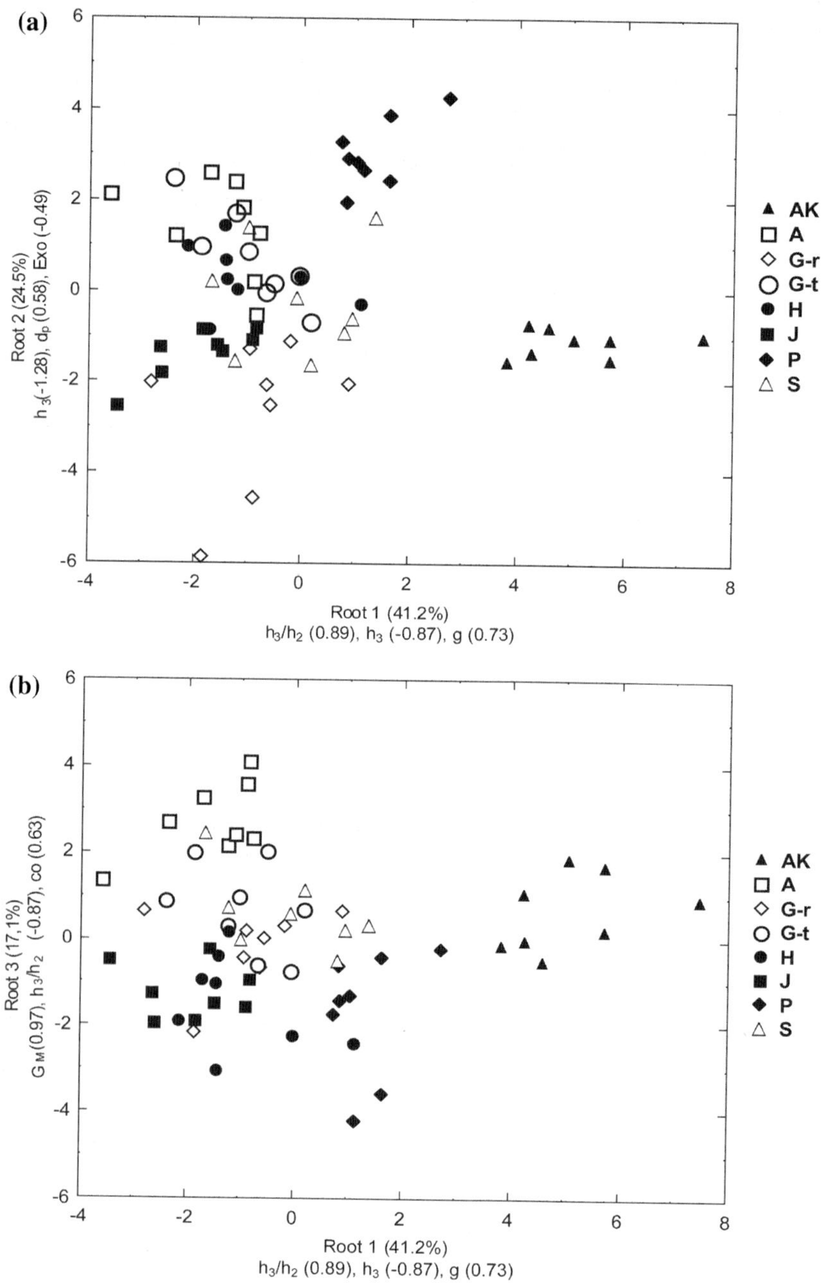

Fig. 11 Plots of the 1st discriminant function (root) against the 2nd (**a**) and of the 1st discriminant function against the 3rd (**b**) depicting discrimination by 12 continuous morphological variables of 64 females belonging to the eight populations (codes for population names as in Table 1, codes for the morphological traits cf. text). Percentages for the explained variance and standardised coefficients of the traits which loaded most heavily the first three roots are added

in limnic sediments deposited during the Upper Miocene in Bavaria, near Regensburg (Lutz, 1965) and it is expected to occur also in the lacustrine environment of the Vienna Basin (current investigations carried on by K. Minati and DLD at the Limnological Institute, Mondsee). Other fossil species potentially belonging to the *P. eremita* lineage were recorded from Miocene deposits in Slovakia (Pipik & Bodergat, 2004) and Pliocene sediments in Turkey (Freels, 1980).

Table 6 Standardised coefficients for the first three roots (canonical functions) of the multiple group discriminant analysis of biometric data collected on females belonging to the *P. eremita* species-group from eight studied populations

Variable	Root 1	Root 2	Root 3
co/A2EI	–0.390	–0.062	0.631
Exo/A2EI	0.468	–0.491	–0.425
G_1/A2EII	0.337	0.629	–0.919
G_3/A2EII	0.108	–0.253	0.079
G_M/A2EII	0.099	0.286	0.970
G_M/G_2	–0.095	–0.418	–0.559
d_p/L7EI	0.219	0.580	–0.599
G/L7EI	0.731	0.115	–0.008
H_1/L7EI	–0.302	0.240	0.133
H_3/L7EI	–0.869	–1.282	0.439
H_3/h_2	0.895	0.286	–0.870
G_p/CRm	0.209	0.378	–0.257
Eigenvalue	5.263	3.134	2.181
Cumulative proportion	0.412	0.657	0.828

P. roaixensis and *P. ratisbonensis* display high triangular valves with well developed umbone and the maximal carapace height represening about 66–68 % of the length. We hypothesise that the present-day stygobitic species living in Central and/or Eastern Europe have as an ancestor either *P. ratisbonensis* or a closely similar species. An additional argument for the idea that species with the triangular valves represent more primitive state of the present day subterranean *P. eremita*-like forms is also the documentation of Danielopol (1978) on the developmental trajectory of the population of Jiblea. The juvenile stages display a well developed umbone and triangular carapace sometimes stronger developed than that of the adults.

If one compares the height/length (h/l) ratio of the left valve of the species belonging to the *Pseudocandona eremita* species-group that live in karstic and interstitial habitats, the clear difference between species of these habitats can be noticed. All the seven interstial dwelling species display the h/l ratios of 51–60% (five species have the ratio between 51 and 57%), while four out of all the six karstic dwelling species exhibit the h/l values above 60%, e.g. the h/l ratio in *P. trigonella* (Klie) and *P. pretneri* Danielopol amounts to 67 % (cf. Namiotko & Danielopol, 2004 and data herein). We consider therefore that the more flat carapace observed in various populations of the interstitial species belonging to the *P. eremita* lineage represent an adaptive trait to life in porous systems of sandy-gravel sediments. Note that this inference is based on the view of one of us (D.L.D.) that the delineation of adaptation of crustaceans, like ostracods, to the aquatic subterranean environment is posible only through neutral correlation arguments (for discussion on this topic cf. Danielopol et al., 1994), while for the difference between adaptation defined through correlation versus causation arguments see *inter alia* Maynard Smith (1987) and Reeve & Sherman (1993). The information presented here should stimulate students to test scenarios related to the generation of adaptive traits with additional field and laboratory observations. One should

Table 7 Matrix of D^2 Mahalanobis distances listed above the diagonal and of the F values (degrees of freedom 12, 45) together with probability levels listed below the diagonal between the females belonging to the *P. eremita* species-group from eight studied populations

	AK	A	G-r	G-t	H	J	P	S
AK	×	57.185	40.438	48.195	45.615	53.752	37.887	29.777
A	15.32***	×	26.995	21.647	21.123	25.257	31.569	11.506
G-r	10.83***	7.23***	×	23.283	13.264	6.785	39.950	10.196
G-t	12.91***	5.80***	6.24***	×	16.782	19.177	27.694	15.613
H	12.22***	5.66***	3.55***	4.50***	×	7.005	15.605	7.449
J	14.40***	6.77***	1.82 n.s	5.14***	1.88 n.s.	×	31.177	10.624
P	10.15***	8.46***	10.70***	7.42***	4.18***	8.35***	×	19.683
S	7.98***	3.08**	2.73**	4.18***	2.00*	2.85**	5.27***	×
Class	100.0	87.5	87.5	100.0	100.0	87.5	100.0	87.5

*** $P < 0.001$; ** $P < 0.01$; * $P < 0.05$; n.s. = not significant

The bottom row (Class) indicates the percent of females that were correctly classified in each population by the classification functions (codes for populations names as in Table 1)

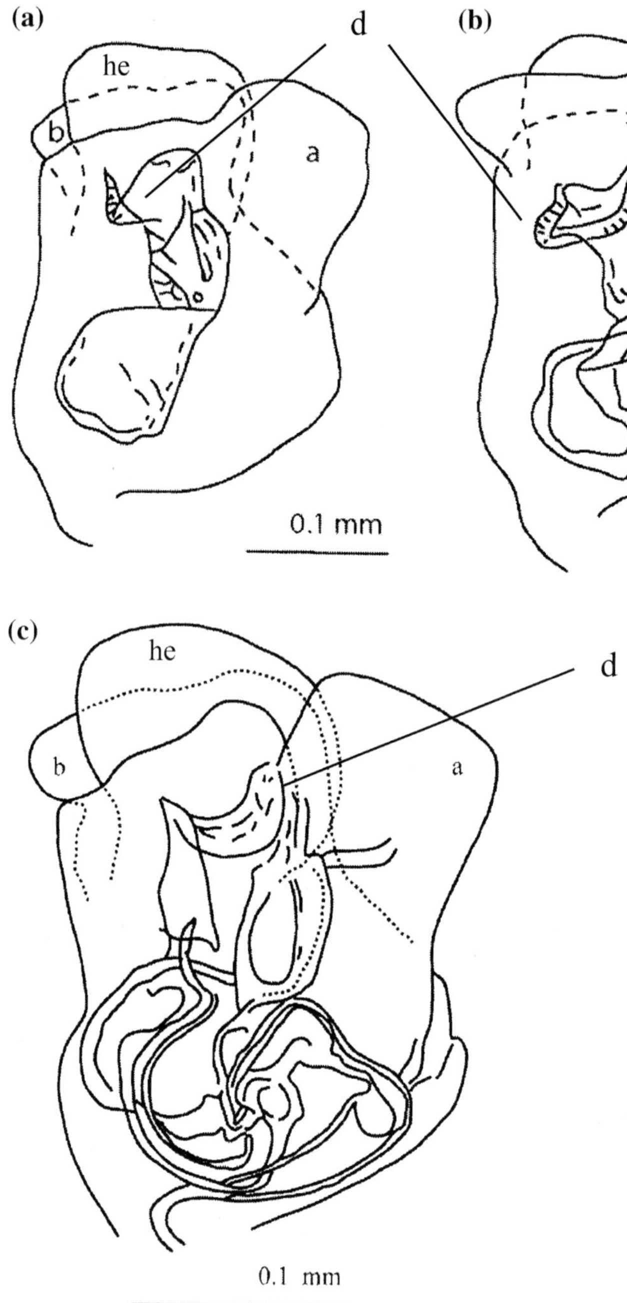

Fig. 12 Hemipenis with details of the central m process (d) and distal lobes a, b, h: (**a**) Ada-Kaleh, lateral sight, (**b**) Şuncuiuş, medial sight, (**c**) Jiblea, lateral sight

consider that if natural selection acts for increased success of life in reduced inter-granular space then individuals with more elongated and/or round-shape carapaces, having the lower h/l ratios, like the morphotypes observed in the A, S, P, G-r and H populations, are at a premium and these morphotypes should be considered adaptive solutions to life in micro-porous sediments (interstitial habitats). Bearing in mind the large distances, which separate geographically these populations one from the other, it is very probable that the dorsally more round-shape carapace

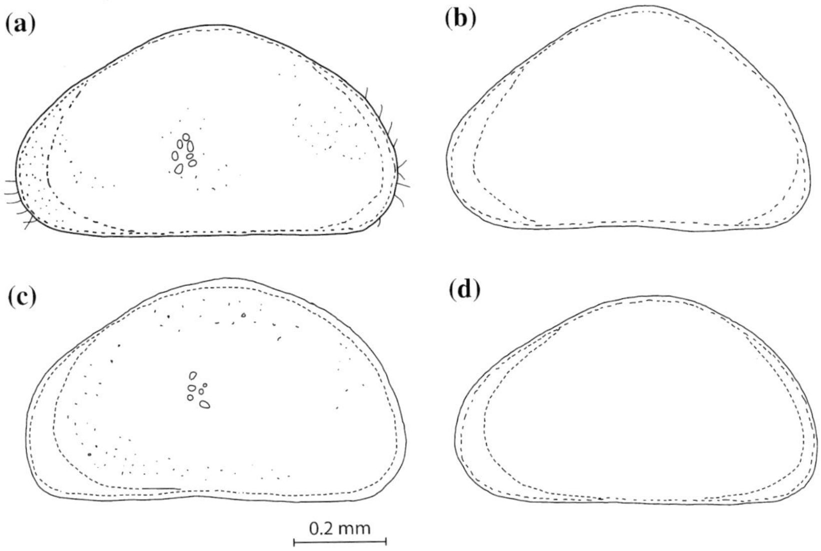

Fig. 13 *Pseudocandona danubialis* sp. nov.: (**a**, **b**) left valves in external view of the holotype (**a**) and allotype (**b**); *Pseudocandona transylvanica* sp. nov.: (**c**, **d**) left valves in external view of the holotype (**c**) and allotype (**d**)

developed independently in a convergent way. Following this line of reasoning, the Ada-Kaleh population with triangular carapace shape has to be viewed as evolutionary less advanced, compared to the other populations from Romania here discussed. Because of the high variability of the carapace shape of the three bisexual populations (AK, J and S) we expect that evolutionary changes will continue the transformation of the carapace into a more flat shape, similar to those occurring in the other populations studied.

The differences in the morphology of the male copulatory organ of the three bisexual populations AK, J and S, point out the existence of two types of morphological traits, those related to mate recognition during the pre-mate act and those which seem to have no adaptive value but apparently are genetically correlated with the former ones. The distal segment of the m process which is slightly crenulated and displays different shapes in the three species (Fig. 12) is in our opinion an adaptive trait as it allows conveying sensorial signals to the female during the pre-mating phase. Such cases were observed at various Candoninae species studied at the "Laboratoire Souterrain" of the C.N.R.S. in Southern France (Danielopol, 1980b). The lobe b of the hemipenis of the Candoninae we studied (Danielopol, 1980b) is neither involved in the pre-mating act nor in the copulatory act, however its distal part differs in the three species, *P. serbani*, *P. danubialis* sp. nov. and *P. transylvanica* sp. nov. We suspect that these morphological differences are due to the genetic correlation with the sexually selected traits of the m process, discussed previously.

Considering the antennal distal claws and that of the walking leg it is apparent that the most elongated phenotypes (expressed as mean values) occur in the bisexual population of AK (Tables 4, 5). The elongated claws on the ambulatory limbs of interstitial dwelling Candoninae help the movement within the inter-granular space, respectively animals can fix well granular structures (Danielopol, laboratory observations). However, elongated ambulatory limbs represent also adaptive traits for animals living in caves on fine grain sediments, like some collembols (Christiansen, 2005). Therefore the AK population could be considered the most advanced population in respect to the development of such phreatomorphic traits as the elongation of ambulatory claws.

The minor differences in the length of the other morphological setae considered to belong to the mechano-sensorial module (Tables 4, 5) do not offer information on a possible trend in the adaptive evolution related to the subterranean environment. Such morphological differences could be the product of neutral mutations too, which potentially occur during regressive evolutionary processes of stygobites (Wilkens, 2005).

Proposal for a new taxonomy of the *Pseudocandona eremita* species group

The taxonomic status of many stygobitic ostracod populations from Central and Eastern Europe that were attributed to *Pseudocandona eremita* (Vejdovský) by various authors (review in Meisch, 2000 and Namiotko & Danielopol, 2004) is problematic because of the incomplete description of the specimens from Prague (the type locality of this species) and/or from Prague surrounding area (cf. Vejdovský, 1882; Vávra, 1891; Jančařík, 1952; Absolon 1978). It is accepted by specialists like W. Klie (1940), H. Löffler (1963), D. L. Danielopol (1982) or I. Karanovic (pers. comm. 2005 to DLD) that *P. eremita* is a morphologically plastic species widely spread through the subterranean waters of Central and Eastern Europe. Hence species like *P. serbani* Danielopol, 1982 or *P. szoecsi* (Farkas, 1958) could be assimilated to *P. eremita* (*sensu lato*).

The present study challenges in part this view using theoretical and empirical arguments. We started from the idea that a species is a lineage segment within a larger species group, or phylogenetic clade, having its existence in time and space (*viz.* de Queiroz, 1998). For pragmatic reasons, for instance when we have to recognise routinely ostracods during ecological projects, we treat species as morphological entities which can be recognised by unique characteristics even when one has a single specimen bearing the diagnostic trait. The clear differences in the structure of the m process and the lobe b of the hemipenis of the males belonging to Ada-Kaleh, Jiblea and Şuncuiuş allowed us unambiguously to recognise besides *P. serbani*, two new species, *P. danubialis* sp. nov. and *P. transylvanica* sp. nov. These species have also distinct geographic distributions. *Pseudocandona serbani* occurs in the eastern sediment accumulation area of the Dacic basin while *P. danubialis* sp. nov. in the western sediment accumulation zone. Both sedimentary areas developed during the Pliocene (cf. Jipa, 1997). *Pseudocandona transylvanica* sp. nov. belongs biogeographically to the intra-Carpathian region.

An unsolved problem is the taxonomic status of the five parthenogenetic populations from Aştileu, Gilău, Hotarele and Pietroşani for which we could not find any diagnostic characters which should allow distinguishing them unambiguously from the other populations, like those of Şuncuiuş or Jiblea. We propose to assign temporarily (it means until one could find better discriminant criteria) to these populations the name of *P. eremita* (Vejdovský). This latter taxon, has to be characterised in a loosely way as *Pseudocandona eremita (sensu lato)*. We consider it a metaspecies, a concept developed for phylogenetic systematics (Donoghue, 1985) but useful in this context too. The advantage of this solution is the possibility to assign a name to the *incertae sedis* Recent and fossil populations resembling *P. eremita*.

Conclusions

(1) The present contribution allows refining the understanding of the evolutionary aspects of the *P. eremita* lineage. We hypothesise that from an epigean ancestor with a highly triangular carapace shape, a trend towards a more flat (dorsally rounded) carapace (with an umbone only slightly developed or totally absent) developed in the case of subterranean species which specialised to micro-porous interstitial life. In this respect it differs from the karstic dwelling species which predominantly display a highly triangular carapace shape.

(2) Our study offers new perspectives for the taxonomy of the species-group *P. eremita*. The best diagnostic traits are offered by the male copulatory organ and based on these, two new species *P. danubialis* sp. nov. and *P. transylvanica* sp. nov. could be recognised in Romania besides *P. eremita* and *P. serbani*.

(3) For the parthenogenetic populations here studied where no clear discriminant diagnostic trait could be found a solution of compromise is proposed, i.e. to assign them

(temporarily) to *Pseudocandona eremita* (Vejdovský, 1882) (*sensu lato*).

(4) Some of the species of the *P. eremita* group recorded in interstitial habitats apparently have circumscribed geographic distributions similar to the species existing in karstified systems.

(5) Additional investigations, using refined morphological, (palaeo-)ecological and possibly genetic approaches should allow better understanding of the evolutionary pathways toward the diversification of the group and subsequently an improvement of the taxonomy of the various parthenogenetic populations as those now assigned to the metaspecies *P. eremita*.

Acknowledgements We benefited from the scientific exchange programmes between the Austrian Academy of Sciences, the Romanian Academy of Sciences and the Polish Academy of Sciences from which financial support for repeated stays at the Institute of Limnology (Mondsee) was obtained for two of us (SI and TN). This work was supported by grants no. 446/2004 from the National University Research Council (Romania) and no. 78/2004 from the Romanian Academy of Sciences to SI. One of us (DLD) profited from the financial support of the Austrian Science Foundation (Grant FWF P17738-BO3). This work was also partly supported by EU Marie Curie Research Training Network (contract no MRTN-CT-2004-512492). The late D. Dancău and E. Şerban (Bucharest) helped DLD during 1967 to sample most of the material presented here. O. T. Moldovan (Cluj) and J. Notenboom (Amsterdam) helped SI during her sampling programme. Discussion with various colleagues, J. Linhart (Salzburg), I. Tabacaru (Bucharest), P. Marmonier (Rennes), I. Karanovic (Perth), K. Minati (Mondsee), R. Pipík (Banska Bystrica), A. Baltanás (Madrid) and the late V. Pokorný (Prague), helped the clarification of some of the issues expressed in this note. We acknowledge also the pertinent comments of two anonymous reviewers which helped to clarify several issues in the MS. Finally, we are indebted to R. Matzke-Karasz, the guest-editor of this volume, for her patience and encouragement to complete this MS.

Note After we submitted the manuscript to Hydrobiologia, we received from Dr. I. Karanovic (14.09.2006) her publication "On the genus *Typhlocypris* Vejdovský, 1882 (Crustacea: Ostracoda: Candoninae), with description of two new species" issued December 2005 in the journal Systematics and Biodiversity, vol. 3(4): 375–406. We consider unacceptable to put in synonymy species which can not be compared with the incomplete description of *Typhlocypris eremita* (Vejdovský, 1882) for which no material from the type locality was studied by this author. Dr. I. Karanovic put in synonymy *Ps. serbani* Danielopol 1982 with *T. eremita*. The drawing of her figure 41 for the details of the hemipenis is not accurate and does not capture the most important aspect, respectively the crenulated distal part of the m process which is a diagnostic character as discussed by Danielopol (1982) and in this contribution! Karanovic (2005, p. 392) gives the following unacceptable argument for the synonymy: "the Romanian populations have exactly the same carapace and hemipenis morphology as *T. eremita*." This statement is incorrect as already Danielopol (1982) pointed out and figured differences between the hemipenis of *Ps. serbani* from Jiblea and a population from Suncuius assigned at that time to *Ps. eremita*. Moreover Karanovic does not specify which male description of *T. eremita* she considered as she did not study any males of this species from Prague area. We noticed other inaccuracies too in literature quotations and/or in the limbs description.

References

Absolon, A., 1978. Die Gattung *Candona* (Ostracoda) im Quartär von Europa. Rozpravy Československé Akademie Věd, Řada Matematických a Přírodnich Věd 88: 1–76.

Baltanás, A., W. Brauneis, D. L. Danielopol & J. Linhart, 2003. Morphometric methods for applied ostracodology: tools for outline analysis of non-marine ostracodes. In Park, L. E. & A. J. Smith (eds), Bridging the Gap—Trends in the ostracode biological and geological sciences. The Paleontological Society Papers 9: 101–118.

Brauneis, W., J. Linhart & D. L. Danielopol, 2005a. Morphomatica, Computer Programme versions 1.3–1.5. Institut für Limnologie, Österreichische Akademie der Wissenschaften, Mondsee, www.oeaw.ac.at/limno/research.

Brauneis, W., A. Strake, J. Linhart, D. L. Danielopol & A. Baltanás, 2005b. Morphomatica (Version 1.5.); User Manual/Tutorial. Institut für Limnologie, Österreichische Akademie der Wissenschaften, Mondsee, www.oeaw.ac.at/limno/research.

Carbonnel, G., 1969. Les Ostracodes du Miocène rhodanien. Systématique, biostratigraphie, écologie, paléoécologie. Documents des Laboratoires de Géologie de la Faculté des Sciences de Lyon 32: 1–469.

Christiansen, K., 2005. Morphological adaptations. In Culver, D. C. & W. B. White (eds), Encyclopedia of Caves. Elsevier, Academic Press, San Diego, 386–397.

Ciampaglio, C. N., M. Kemp & D. W. McShea, 2001. Detecting changes in morphospace occupation pattern in the fossil record: characterisation and analysis of measure of disparity. Paleobiology 27: 695–715.

Clarke, K. R. & R. N. Gorley, 2001. Primer v5: Computer Programme and User Manual/Tutorial. Plymouth Marine Laboratory, Plymouth.

Clarke, K. R. & R. M. Warwick, 2001. Change in Marine Communities: An Approach to Statistical Analysis and Interpretation. Plymouth Marine Laboratory, Plymouth.

Coineau, N., 2000. Adaptations to interstitial groundwater life. In Wilkens, H., D. C. Culver & W. F. Humphreys (eds), Ecosystems of the World, Subterranean Ecosystems, Elsevier Publ., Amsterdam, 189–210.

Danielopol, D. L., 1969. Recherches sur la morphologie de l'organe copulateur mâle chez quelques ostracodes du genre *Candona* Baird (Fam. Cyprididae Baird). In Neale, J. W. (ed.), The Taxonomy, Morphology and Ecology of Recent Ostracoda. Olivier & Boyd, Edinburgh, 136–153.

Danielopol, D. L., 1978. Über Herkunft und Morphologie der süßwasserhypogäischen Candoninae (Crustacea, Ostracoda). Sitzungsberichte der Österreichischen Akademie der Wissenschaften, Mathematisch-Naturwissenschaftliche Klasse, Abteilung I, 187: 1–162.

Danielopol, D. L., 1980a. An essay to assess the age of the freshwater interstitial ostracods of Europe. Bijdragen tot de Dierkunde 50: 243–291.

Danielopol, D. L., 1980b. Sur la biologie de quelques Ostracodes Candoninae épigés et hypogés d'Europe. Bulletin du Museum National d'Histoire Naturelle de Paris 4, 2: 471–506.

Danielopol, D. L., 1982. Nouvelles donnée sur les Candoninae (Ostracoda) hypogés de Roumanie et Yougoslavie. Bulletin du Musée National d' Histoire Naturelle de Paris 4, 4: 369–396.

Danielopol, D. L., M. Creuzé des Châtelliers, F. Mösslacher, P. Pospisil & R. Popa, 1994. Adaptation of Crustacea to interstitial habitats: a practical agenda for ecological studies. In Gibert, J., D. L. Danielopol & J. A. Stanford (eds), Groundwater Ecology. Academic Press, San Diego, 217–243.

de Queiroz, K., 1998. The general lineage concept of species, species criteria, and the process of speciation. In Howard, D. J. & S. H. Berlocher (eds), Endless Forms, Species and Speciation. Oxford University Press, New York, 57–75.

Donoghue, M., 1985. A critique of the biological species concept and recommendations for a phylogenetic alternative. The Bryologist 88: 172–181.

Farkas, H. K., 1958. *Candona szöcsi* n. sp., eine neue Ostracoden-Art der *rostrata*- Gruppe aus der ungarischen Brunnen-Fauna. Zoologischer Anzeiger 160: 110–112.

Freels, D., 1980. Limnische Ostrakoden aus Jungtertiär und Quartär der Türkei. Geologisches Jahrbuch B 39: 3–169.

Gibert, J. & D. C. Culver, 2005. Diversity patterns in Europe. In Culver, D. C. & W. B. White (eds), Encyclopedia of Caves. Elsevier, Academic Press, San Diego, 196–201.

Hartmann, G., 1973. Zum gegenwärtigen Stand der Erforschung der Ostracoden interstitieller Systeme. Annales de Spéléologie 28: 417–426.

Jančařík, A., 1952. Novy nalez lasturnatky *Candona eremita* Vejd 1882. Sbornik Klubu Prirodovedeckeho v Brne 29: 232–235.

Jipa, D., 1997. Late Neogene-Quaternary evolution of Dacian Basin (Romania) an analysis of sediment thickness pattern. Geo-Eco-Marina 2: 127–134.

Klie, W., 1940. Zur Kenntnis der Ostracodenarten *Candona eremita* (Vejdowsky) und *Candona reducta* Alm. Mitteilungen über Höhlen- und Karstforschung 1: 24–29.

Krebs, C. J., 1999. Ecological Methodology. Addison Wesley Longman, Menlo Park, California.

Löffler, H., 1963. Die Ostrakodenfauna Österreichs. Sitzungsberichte der Österreichischen Akademie der Wissenschaften, Mathematisch-Naturwissenschaftliche Klasse, Abteilung I 72: 199–211.

Lutz, A.-K., 1965. Jungtertiäre Süßwasser-Ostracoden aus Süddeutschland. Geologisches Jahrbuch 82: 271–330.

Manly, B. F. J., 1998. Multivariate Statistical Methods, A Primer. Chapman & Hall, London.

Maynard Smith, J., 1987. Reply to Sober. In Dupré, J. (ed.), The latest on the best. Essays on evolution and optimality. The MIT Press, Cambridge, MA, 147–149.

Meisch, C., 2000. Freshwater Ostracoda of Western and Central Europe. In Schwoerbel, J. & P. Zwick (eds), Sü*ß*wasserfauna von Mitteleuropa 8/3, Spektrum Akademischer Verlag, Heidelberg, Berlin.

Namiotko, T. & D. L. Danielopol, 2004. Review of the *eremita* species-group of the genus *Pseudocandona* Kaufmann (Ostracoda, Crustacea), with the description of a new species. Revista Española de Micropaleontología 36: 117–134.

Pipík, R. & A.-M. Bodergat, 2004. Cyprididae (Ostracoda) du miocène supérieur du Bassin de Turiec (Slovaquie): Taxonomie et paléoécologie. Revue de Micropaléontologie 47: 225–242.

Pokorný, V., 1986. Freshwater ostracodes from the Lower Miocene of Tuchořice (Bohemia, Czechoslovakia). Acta Universitatis Carolinae-Geologica 3: 281–315.

Reeve, H. K. & P. W. Sherman, 1993. Adaptation and the goals of evolutionary research. Quarterly Review of Biology 68: 1–32.

Rohlf, F. J., 2001. Tps-dig, Program version 1.43. Department of Ecology and Evolution, State University of New York, Stony Brook, NY 8: 129–132.

StatSoft, Inc., 1997. STATISTICA for Windows, Computer Programme and Manual. StatSoft, Tulsa OK.

Vávra, W., 1891. Monographie der Ostracoden Böhmens. Archiv der naturwissenschaftlichen Landesdurchforschung von Böhmen 8: 1–116.

Vejdovský, F., 1882. Thierische Organismen der Brunnenwässer von Prag. Prag.

Wilkens, H., 2005. Neutral mutation. In Culver, D. C. & W. B. White (eds), Encyclopedia of Caves. Elsevier, Academic Press, San Diego, 411–415.

Winther, R. G., 2001. Varieties of modules, kinds, levels, origins and behaviours. Journal of Experimental Zoology 291: 116–129.

Hydrobiologia (2007) 585:181–200
DOI 10.1007/s10750-007-0637-2

OSTRACODA (ISO15)

On the origin of the putative furca of the Ostracoda (Crustacea)

Claude Meisch

Abstract The posterior end of body of the extant ostracods exhibits a pair of variously shaped appendages, commonly designated as furca(e), uropods or caudal rami, used for feeding and/or locomotion. It is here shown that the so-called furca of all extant ostracods has evolved from the (probably epipodal) vibratory plates of a pair of uropods.

The transformation comprised the following steps: (a) complete reduction of the uropodal protopodite and endopodite; (b) sclerotisation of the lateral walls of the vibratory plates; (c) transformation of the branchial filaments into spines and/or claws; (d) re-orientation of the plates from posterodorsal to posteroventral. These modifications are suggested to have evolved in parallel with a change in function, from respiratory to locomotory and/or feeding.

The most primitive condition, reminiscent of the ancestral state of character, is seen in the Platycopida: the 'furca' still appears similar in shape to the vibratory plates of the pair of sixth limbs. In the Podocopida the uropodal plates have been modified into plate-like, more often into rod-shaped rami mainly used for locomotion. In both the Platycopida and Podocopida the anus has remained in its original place, posterior to the 'furcal' plates or rami.

In the Myodocopida and Halocyprida the uropodal vibratory plates are transformed into heavily developed lamellae bearing sturdy spines. They are activated by a complex apparatus of muscles and sclerites, the development of which necessitated the displacement of the anus from the end of the body towards its present place, anterior to the 'furca'.

The furca of the Ostracoda being not a 'true' furca, a change in terminology is proposed: uropodal plates or lamellae in the Platycopida, Palaeocopida and Myodocopida/Halocyprida; uropodal rami in the Podocopida. The so-called furcae of the Ostracoda being homologous structures, it is concluded that all extant ostracods belong to a monophyletic lineage.

Keywords Crustacea · Ostracoda · Furca · Uropod · Vibratory plates · Evolution

Guest editors: R. Matzke-Karasz, K. Martens & M. Schudack
Ostracodology – Linking Bio- and Geosciences

C. Meisch (✉)
National Natural History Museum of Luxembourg, 25, rue Munster, 2160 Luxembourg, Luxembourg
e-mail: claude.meisch@education.lu

Introduction

Ostracods are one of the most diverse groups of living crustaceans, with approximately 8,000 described species (for references, see caption of

Table 1 Classification of living Ostracoda (after Horne et al., 2002; Horne, 2005)

Class Ostracoda	
Subclass Myodocopa	ca. 750 species
Order Myodocopida	
Superfamily Cypridinoidea and two other superfamilies	
Order Halocyprida	
Three superfamilies	
Subclass Podocopa	
Order Platycopida	ca. 60 species
Superfam. Cytherelloidea	
Order Podocopida	>7,000 species
Eight superfamilies	
Order Palaeocopida	
Superfam. Puncioidea	One species, *Manawa staceyi*

Species numbers are from Morin & Cohen (1991), Cohen et al. (1998) and Cohen (2005) (Oct. 10 2005, informal communication to Ostracon, the discussion list on the Ostracoda on the Internet) and Kempf (1986, 1995) for the Cytherelloidea

Table 1). Although their classification is still the subject of debate (Martens, 1992; Cohen et al., 1998; Park & Ricketts, 2003; Horne, 2005; Liebau, 2005), modern systematists generally recognize extant ostracods to belong to five lineages classified as orders (Table 1).

Ostracods have up to eight pairs of limbs in the adult stage, including male copulatory appendages. In addition, their posterior body region generally exhibits a pair of conspicuous, variously shaped structures commonly designated as furca (e.g., Müller, 1894; Klie, 1938; Hartmann, 1967; Maddocks, 1982; Parker, 1997) or caudal rami (Smith & Martens, 2000; Tsukagoshi & Parker, 2000), sometimes also, but incorrectly so, as furcae. The furcal structures are generally not considered to be limbs (Schram, 1986). (For a review of the ostracod appendages in general, the furca included, see Horne et al., 2002).

The 'furca' either consists of (a) a pair of unsegmented, rod-shaped rami bearing claws and setae used for crawling and/or digging (Podocopida) (Fig. 1); (b) a pair of membranous plates bearing terminal filaments (Platycopida). (Fig. 4a–c); or (c) a pair of lamellae carrying sturdy spines, used for feeding and/or locomotion (Myodocopida) (Fig. 5). The 'furca' is partly, rarely totally reduced in a number of taxa (e.g., Cytheroidea; Cypridopsinae of Cyprididae; some Notodromadidae).

In recent years a number of attempts have been made to homologise cephalic and postcephalic limbs throughout the Ostracoda and also limbs of ostracods with those of other crustaceans (Boxshall, 1997, 2004; Cohen et al., 1998; Horne, 2005; Newman, 2005). The so-called furca, however, was excluded from these analyses. Obviously, both the great variation in structure and the varying position of the 'furca' in relation to the anus within the different ostracod lineages have defied any attempt of homologisation.

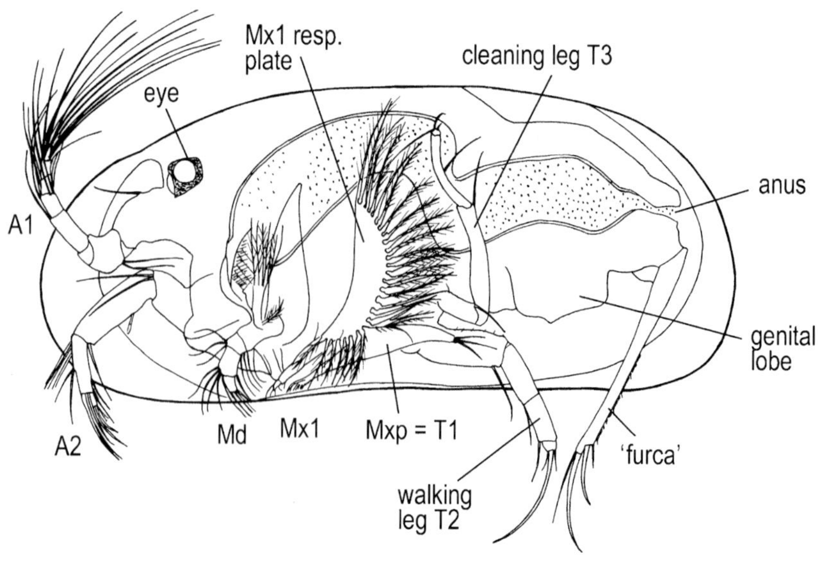

Fig. 1 Morphology of *Herpetocypris reptans* (Baird) (Podocopida, Cypridoidea, Cyprididae) showing the position of the 'furca' in relation with the remaining appendages. Female, left valve removed. Redrawn from Rome, 1947. A1, antennula; A2, antenna; Md, mandibula; Mx1, maxillula; Mxp = T1, maxilliped, first thoracopod; T2, second thoracopod; T3, third thoracopod

Here I intend to show that:

(1) the posteriormost pair of appendages of the Ostracoda is not a 'true' furca;
(2) despite the conspicuous variation in structure and position relative to the anus, the so-called furca of all living Ostracoda derives from the (probably epipodal) vibratory plates of a pair of uropods.

The evidence in favour of this hypothesis, which is here designated as the vibratory plate hypothesis, or VP hypothesis, is mainly taken from the comparative study of the morphology and structure of the 'furca' throughout the extant Ostracoda. Additionally, evidence is sought from studies of the larval development of representatives of the extant lineages and the analysis of fossil ostracods with preserved appendages.

The hypothesis presented here was foreshadowed by Hartmann (1967: 217, 220) who briefly considered the 'furcal' plates of the Platycopida to express the original condition from which the lamellar 'furca' of the Myodocopida and the 'furcal' rami of the Podocopida evolved.

For a better understanding of the following it is useful to give a brief description of the posterior end of the ostracod body and to provide definitions of the telson, the furcal rami and the uropods.

The posteriormost region of the ostracod body

Bowman (1971) discussed the presence of both the telson and the furca in the Ostracoda and the Crustacea in general. Relying on the position of the anus and the presence of an abdominal 'process' posterior to the 'furca' in a number of ostracod species, Bowman (op. cit.) suggested that:

(a) the furcal rami of the Podocopida are actually not a furca but "could well be interpreted as vestigial nonarticulated uropods";
(b) the furcal rami of the Myodocopida "may be a deeply incised telson" (pers. comm. of L. S. Kornicker to T. E. Bowman; see also Kornicker & Sohn, 1976).

Thus, following both Bowman (1971) and Kornicker & Sohn (1976) the furca of the Myodocopida and Halocyprida on the one side and that of the Platycopida and Podocopida on the other side would not be homologous structures. It should be stressed that the position of the anus in relation to the 'furca' was essential to that conclusion.

In his reply to Bowman (op. cit.), Schminke (1976) pointed out that the anus may occupy various positions in the Crustacea, the posterior end of the telson included, and that in general 'the position of the anus is not a useful and reliable character when trying to interpret the caudal region of various Crustacea'.

The telson (Fig. 2a)

The telson is classically defined as the unsegmental terminal lobe of the trunk that remains after segmentation has been completed (Moore & McCormick, 1969; Bowman, 1971; Schminke, 1976; McLaughlin, 1980; Hessler, 1982; Schram, 1986; Gruner & Scholtz, 2004). Following Schminke (1976) all Crustacea have a telson, at least in the embryological stage.

The telson being not a 'true' segment, it follows that it is devoid of limbs, paired ganglia, and mesodermal somites (Schminke, 1976).

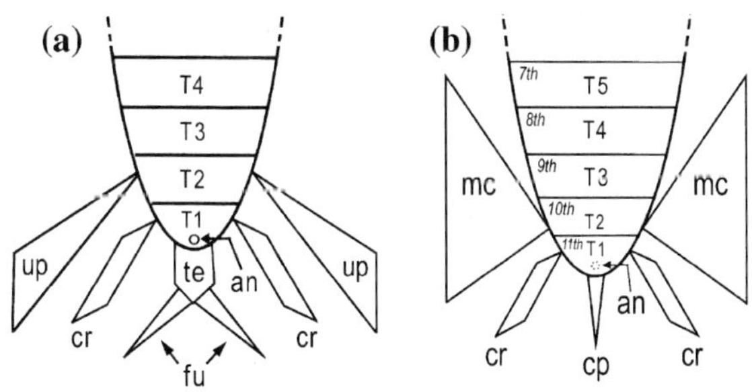

Fig. 2 Diagrams of the posterior end of body of a generalised (**a**) crustacean and (**b**) podocopid ostracod. The body plan of the Podocopida was shown to comprise 11 trunk segments (Tsukagoshi & Parker, 2000). Designation of the 'furca', caudal rami and uropods follows Schram (1986). Segments are numbered from posterior to anterior. T1 is the anal segment. The caudal process may be a telsonic structure. an, anus; cd, caudal ramus; cp, caudal process; fu, 'true' furca; te, telson; up, uropod; mc, male copulatory organ. Slightly modified from Tsukagoshi & Parker (2000)

The situation, however, becomes often more complex, when one or more of the posteriormost abdominal segments have fused with the telson, forming a complex often designated as pleotelson (e.g., McLaughlin, 1980). In that case, the limbs of that (those) abdominal segment(s) seem to originate from the telson (Schminke, 1976).

The furca (Fig. 2a)

By definition, the 'true' furca of crustaceans (also designated as caudal furca) develops from the telson. It usually consists of two rami (called furcal or caudal rami) that are directed posteriorly (Schram, 1986). As the telson is not a 'true' segment, the furcal rami have no internal musculature (Bowman, 1971; Schminke, 1976). In fact, the furca consists of two 'simple' posterior extensions of the telson.

No 'true' furca has so far been identified in ostracods. In many Podocopida, the posterior end of body bears a small 'caudal process' or 'caudal seta' which may be interpreted as being a (reduced) telsonic structure (Tsukagoshi & Parker, 2000; also extensive observations by the present author).

The position of the anus (Fig. 2)

In the course of evolution the anus of various crustaceans has migrated from its primitive position on the posteriormost somite (the anal somite) to the ventral surface of the telson (Bowman, 1971; Schram, 1986) and even the posterior end of the telson (Schminke, 1976). The latter author stressed that "the position of the anus may change within the same animal in the course of its development" and concluded that "the position of the anus is not an important feature for the definition of the telson".

Uropods and caudal rami (Fig. 2)

In general, the specialised limbs of the posteriormost crustacean body segment(s) are defined as uropods (e.g., Bowman, 1971; McLaughlin, 1980; review in Martin, 2005, 2006).

However, Schram (1986), relying on Bowman (1971), provided definitions of the uropods and caudal rami in relation with the position of the anus. Thus, uropods are a pair of "specialised appendages of the segment just anterior to the anal segment"; the appendage-like structures arising from the anal segment or near the base of the telson are designated as caudal rami, which are considered "analogs of uropods" (Schram, 1986).

No uropods have so far been clearly identified in ostracods, except by Bowman (1971; see above), Athersuch et al. (1989, following Bowman's view), and Meisch (1999, 2000). Those authors recognized the posteriormost pair of appendages of the Podocopida not to be a 'true' furca and therefore suggested them to be designated as uropods. Additionally, Swanson (1989) tentatively designated one of the abdominal pairs of appendages (but not the 'furca') of the single living punciid species (*Manawa staceyi* Swanson) as uropods.

Methods

For definitions of the crustacean appendages I rely on the extensive glossaries provided by Moore & McCormick (1969) and McLaughlin (1980) and, above all, the comprehensive compilation of definitions presented on the Web by Martin (2005, 2006).

The morphological data provided here derive from both the author's research on ostracods and the extensive study of the literature. The illustrations are redrawn from the literature. Their accuracy was carefully checked against those of various other authors.

In most ostracods, the 'furca' is similarly built in the two sexes, there being no sexual dimorphism. The main exceptions are found in the Cypridopsinae (Cyprididae, Cypridoidea), in which the 'furca' is totally reduced in the males, and the Cytheroidea, in which it is incorporated into the penis. Unless expressly mentioned, only the 'furca' of the females is considered.

The species selected for the morphological and anatomical parts of the present study have been selected as follows.

(1) Platycopida. *Cytherella sordida* G. W. Müller, 1894 is selected as a representative of its order. Müller's (1894) original drawings accurately represent the entire posterior abdomen of both sexes, showing the details of the morphology of the 'furca' and the preceding sixth pair of appendages as well.
(2) Podocopida. For seven of the eight superfamilies of this order at least one representative has been selected and its 'furca' is illustrated.
(3) Palaeocopida. The single known species of this order is so far only known from the original description and illustrations, from which all data of the present study are taken.
(4) Myodocopida and Halocyprida. The data for the Myodocopida are taken from the very detailed study of the 'furcal' apparatus of *Vargula* sp. (Cypridinidae) by Parker (1997). No similarly detailed study has been published for the Halocyprida, in which, however, the 'furca' appears similar in morphology to that of the Myodocopida.

Larval development

The data used here rely on the detailed studies of the ontogenic development of the following species of Podocopida and Myodocopida:

(a) a representative of the Bairdioidea: *Neonesidea oligodentata* (Kajiyama, 1913);
(b) two representatives of the Cytheroidea: *Loxoconcha japonica* Ishizaki, 1968 and *Uncinocythere occidentalis* (Kozloff & Whitman, 1954);
(c) several representatives of the Cytheroidea Bythocytheridae;
(d) one species of the Terrestricytheroidea: *Terrestricythere elisabethae* Horne et al., 2004;
(e) a representative of the Cypridoidea: *Eucypris virens* (Jurine, 1820);
(f) one species of Myodocopida: *Skogsbergia lerneri* (Kornicker, 1958).

For (a) and (b) see Smith & Kamiya (2002, 2003, 2005) for (c) Schornikov (1988) for (d) Horne et al. (2004) for (e) Smith & Martens (2000) and for (f) Cohen (1983).

Fossil ostracods with soft parts

Only the oldest records of fossils with preserved appendages, the 'furca' included, are considered here.

Results: morphological, developmental and palaeontological data

Platycopida (Figs. 3, 4)

This order, which is exclusively marine, comprises one single extant lineage, the Cytherelloidea, with the single family Cytherellidae (Maddocks, 1982; Horne et al., 2002; Horne, 2005). The 'furca' of the platycopids consists of a pair of lamellar

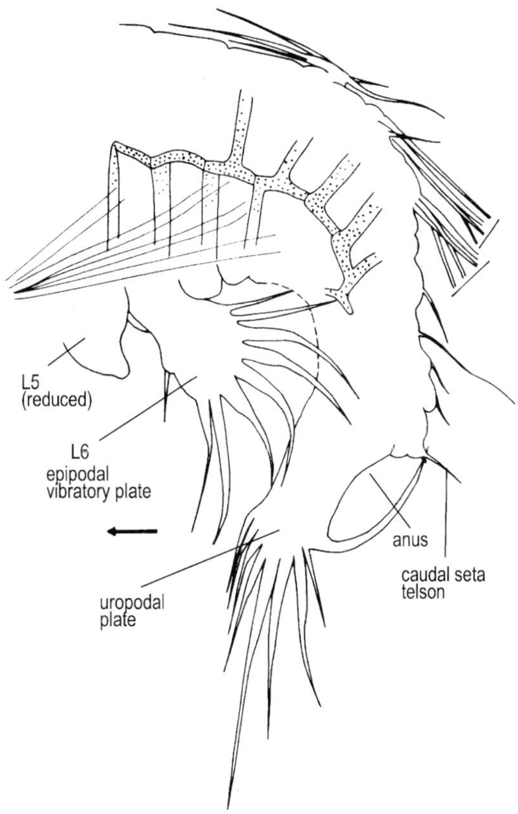

Fig. 3 *Cytherella sordida* (Platycopida, Cytherelloidea). Female. Limbs L5, L6 and posterior end of body. Slightly modified from Müller (1894)

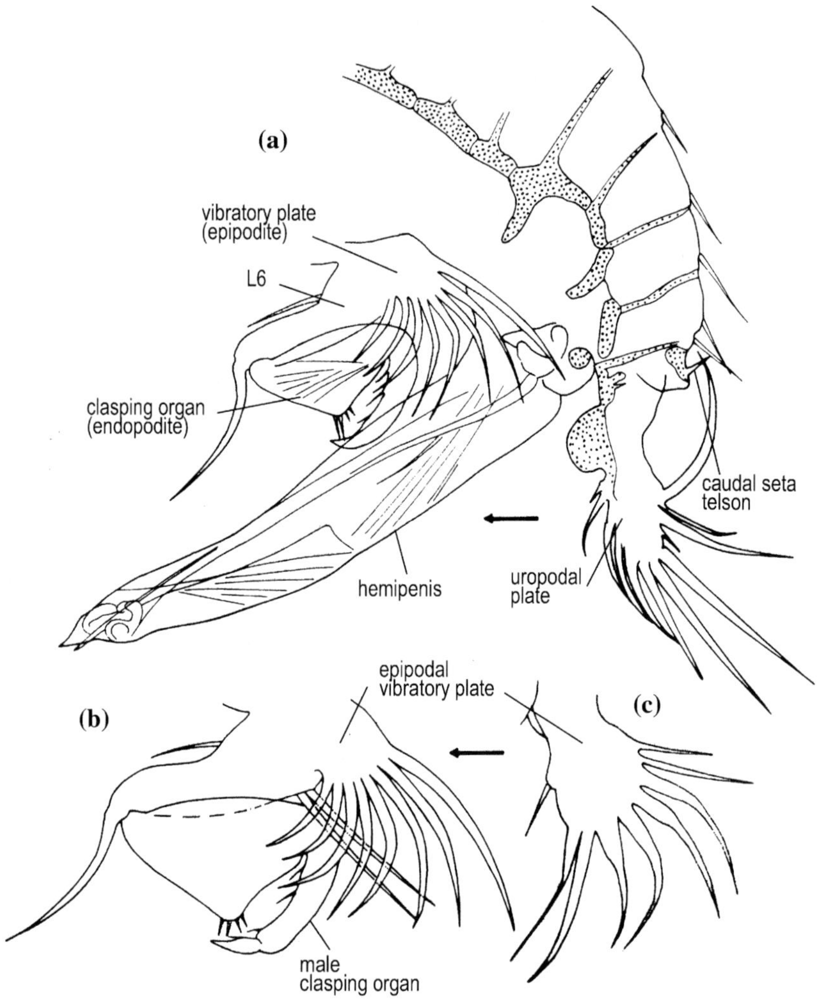

Fig. 4 *Cytherella sordida* (Platycopida, Cytherelloidea). (**a**) end of male body, (**b**) male sixth limb (L6) with clasping organ, (**c**) female sixth limb (L6): the endopodite is totally reduced. Not to scale. Redrawn from Müller (1894)

plates that bear sturdy marginal spines (Müller, 1894; Sars, 1923; Maddocks, 1982; Horne et al., 2002). The plates do not contain muscles (Müller, 1894). The anus of the cytherelloids is situated dorsal (posterior) to the 'furca' (Kornicker, 1975; Maddocks, 1982). The small abdominal process with the caudal seta situated above the anus is here considered the telson (or a remnant of the telson).

Myodocopida and Halocyprida (Fig. 5)

Each of these two orders comprises three extant, exclusively marine lineages classified as superfamilies (Horne et al., 2002; Horne, 2005).

The Cypridinidae, the 'furca' of which has recently been studied in great detail (Parker, 1997), are taken as representatives of the entire Myodocopa. Cypridinidae is the single living family of Cypridinoidea and is considered the most primitive group of myodocopans in appendage morphology (Maddocks, 1982). The 'furca' appears to be very similar in the remaining myodocopan lineages (e.g., Hartmann, 1967; Angel, 1993). It consists of two movable lamellae (also called plates) connected to basal sclerites situated within the abdomen. The two plates are activated by powerful muscles attached to the sclerites and the base of the lamellae themselves. No muscles penetrate into the lamellae. (All details of this paragraph from Parker, 1997).

The anus of both the Myodocopida and Halocyprida is situated anteroventral to the 'furca' (review and original data and illustrations in

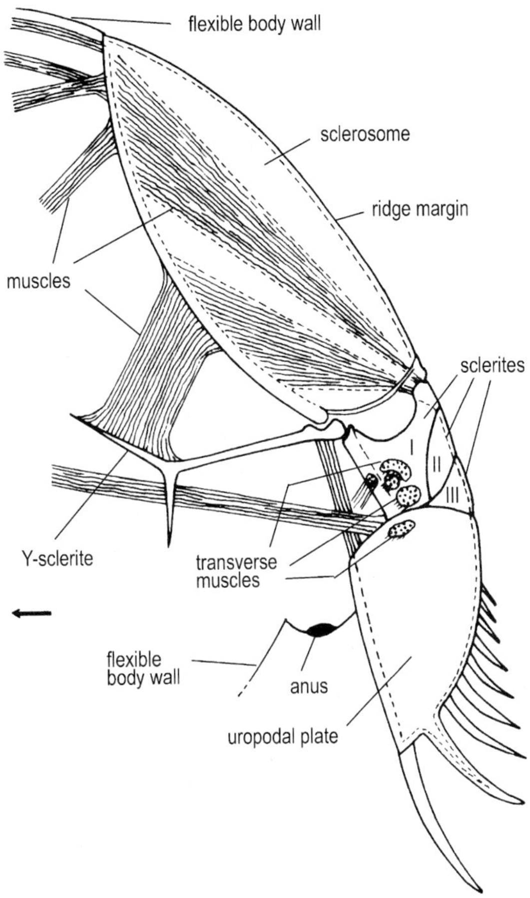

Fig. 5 *Vargula* sp. (Myodocopida, Cypridinidae). Male. Posterior end of body. Redrawn from Parker (1997)

Kornicker, 1975: 81–82; Parker, 1997). The posteriormost abdomen of the Myodocopa shows not trace of segmentation (Parker, 1997).

Podocopida (Figs. 6, 7)

It is in this group that the 'furca' appears most diverse in shape and structure. The bases of the rami are connected to strong muscles attached to chitinous supports located within the body. When the carapace closes, the rami are recurved antero-ventrally to be placed between the protopodites of the walking legs. There is a general consensus that the Podocopida comprise eight lineages classified as superfamilies (e.g., Horne et al., 2002, 2004, 2005), the 'furcae' of which are briefly described hereafter.

In the Bairdioidea (Fig. 6a), which are probably the most primitive podocopids (Maddocks, 1969, 1982; Horne et al., 2002), the rod-shaped ramus bears up to 7 setae, spines and/or claws (Maddocks, 1969).

In the Pontocypridoidea (Fig. 6d, e), the ramus is usually lamella-like and bears up to 7 setae, spines and/or claws (e.g., Athersuch et al., 1989; Maddocks, 1991; Wouters, 1997).

In the Cypridoidea (Fig. 7b, d–f), the ramus is rod-shaped, laterally compressed; it usually bears two distal claws, one distal seta and one (rarely more) posterior seta. In a number of cypridoid genera the ramus is remarkably thin, as a result of partial reduction (e.g., Cyprettinae), in still others it is reduced to a flagellum (Fig. 7e, f) (e.g., Cypridopsinae of Cyprididae; the furca is totally reduced in the males, sometimes also in the females). (For more illustrations, see Meisch, 2000).

The 'furca' of the Sigillioidea (Fig. 6b, c) is similar to that of the cypridoids (e.g., McKenzie, 1967; Maddocks, 1972; Wouters, 1988).

In the Macrocypridoidea (Fig. 7c), the 'furcal rami' are reduced and often basally fused (Maddocks, 1990).

In the three remaining podocopid lineages, the Darwinuloidea, Cytheroidea and Terrestricytheroidea, the 'furca' is either strongly or totally reduced (Fig. 7a, g). In the male cytheroids the rami are incorporated into the (hemi)penes (e.g., Meisch, 2000: Fig. 14B).

In the Podocopida the anus is situated posterior to the 'furca' (Horne et al., 2002; also extensive observations by the present author). The posteriormost region of the abdomen shows conspicuous traces of segmentation in at least one order, the Sigillioidea (Fig. 8d).

Palaeocopida

Manawa staceyi Swanson, 1989 is the only known extant species of this order, which was most diverse in the Palaeozoic. The appendages of that species, which is less than 1 mm in length, are somewhat imperfectly known. The 'furca' is well developed, it consists of two unsegmented plates with four filaments and several setae. Overall, it appears very similar to that of the cytherelloids.

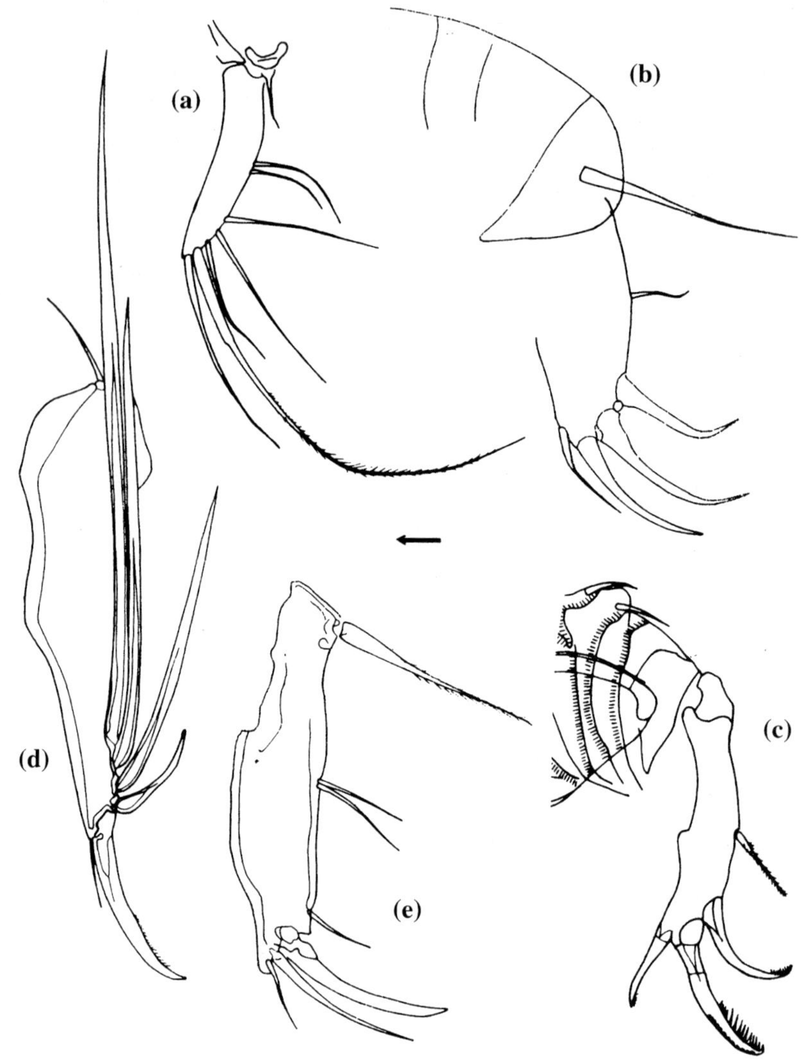

Fig. 6 Uropodal rami of (**a**) *Bairdoppilata craticericola* Maddocks (Bairdioidea), (**b**) *Saipanetta bensoni* Maddocks (Sigillioidea), (**c**) *Saipanella cloudi* McKenzie (Sigillioidea), (**d**) *Peripontocypris magnafurcata* Wouters (Pontocypridoidea), (**e**) *Pontocypris mytiloides* (Norman) (Pontocypridoidea). Not to scale. (**a**) redrawn from Maddocks (1969) (**b**) from Maddocks (1972) (**c**) from McKenzie (1967) (**d**) from Wouters (1997) (**e**) from Athersuch et al. (1989)

The position of the anus is not known with certainty. Swanson (1989: 239, 244) found "fine faecal (?) material between (furcal) lamellae", which observation led him to suggest that the anus possibly lies between the lamellae, but also to stress that "the evidence is equivocal". The posterior extremity of the body shows distinct traces of segmentation. All data of this paragraph from Swanson (1989).

Larval development

The larval development of a number species belonging to four of the eight superfamilies of the Podocopida has been studied in great detail in recent years (see list of species and references in Methods). In all these species, the 'furcal' *anlage*, throughout the different larval stages, develops into the final adult ramus without showing any trace of segmentation. Of these results, that of the bairdioid representative is stressed here, as bairdioids are considered the 'most primitive' living podocopid lineage. See also point (4) of the Discussion below.

In the single myodocopid species so far studied in detail, the 'furca' appears as a pair of unsegmented plates with claw-like filaments in the first instar and then progressively develops into the adult 'furcal' plates throughout the subsequent moults (Cohen, 1983: Fig. 9).

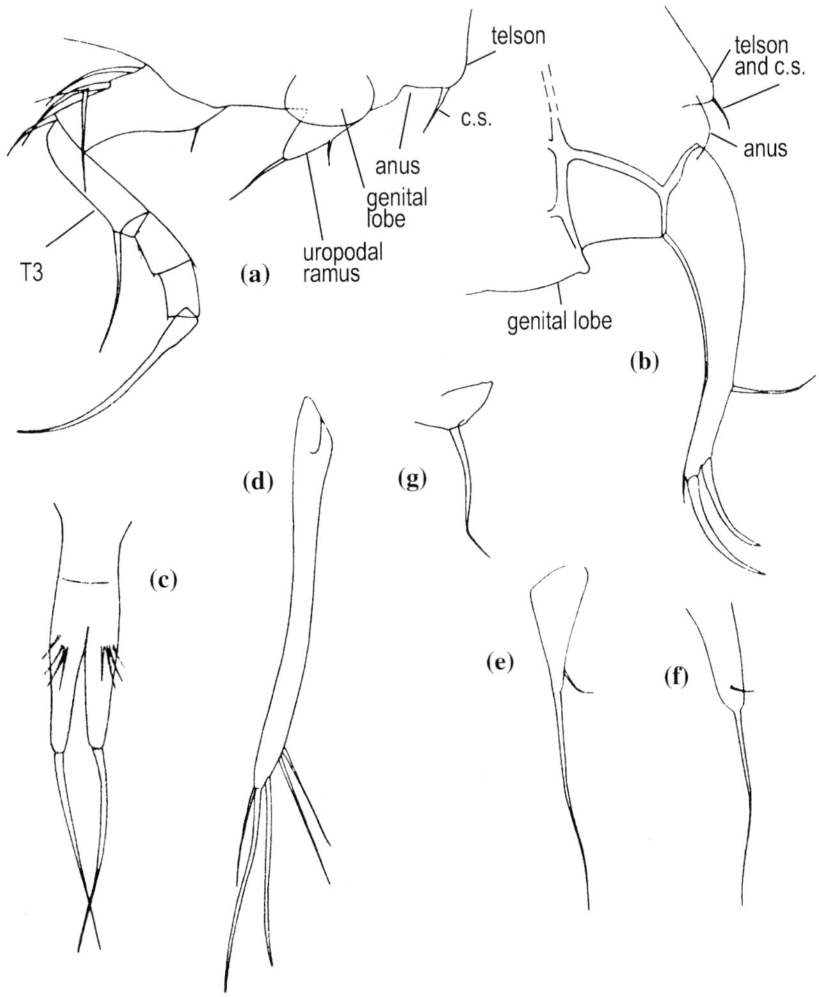

Fig. 7 Uropodal rami of (**a**) *Limnocythere inopinata* (Baird) (Cytheroidea), (**b**) *Candona lindneri* Petkovski (Cypridoidea), (**c**) *Macropyxis arta* Maddocks (Macrocypridoidea), (**d**) *Hungarocypris madaraszi* (Örley), (**e**) *Cypridopsis vidua* (Müller), (**f**) *Plesiocypridopsis newtoni* (Br. & Rob.), (**g**) *Penthesilenula brasiliensis* (Pinto & Kotzian) (Darwinuloidea). (**b**) and (**d**)–(**f**), Cypridoidea. cs, caudal seta. Not to scale. (**b**) redrawn from Petkovski (1969) (**c**) from Maddocks (1990), (**d**) from Bronshtein (1947) (**a**), (**e**) from Meisch (2000) (**g**) from Rossetti & Martens (1998)

Fossil ostracods with preserved appendages

Ostracods have a remarkably long and diverse fossil record, ranging from at least the Ordovician (Hou et al., 1996; Horne et al., 2002). Although only fossil carapaces and valves are usually found, very occasionally ostracods are preserved with their appendages and soft parts intact. Fossils with preserved soft parts have been described for both the Podocopida and Myodocopida, but are so far missing for the Platycopida (review in Smith, 2000; Siveter et al., 2003).

Podocopids with well preserved appendages have been described from Cretaceous deposits in Brazil, allowing the reconstruction of the appendages not only of the adults but also of the last three instars (Smith, 2000). Both the structure and chaetotaxy of the appendages are surprisingly similar to those of the present-day members of the Cypridoidea Cyprididae. In particular, the 'furca' is almost identical to that of the extant Cyprinotinae, the subfamily to which these fossils were assigned (Smith, 2000).

Fossil Myodocopida with preserved appendages, the 'furca' included, were reported from Lower Triassic and Upper Jurassic deposits at Spitsbergen and in Russia, respectively (Dzik, 1978; Weitschat, 1983). The appendages of these fossils, the 'furca' included, are remarkably similar to those of the living representatives of the myodocopids.

More recently, remarkable myodocopid fossils were reported from the Silurian in England (Siveter et al., 2003). The appendages of these

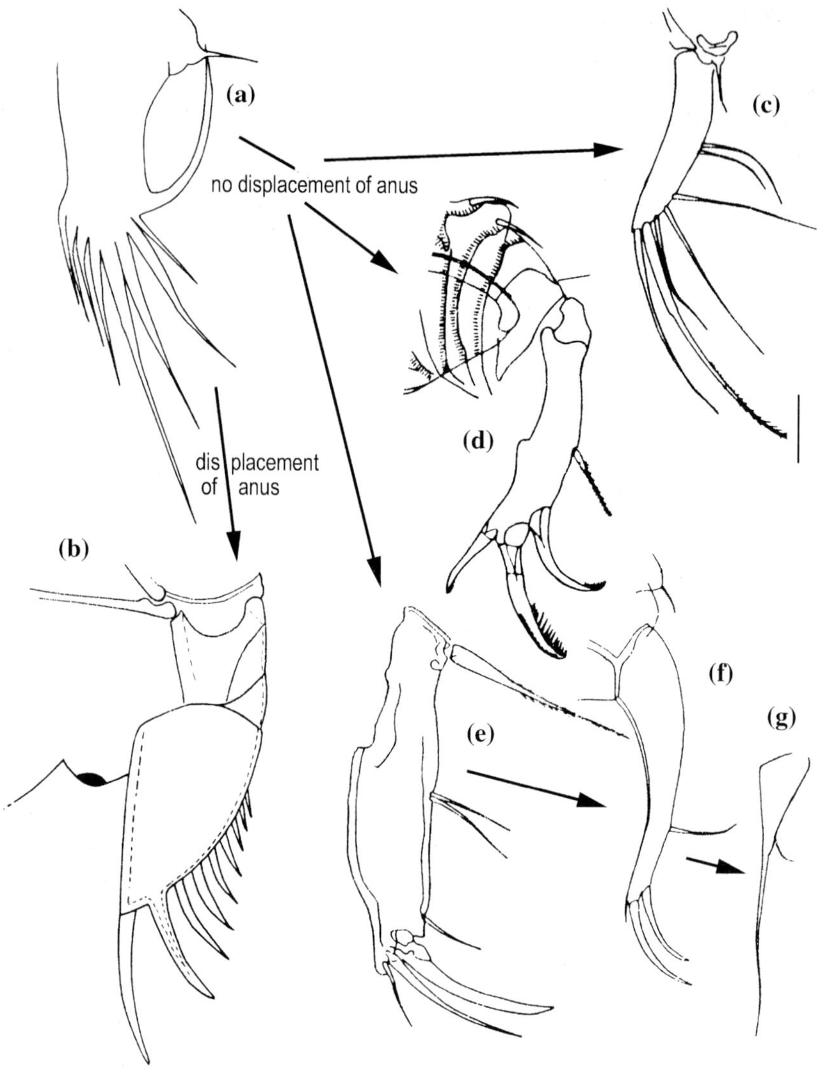

Fig. 8 Tentative evolutionary pathways of the uropodal plates and rami. (**a**) platycopid uropodal plate, (**b**) myodocopid plate, (**c**) bairdioid ramus, (**d**) sigillioid ramus, (**e**) pontocypridoid ramus, (**f**) and (**g**) cypridoid rami. Same species as in Figs. 1–5. Not to scale

fossils too, the lamellar 'furcae' included, appear strikingly similar to those of the modern representatives of their order.

From these data it appears that the morphology of the 'furca', and also that of the remaining appendages, has been conserved for over at least around 110 million years in the Podocopida and, very surprisingly, over at least around 430 million years in the Myodocopida.

Results: the vibratory plate hypothesis

The so-called 'furca' of the extant Ostracoda is here shown to have evolved from the vibratory plates of a pair of uropods.

Vibratory plates

Vibratory plates, also designated as respiratory or branchial plates, are common in extant ostracods. They are generally interpreted as epipodites (Müller, 1894; Hartmann, 1966, 1967; Maddocks, 1982; Cohen et al., 1998) (but see Horne, 2005; discussed under point 8 of the Discussion below). Vibratory plates clearly fulfil a vital respiratory function in ostracods. Their body being completely enclosed within two valves, the water that stagnates within the carapace is rapidly impoverished in oxygen and, in parallel, enriched in carbon dioxide. Against this, the vibratory plates generate a continuous water current across the body for respiration.

This certainly explains why vibratory plates are extremely common in ostracods, being found on both cephalic (mandibles and maxillules) and trunk appendages. The platycopids retain what is here believed the most primitive condition, i.e. relatively small vibratory plates on the mandibles and well developed ones on the maxillules, the fifth, and sixth limbs (e.g., Müller, 1894; Maddocks & Iliffe, 1986; Horne et al., 2002).

The musculature of vibratory plates has been relatively poorly studied. Vibratory plates are connected to muscles attached at the base of the plates (Rome, 1947 for *Herpetocypris reptans*); the large Mx1-plates of Cyprididae (Podocopida) additionally have two muscles that penetrate into the plates, being attached to the inner walls of the latter (Claus, 1892; Kesling, 1951; Horne, 2005). Muscles are also found in the vibratory plates of L5 and L6 (fifth and sixth limbs) of various Myodocopida (Cohen et al., 1998).

It is important to notice that the uropodal plates of the platycopid cytherellids (Figs. 3, 4), which are here suggested to display the primitive condition for the evolution of the uropodal plates and rami of the Podocopida, Myodocopida and Halocyprida, are devoid of inner muscles (Müller, 1894; Sars, 1923).

Cytherellidae (Platycopida)

It is here suggested that the Cytherellidae, the single extant family of the Platycopida, hold the key to the understanding of the origin and evolution of the uropodal plates and rami of the Ostracoda. The Platycopida have indeed retained an (at least relatively) primitive postero-abdominal morphology that permits the recognition of:

(a) the original condition, i.e. an ancestral uropod reduced to an epipodal vibratory plate and an anus located posteriorly to the uropodal plates (Figs. 3, 4);
(b) the evolutionary steps that have led to the morphology of the uropodal plates and rami of the extant Myodocopida/Halocyprida and Podocopida, respectively.

The VP hypothesis holds that the platycopid uropodal plates (the so-called furca) have evolved from a pair of originally 'complete' ancestral abdominal appendages (uropods) which each originally comprised a protopodite, an endopodite, and a (probably epipodal) vibratory plate. The evolutionary transformation into the uropodal plates seen in the present-day platycopids comprised the following steps:

(a) total reduction of the endopodite and protopodite;
(b) sclerotization of the lateral walls of the plates;
(c) transformation of the marginal 'respiratory filaments' into spines and/or claws;
(d) fusion of the uropodal body-segment with the telson and displacement of the two uropodal plates towards the posterior end of body;
(e) re-orientation of the epipodal plates from posterodorsal to posteroventral;
(f) development of powerful muscles within the posterior end of body that act either on the chitinous attachments (sclerites) and/or the base of the uropodal plates.

These modifications are suggested to have evolved in parallel with a change in function, from respiratory to locomotory and/or feeding.

A strong support for the transformation described above comes from a look at the male and above all the female sixth limb (L6) of *Cytherella sordida* and its relatives. That limb has indeed very obviously undergone an evolutionary transformation which appears very similar to that of the uropod.

In the male of *Cytherella sordida*, L6 consists of a protopodite, an endopodite that is transformed into a clasping organ used during copulation, and a well-developed vibratory plate directed posteroventrally (Fig. 4a, b). In the female, however, only the vibratory plate is preserved, the (unneeded) endopodite having been completely reduced (Fig. 3); by contrast, the epipodal plate is similar in development and orientation to that of the male (compare Figs. 3 and 4b, c). For more cytherellid species see for instance Maddocks & Iliffe (1986) and Horne et al. (2002).

It might be objected that the plates of L6 and the uropod ('furca') appear only superficially vibratory plate-like and that they actually could have evolved from a modified protopodite or any other part of an abdominal appendage. This hypothesis is easily rejected here for L6 and, by analogy, also for the uropodal plate. Both the position and structure of the L6-plate of the male of *Cytherella sordida* clearly point towards its origin from a vibratory plate. Indeed, the male L6-plate cannot derive from a protopodite nor an endopodite, as both of these are present and well developed in the male (with the endopodite being modified into a clasping organ) (Fig. 4a, b).

Myodocopida (Fig. 5)

The myodocopid uropodal lamellae appear similar in structure to those of the cytherelloids. They are here hypothesized to have derived from platycopid or platycopid-like vibratory plates.

The position of the myodocopid/halocyprid anus, situated ventral (anterior) to the uropodal plates, needs to be explained. It is suggested here that in the course of evolution the anus of myodocopans has been displaced from the original position (posterior to the uropodal plates) towards the antero-ventral position it holds in modern forms (Fig. 5).

The secondary displacement of the anus is easily explained by functional needs. Indeed, in the course of evolution, the myodocopid/halocyprid uropodal plates have become extremely powerful tools for feeding and movement (e.g., Hartmann, 1967; Cohen, 1983; review in Parker, 1997). They therefore need to be activated by strong muscles attached to well developed sclerotized plates, of which the so-called sclerosome (Fig. 5) is the most important (Parker, 1997). It is suggested here that the development of these heavily sclerotized structures, mainly the sclerosome, at the posterior end of body, where the anus was originally situated, necessitated the displacement of the anus to its present, antero-ventral position. The complex apparatus of muscles and sclerites connected to the uropodal plates is suggested here to have originated from two or more abdominal segments. The development of this apparatus is suggested to have occurred in parallel to the fusion of the corresponding abdominal segments. This view is supported by the complete absence of any trace of segmentation of the posterior extremity of the myodocopid abdomen (Fig. 5). The fusion of segments is suggested to have facilitated the ventral displacement of the anus from the rear end of body towards its position seen in the modern representatives of the subclass.

Podocopida (Figs. 6, 7)

Podocopids show the most derived condition: the platycopid-like plates, which lack muscles, have been transformed into laterally flattened rami bearing usually three distal claws and/or setae and 1–4 posterior setae or claws.

The transformation of platycopid-like plates into the podocopid rami comprised the following steps:

(a) transformation of the 'respiratory filaments' into claws and/or setae; the maximum number of claws/setae is found on the rami of the bairdioids; as already mentioned above, the bairdioids, on the base of other evidence, are considered the most primitive podocopids;
(b) reduction of the number of claws/setae in the non-bairdioid lineages of Podocopida;
(c) overall reduction of the entire uropodal rami in a number of podocopid lineages (e.g., Macrocypridoidea, Cytheroidea, Cypridopsinae of Cypridoidea).

This somewhat surprising transformation is supported by

(a) the still lamella-like appearance of the rami found in some podocopids, such as for instance the pontocypridoid genus *Pontocypris* Wouters (Fig. 6d). This genus is characterized by a number of other unique features, which may similarly be interpreted as plesiomorphic (Wouters, 1997);
(b) the absence of any segmentation of the rami in all known podocopids;
(c) the complete absence of muscles within the rami; the powerful muscles acting on the rami are attached to the base of the rami and their chitinous supports.

Discussion

The 'furca' of the Ostracoda is not a true furca

The 'furcal' plates and rami of ostracods are connected to powerful muscles which are attached to chitinous sclerites and rod-shaped supports (for the myodocopids see Fig. 5; for the podocopids see Rome, 1947, 1969). Both those muscles and chitinous structures have necessarily evolved from one or more 'true' somites or body segments, not from the telson, which lacks any mesodermal tissue. Furthermore, the 'furcal' rami of the podocopids usually bear articulated distal claws (Figs. 6, 7), which develop from setae or setae-like structures of 'true' appendages, not from telsonic structures.

It might be objected that the telson could have fused with one or more of the posteriormost abdominal segments which would have contributed the muscles to a true 'furca'. However, it appears highly implausible that a complex muscular and chitinous apparatus would have engaged into an elaborate connection with a telsonic structure which embryologically does not depend on mesodermal tissue. Furthermore, the development, from a telsonic structure, of articulated distal claws similar to those of true appendages is highly improbable.

From all this it is here concluded that the 'furca' of the Ostracoda does not originate from the telson and is therefore not a 'true' furca as defined above.

The putative furca is not derived from the endopodites of a pair of uropods

The putative furca of many podocopids appears somewhat similar in structure to the endopodite of a thoracic limb, with two terminal claws and one terminal seta (e.g., Figs. 6e, 7b) and could therefore have been derived from the endopodite of an abdominal limb. This hypothesis is rejected here for the following reasons:

(a) the 'furcal' rami of the Podocopida never show any trace of segmentation (Klie, 1938; Hartmann, 1966; also extensive observations of the present author); although fusion of endopodal segments is common in ostracods (see for instance the 4-segmented thoracic endopodites of Cytheroidea), the fusion is always restricted to a limited number of segments;

(b) the podocopid 'furcal' rami are devoid of any muscles (e.g., Rome, 1947 for *Herpetocypris reptans*; Hartmann, 1967; also extensive observations of the present author). Muscles, even unstained, are usually easily recognised within the podomeres of ostracod appendages. It appears therefore highly improbable that muscles would have escaped observations by so many authors.

For the same reasons, the derivation of the lamella-like structure of the locomotory platycopid and myodocopan 'furca' from the endopodites of a pair of abdominal limbs appears even more implausible.

The putative furca is not derived from the protopodites of a pair of uropods

The podocopid 'furcal' ramus looks somewhat similar to a limb (uropod) protopodite, the endopodite, exopodite and epipodite of which would have been reduced in the course of evolution. This hypothesis is rejected here for the following reason: the absence of muscles within the modified protopodite; indeed, it appears very implausible that these muscles would have been totally reduced in an appendage with an important locomotory function; the evolutionary scenario compatible with a protopodite origin would be that first any locomotory function of the protopodite (or any function necessitating internal muscles) would have been lost, followed by the reduction of all muscles, but that the locomotory function was later acquired again; this scenario is here considered implausible (and unparsimonious).

Larval development of the Podocopida and Myodocopida

The study of the larval development may reveal the 'true' nature, i.e. the evolutionary origin of at least part of the appendages of crustaceans in general, and ostracods in particular, as illustrated by the recent discussion on the nature of the fifth limb (L5) of cypridids, and cypridoids in general: the adult L5 appears maxilla-like in shape and function, and for decades it was therefore (and also for other 'evidence') regarded as homologous to the maxilla of the other crustaceans. However, the L5 of the instar A-4 of *Eucypris virens* is a 4-segmented leg similar to an adult thoracopod, to be progressively reduced and modified, throughout the subsequent moults, into the maxilla-like adult appendage (Smith & Martens, 2000). From this and other evidence unrelated to postembryonic development, it was concluded that the L5 of the adult Cyprididae (and Cypridoidea in general) represents the first thoracopod that had been modified into a maxilliped (Smith & Martens, 2000; see also Meisch, 1996, 2000; Meisch et al., in prep.).

A similar reasoning may be applied to the uropodal appendage.

It is stressed here that in the five podocopid species mentioned above, the uropodal *anlage*, throughout the different larval stages, develops into the final adult ramus without showing any trace of segmentation (see Results), as is expected for a structure originating from an epipodal plate. Of these results, that of the bairdioid representative is stressed here, as bairdioids are considered the most primitive living podocopid lineage (see above).

In at least a number of Cytheroidea and also in Terrestricytheroidea the 'furcal' appendage changes in morphology and function during ontogeny. For example, in the early instars A-7 to A-5 of *Terrestricythere elisabethae* Horne et al., 2004, the 'furca' consists of two broad plate-like rami, each bearing 3 or 4 claws, to become progressively reduced in the subsequent instars. This transformation occurs in parallel with a change in function: the 'furca' in instars A-7 to A-5 is used as a locomotory structure. In the subsequent stages, when first the fifth and then the sixth limbs (L5 and L6) have become functional legs, the 'furcal' rami of the female are progressively reduced into two small claws, those of the male into three setae. (All data from this paragraph from Horne et al., 2004).

The latter data are here suggested to support the rejection of the possible origin of the podocopid 'furcal' rami from the endopodites of a pair of abdominal appendages (uropods). Indeed, if this would be true, then, in those species in which the 'furcal rami' have a locomotory function in the early instars, they would be expected to show up as segmented larval rami; the segmentation would then disappear in parallel to the reduction of the endopodal ramus and together with the loss of its locomotory function, similarly to what occurs in the L5 of the Cypridoidea (see above).

In the single myodocopid species so far studied in detail, the uropodal appendage is an unsegmented plate throughout the entire larval development (see Results).

It is here concluded that the results obtained from the ontogeny of both the Podocopida and Myodocopida support, or at least do not contradict, the vibratory plate-origin of the uropodal rami and plates (see also the remark on the possible falsification of the VP hypothesis below).

Fossils with soft parts

No information regarding the origin of the 'furca' can be retrieved from the study of fossil ostracods. Indeed, the oldest known fossils with preserved appendages already possess 'furcal' structures that are similar to those of the modern representatives of their lineage.

The position of the anus

The VP hypothesis provides an explanation for the so far unsolved problem of the position of the anus in the three extant ostracod lineages. In both the Platycopida and Podocopida the anus has remained in its original place, posterior to the uropodal lamellae, as there was no functional need for a displacement. This view is supported by the presence of distinct traces of segmentation of the posterior abdomen in at least a number of Podocopida; the anus is situated on the posteriormost segment (somite), which has fused with the telson (Fig. 2b) (Tsukagoshi & Parker, 2000).

In the Myodocopida/Halocyprida the anus has been secondarily displaced for functional reasons (see above). The following two possibilities may be considered: (a) the anus has remained on the original (anal) segment but has shifted to the anterior end of that segment, anteriorly to the uropodal lamellae; (b) the anus has shifted from the anal segment to the segment situated in front of the latter; that displacement could have been made possible, or at least facilitated, by the fusion of two (or more) of the terminal body segments. That fusion is supported by the complete absence of any trace of segmentation in the posterior end of the myodocopan body (Fig. 5) (Parker, 1997).

Alternatively, one might consider the position of the myodocopan anus, anterior to the uropodal plates, to be the primitive condition, while the platycopids and podocopids would display the derived condition, their anus having been displaced towards the rear end of body.

In that case, the myodocopans, which possess the most derived uropodal condition (fusion of segments with total reduction of segmentation, a complex apparatus of muscles and sclerites), would have retained the primitive anal condition. By contrast, the platycopids, which possess a far less complex uropodal apparatus (preserved segmentation of the end of body, moderately transformed uropodal vibratory plates), would have acquired the derived anal condition, the anus having been shifted towards the rear end of body. This scenario is considered implausible and is therefore rejected here.

Evolutionary pathways (Fig. 8)

The two main evolutionary pathways leading from platycopid-like uropodal plates towards those of the myodocopans on the one side and the podocopids on the other side are tentatively illustrated in Fig. 8. The pathway towards the myodocopids/halocyprids necessitated the displacement of the anus, while in the pathway towards the podocopids the anus remained in its original place. It should be stressed that these pathways do not express phylogenetic relationships.

Monophyly versus polyphyly of the Ostracoda

Recently, the monophyly of the Myodocopa and Podocopa has been questioned on the base of both morphological (Horne, 2005; Horne et al., 2005) and molecular data (references in Horne et al., 2005). However, no clear support for the possible polyphyletic origin of the myodocopans and podocopans was found (Horne et al., 2005), except by Horne (2005). From the detailed study of the comparative morphology and musculature of the branchial (vibratory) plates in relation with the protopodites, the latter author (Horne, 2005) concluded that the vibratory plates in myodocopan ostracods are epipodites arising from the coxa or precoxa; by contrast, his study led him to "regard it as more probable that the branchial plates of podocopans are modified exopodites"; therefore the branchial plates of the two lineages would be the results of convergent evolution, being homoeomorphic rather than homologous structures. Although the 'furca' was excluded from that study, most plausibly the vibratory plates associated with the uropods originated from the same structures (epipodites or exopodites) as those of the remaining limbs in each of the two lineages (Podocopa and Myodocopa). From his results, Horne (op. cit.) logically inferred the Ostracoda to be probably a polyphyletic group.

The extensive discussion of Horne's (2005) findings is beyond the scope of the present work. However, it must be pointed out that, to be valid for both the podocopans and myodocopans, the VP hypothesis requires the uropodal rami and plates to derive from homologous structures, i.e. either from epipodites or exopodites in both lineages.

By contrast, the VP hypothesis supports the monophyly of the extant Ostracoda. Indeed, it appears highly improbable that the complex modifications described here would have evolved independently in separate lineages. See Cohen et al. (1998) and Park & Ricketts (2003) for independent support of the monophyly of the ostracods.

Phylogeny of the extant Ostracoda

The VP hypothesis throws new light on the phylogeny of the extant ostracod lineages.

First, it confirms the basal position of the Platycopida. Their moderately transformed uropodal plates, which appear still similar to branchial plates and probably still fulfil that function, are, according to the VP hypothesis, plesiomorphic in relation to those of the podocopids and myodocopans. This adds another plesiomorphic character to those already known in the platycopid appendages, which are: (a) a distinctly biramous antenna (exopodite and endopodite almost similarly developed); (b) vibratory plates present on all postantennal appendages (the mandibule, maxillule, and the male fifth and sixth limbs; (c) a body with 11 recognisable trunk segments, telson included (Schulz, 1976; Tsukagoshi & Parker, 2000). Also, the valves display a number of plesiomorphic characters, which led Jaanusson (1985) to characterize the extant platycopids as 'living fossils... [with] arrested evolution'. However, the platycopids may not be regarded as having retained an overall primitive condition. As all animals, they are best characterised as mosaics of both plesiomorphic and apomorphic (derived) characters. The most conspicuous of the latter are (a) well developed filter combs on the mandibulae and maxillulae; (b) the absence of any walking legs, due to the partial reduction of the limbs L5 and L6, and the loss of the subsequent limb (L7).

Second, the enigmatic Puncioidea: the uropodal plates of *Manawa staceyi* appear similar to those of the platycopids and hence must be considered plesiomorphic too. The position of the anus remains imperfectly known (see above). As the puncioids have retained relatively primitive uropodal plates and, unlike the Myodocopida, do not have chitinous plates at the rear end of body, the VP hypothesis rather predicts the puncioidean anus to be located posterior to the uropodal lamellae. However, if, as presumed by Swanson (1989), the puncioid anus will be confirmed to be situated between the uropodal plates, this could be interpreted as expressing the transition between the Platycopida/Podocopida on the one and the Myodocopida on the other side and would provide a further and major support of the VP hypothesis.

Interestingly, the puncioids, which are now generally considered to belong to the palaeocopids (Horne et al., 2002; Horne, 2005), were originally assigned to the Platycopida (Swanson, 1989), to be again placed within the Platycopida in a recently published classification of the Crustacea, the Ostracoda included (Martin & Davis, 2001; the puncioid 'decision' is said to follow a personal communication by R. Whatley).

Terminology: uropods versus caudal rami

Recently, Tsukagoshi & Parker (2000; Fig. 11) recognized the 'furca' of the Podocopida to depend on the anal segment. Applying Schram's (1986) definitions (see Introduction), they consequently designated the 'furcal' rami as 'caudal rami' (Fig. 2b).

However, designations of appendages in relation with the position of the anus makes the distinction of caudal rami and uropods (following Schram, 1986, the latter originate from the segment anterior to the anal segment) at least uneasy in those cases where the anus has shifted posteriorly (to the telson) or anteriorly, as suggested here for the Myodocopa.

Furthermore, although the term 'caudal rami' does not necessarily imply a connection with or a developmental origin from the telson (Schram, 1986), it is nevertheless generally considered synonymous with the terms 'furcal rami', 'caudal furca', or 'furca' (e.g., McLaughlin, 1980; Schram, 1986; Martin, 2005).

Therefore, in order to avoid any confusion in limb terminology (furca versus caudal rami and uropods) and homology it is here suggested (a) to follow those authors who adopt a broader definition of the term 'uropod' (e.g., McLaughlin, 1980): any of the modified posteriormost pair of abdominal limbs; (b) to designate the 'furcal' rami and plates of the Ostracoda as modified uropods. In particular, the following designations are proposed.

As in the Platycopida, Palaeocopida and Myodocopida/Halocyprida the plates have retained the original lamellar shape, they are here suggested to be designated as uropodal plates or uropodal lamellae. In the Podocopida, the uropodal vibratory plates have been modified into rami and are therefore suggested to be designated as uropodal rami.

Evolutionary potential of the crustacean epipodites

The transformation of a pair of vibratory plates into a pair of variously shaped structures adapted for locomotion and/or feeding demonstrates the powerful evolutionary potentiality of the epipodites of ostracods. And ostracods are not the only crustaceans in that case. There is currently strong evidence from both morphological and genetic data that the wings of insects, which are now thought to descend from an early crustacean lineage, originated from the modification of a pair of epipodites (references in Boxshall, 2004). In this context, it is interesting to mention the possible phylogenetic relationship between Ostracoda and Insecta (Newman, 2005).

Falsification of the vibratory plate hypothesis

The VP hypothesis can be falsified by any of the following observations in ostracods:

(a) the finding of a segmentation or trace of segmentation in the uropodal plates and/or rami of adults and, above all, juveniles;
(b) the finding of any muscles (deeply) penetrating into the uropodal lamellae or rami;
(c) the finding of any phylogenetic relationship incompatible with the evolutionary pathways proposed here.

Open questions

The hypothesis presented here leaves many questions open to future research. For example:

(a) the morphology and anatomy (uropodal muscles and basal sclerites) of the posterior abdomen of the cytherelloids (Platycopida) remain imperfectly known and need to be (re)studied in detail;
(b) similarly, the detailed study of the structure of the posterior end of body of *Manawa staceyi* Swanson, the only living puncioid species, could throw new light on the hypothesis presented here;
(c) the uropodal attachments of the Podocopida could at least partly be homologous with those of the platycopids and myodocopids. Parker (1997: 647) pointed out the similarity between the mode of attachment of a 'muscular tendon' of the uropodal plates of the myodocopid *Vargula* sp. and the extensor muscles of the uropodal rami of the podocopan *Herpetocypris reptans*. This possible homology, if confirmed, would provide further support for the common origin and homology of the myodocopan uropodal plates and the podocopan uropodal rami proposed here.

Conclusions

The so-called furca of the Ostracoda is not derived from the telson and it is therefore not a 'true' furca. It has evolved from the (probably epipodal) vibratory plates of a pair of homologous uropods. Despite their remarkable morphological diversity, the uropodal lamellae and rami in all ostracod lineages are therefore homologous structures. The platycopid Cytherelloidea have retained a postero-abdominal morphology which comes close to the primitive condition: a pair of uropodal plates that remain morphologically similar to vibratory plates; the anus has remained in its primitive position, posterior to the uropodal plates. The plesiomorphic structure of the uropodal ('furcal') plates is an additional primitive (plesiomorphic) character retained by the extant platycopids.

In the myodocopans the uropodal vibratory plates are modified into powerful lamellae with sturdy spines used for locomotion and/or feeding. The evolutionary development of strong sclerites and the sclerosome for attachment of muscles necessitated the anteroventral displacement of the anus.

In the Podocopida, the uropodal plates are transformed into lamella-like, more often into rod-shaped rami that are partly or totally reduced in some lineages. The anus has preserved its original position posterior to the uropodal rami.

The vibratory plate hypothesis has important bearings on the phylogeny and classification of the Ostracoda:

(a) extant ostracods are confirmed to form a monophyletic group. It appears indeed highly improbable that the complex transformation described here could have evolved independently in separate crustacean lineages;
(b) the Platycopida, which have retained the most primitive uropodal plates and posterior end of body, are likely either to be the ancestors of the Podocopida and Myodocopa or at least to occupy a position close to the common ancestors of those orders.

Finally, the following features of the VP hypothesis are stressed here:

(a) it is parsimonious (the same 'furcal' origin for the two main ostracod lineages);
(b) it can be tested with (new) palaeontological, phylogenetical, ontological, molecular, and morphological data and might therefore stimulate future research.

Acknowledgments I am very grateful to David J. Horne and Koen Martens for their critical comments and suggestions, which helped me to improve the manuscript. I thank Karel Wouters for extensive discussions on the morphology of the uropodal rami and plates throughout the Ostracoda. I am grateful to Georges Bechet, director of the National Natural History Museum of Luxembourg, for his constant support of my research activities. The present work is part of the research project ACU/06/01 of the Luxembourg Ministry of Culture, Higher Education and Scientific Research.

References

Angel, M. V., 1993. Marine planktonic ostracods. Synopses of the British Fauna, Vol. 48. E. J. Brill, Leiden.

Athersuch, J., D. J. Horne & J. E. Whittaker, 1989. Marine and Brackish Water Ostracods Synopses of the British Fauna, Vol. 43. E. J. Brill, Leiden.

Bowman, T. E., 1971. The case of the nonubiquitous telson and the fraudulent furca. Crustaceana 21: 165–175.

Boxshall, G., 1997. Comparative limb morphology in major crustacean groups: the coxa-basis joint in postmandibular limbs. In Fortey, R. A. & R. H. Thomas (eds), Arthropod Relationships. Systematics Association Special Volume Series 55, 155–167.

Boxshall, G., 2004. The evolution of arthropod limbs. Biological Reviews 79: 253–300.

Bronshtein, Z. S., 1947. Faune de l'URSS. Crustacés, Vol. 2: no 1: Ostracodes des eaux douces. Zoologicheskiy Institut Akademii Nauk SSSR, n.s. 31 (English translation: Freshwater Ostracoda, Z.S. Bronshtein, 1988, Oconian Press, New Delhi. Reprint published by the U.S. Department of Commerce, National Technical Information Service, Springfield, Virginia).

Claus, C., 1892. Beiträge zur Kenntnis der Süßwassserostracoden. I. Über den Körper- und Gliedmassenbau der Cypriden nebst Bemerkungen über einzelne innere Organe derselben. Arbeiten aus dem zoologischen Institut der Universität Wien und der zoologischen Station in Triest 10: 147–216.

Cohen, A. C., 1983. Rearing and postembryonic development of the myodocopid ostracode *Skogsbergia lerneri* from coral reefs of Belize and the Bahamas. Journal of Crustacean Biology 3: 235–256.

Cohen, A. C., J. W. Martin & L. S. Kornicker, 1998. Homology of Holocene ostracode biramous appendages with those of other crustaceans: the protopod, epipod, exopod and endopod. Lethaia 31: 251–265.

Dzik, J., 1978. A myodocopid ostracode with preserved appendages from the Upper Jurassic of the Volga River region (USSR). Neues Jahrbuch für Geologie und Paläontologie, Monatshefte 7: 393–399.

Gruner, H.-E., & G. Scholtz, 2004. Segmentation, tagmata, and appendages. In Forest, J. & J. C. von Vaupel Klein (eds), The Crustacea, revised and updated from Traité de Zoologie, Vol. 1. Brill, Leiden, Boston, 13–57.

Hartmann, G., 1966. Dr. H.G. Bronns Klassen und Ordnungen des Tierreichs. In Gruner, H. E. (ed.), Fünfter Band: Arthropoda, 1. Abt.: Crustacea, 2. Buch, IV. Teil, 1. Lieferung (1966). Akademische Verlagsgesellschaft Geest & Portig K.-G., Leipzig, 1–216.

Hartmann, G., 1967. Dr. H. G. Bronns Klassen und Ordnungen des Tierreichs. In H. E. Gruner (ed.), Fünfter Band: Arthropoda, 1. Abt.: Crustacea, 2. Buch, IV. Teil, 2. Lieferung (1967). Akademische Verlagsgesellschaft Geest & Portig K.-G., Leipzig, 217–408.

Hessler, R. R., 1982. Evolution within the Crustacea. Part 1: General: Remipedia, Branchiopoda, and Malacostraca. In Abele, L. G. (ed.), The Biology of Crustacea, Vol. 1: Systematics, the Fossil Record, and Biogeography. Academic Press, 150–185.

Horne, D. J., 2005. Homology and homoeomorphy in ostracod limbs. Hydrobiologia 538: 55–80.

Horne, D. J., A. Cohen & K. Martens, 2002. Taxonomy, morphology and biology of Quaternary and living Ostracoda. In Holmes, J. A., A. R. Chivas (eds), The Ostracoda: Applications in Quaternary Research. American Geophysical Union Washington DC, Geophysical Monograph 131, 5–36.

Horne, D. J., I. Schön, R. J. Smith & K. Martens, 2005. What are Ostracoda? A cladistic analysis of the extant superfamilies of the subclasses Myodocopa and Podocopa (Crustacea: Ostracoda). In Koenemann, S.

& R. A. Jenner (eds), Crustacea and Arthropod Relationships. Taylor & Francis, 249–273.

Horne, D. J., R. J. Smith, J. E. Whittaker & J. W. Murray, 2004. The first British record and a new species of the superfamily Terrestricytheroidea (Crustacea, Ostracoda): morphology, ontogeny, lifestyle and phylogeny. Zoological Journal of the Linnean Society 142: 253–288.

Hou, X., D. J. Siveter, M. Williams, D. Walossek & J. Bergström, 1996. An early Cambrian bradoriid arthropod from China with preserved appendages: its bearing on the origin of the Ostracoda. Philosophical Transactions of the Royal Society of London, Series B 351: 1131–1145.

Jaanusson, V., 1985. Functional morphology of the shell in platycope ostracodes – a study of arrested evolution. Lethaia 18: 73–84.

Kempf, E. K., 1986. Index and Bibliography of Marine Ostracoda. Part 1: Index A. Sonderveröffentlichungen des Geologischen Instituts der Universität zu Köln 50.

Kempf, E. K., 1995. Index and Bibliography of Marine Ostracoda. Part 6: Index A Supplement 1. Sonderveröffentlichungen des Geologischen Instituts der Universität zu Köln 100.

Kesling, R. V., 1951. The Morphology of Ostracod Molt Stages. Illinois Biological Monographs 21.

Klie, W., 1938. Krebstiere oder Crustacea III: Ostracoda, Muschelkrebse. Die Tierwelt Deutschlands und der angrenzenden Meeresteile nach ihren Merkmalen und nach ihrer Lebensweise 34.

Kornicker, L. S., 1975. Antarctic Ostracoda (Myodocopina). Smithsonian Contributions to Zoology 163.

Kornicker, L. S. & I. G. Sohn, 1976. Phylogeny, Ontogeny, and Morphology of Living and Fossil Thaumatocypridacea (Myodocopa: Ostracoda). Smithsonian Contributions to Zoology 219.

Liebau, A., 2005. A revised classification of the higher taxa of the Ostracoda (Crustacea). Hydrobiologia 538: 115–137.

Maddocks, R. F., 1969. Revision of Recent Bairdiidae (Ostracoda). United States National Museum Bulletin 295.

Maddocks, R. F., 1972. Two new living species of *Saipanetta* (Ostracoda, Podocopida). Crustaceana 23: 28–42.

Maddocks, R. F., 1982. Ostracoda. In Bliss, D. E. (ed.), The Biology of Crustacea, Vol. 1, Systematics, the Fossil Record, and Biogeography. Academic Press, New York.

Maddocks, R. F., 1990. Living and Fossil Macrocyprididae (Ostracoda). The University of Kansas Paleontological Contributions, Monograph 2.

Maddocks, R. F., 1991. Revision of the family Pontocyprididae (Ostracoda), with new anchialine species and genera from Galapagos Islands. Zoological Journal of the Linnean Society 103: 309–333.

Maddocks, R. F. & T. M Iliffe, 1986. Podocopid Ostracoda of Bermudian caves. Stygologia 2: 26–76.

Martens, K., 1992. On *Namibcypris costata* n. gen., n. sp. (Crustacea, Ostracoda, Candoninae) from a spring in northern Namibia, with the description of a new tribe and a discussion on the classification of the Podocopina. Stygologia 7: 27–42.

Martin, J. W., 2005, 2006. Crustacean Glossary. Website of the Natural History Museum of Los Angeles County: http://crustacea.nhm.org/glossary/browse.html.

Martin, J. W. & G. E. Davis, 2001. An updated classification of the Recent Crustacea, Ostracoda. Natural History Museum of Los Angeles County, Science Series 39, 28–30, 64–65.

McKenzie, K. G., 1967. Saipanellidae: a new family of podocopid Ostracoda. Crustaceana 13: 103–113.

McLaughlin, P. A., 1980. Comparative Morphology of Recent Crustacea. W. H. Freeman and Company, San Francisco.

Meisch, C., 1996. Contribution to the taxonomy of *Pseudocandona* and four related genera, with the description of *Schellencandona* nov. gen., a list of the Candoninae genera, and a key to the European genera of the subfamily (Crustacea, Ostracoda). Bulletin de la Société des naturalistes luxembourgeois 97: 211–237.

Meisch, C., 2000. Freshwater Ostracoda of Western and Central Europe. In Schwoerbel, J. & P. Zwick (eds), Süßwasserfauna von Mitteleuropa 8/3, Spektrum Akademischer Verlag, Heidelberg, Berlin.

Moore, R. C. & L. McCormick, 1969. General features of Crustacea. In Moore, R. C. (ed.), Treatise on Invertebrate Palaeontology, Part R (4), 57–120.

Morin, J. G. & A. C. Cohen, 1991. Bioluminescent displays, courtship, and reproduction in ostracodes. In Bauer, R. T. & J. W. Martin (eds), Crustacean Sexual Biology. Columbia University Press, 1–16.

Müller, G. W., 1894. Die Ostracoden des Golfes von Neapel und der angrenzenden Meeres-Abschnitte. Fauna und Flora des Golfes von Neapel und der angrenzenden Meeres-Abschnitte. Zoologische Station zu Neapel 21.

Newman, W. A., 2005. Origin of the Ostracoda and their maxillopodan and hexapodan affinities. Hydrobiologia 538: 1–21.

Park, L. E. & D. Ricketts, 2003. Evolutionary history of the Ostracoda and the origin of nonmarine faunas. In Park, L. E. & A. J. Smith (eds), Bridging the Gap. Trends in the Ostracode Biological and Geological Sciences. The Paleontological Society Papers 9, 11–36.

Parker, A. R., 1997. Functional morphology of the myodocopine (Ostracoda) furca and sclerotized body plate. Journal of Crustacean Biology 17: 632–653.

Petkovski, T., 1969. Einige neue und bemerkenswerte Candoninae aus dem Ohridsee und einigen anderen Fundorten in Europa. Acta Musei Macedonici Scientiarum Naturalium 11: 81–111.

Rome, D. R., 1947. *Herpetocypris reptans* (Ostracode). Etude morphologique et histologique. 1. Morphologie externe et système nerveux. La Cellule 51: 49–152.

Rome, D. R., 1969. Morphologie de l'attache de la furca et son utilisation en systématique. In Neale, J. W. (ed.), The Taxonomy, Morphology and Ecology of Recent Ostracoda. Proceedings of the 2nd International Symposium on Ostracoda. Oliver and Boyd, Edinburgh, 168–193.

Rossetti, G., & K. Martens, 1998. Taxonomic revision of the Recent and Holocene representatives of the Family Darwinulidae (Crustacea, Ostracoda), with a description of three new genera. Bulletin de l'Institut Royal des Sciences Naturelles de Belgique, Biologie 68: 55–110.

Sars, G. O., 1923. An account of the Crustacea of Norway with short descriptions and figures of all the species. 9. Ostracoda. Parts III & IV, 33–72.

Schminke, H. K., 1976. The ubiquituous telson and the deceptive furca. Crustaceana 30: 292–300.

Schornikov, E. I., 1988. The concept of cyclicity of morphogenesis. In Hanai, T., N. Ikeya, & K. Ishizaki (eds), Evolutionary Biology of Ostracoda: Developments in Palaeontology and Stratigraphy 11. Kodansha, Tokyo & Elsevier, Amsterdam, 195–205.

Schram, F. R., 1986. Crustacea. Oxford University Press.

Schulz, K., 1976. Das Chitinskelett der Podocopida und die Frage der Metamerie dieser Gruppe. Unpublished doctoral dissertation, University of Hamburg.

Siveter, D. J., M. Sutton, D. E. G. Briggs & D. J. Siveter, 2003. An ostracode crustacean with soft parts from the Lower Silurian. Science 302: 1749–1751.

Smith, R. J., 2000. Morphology and ontogeny of Cretaceous ostracods with preserved appendages from Brazil. Palaeontology 43: 63–98.

Smith, R., & T. Kamiya, 2002. The ontogeny of *Neonesidea oligodentata* (Bairdioidea, Ostracoda, Crustacea). Hydrobiologia 489: 245–275.

Smith, R., & T. Kamiya, 2003. The ontogeny of *Loxoconcha japonica* Ishizaki, 1968 (Cytheroidea, Ostracoda, Crustacea). Hydrobiologia 490: 31–52.

Smith, R., & T. Kamiya, 2005. The ontogeny of the entocytherid ostracod *Uncinocythere occidentalis* (Kozloff & Whitman, 1954) Hart, 1962 (Crustacea). Hydrobiologia 538: 217–229.

Smith, R., & K. Martens, 2000. The ontogeny of the cypridid ostracod *Eucypris virens* (Jurine, 1820) (Crustacea, Ostracoda). In Horne, D. J. & K. Martens (eds), Evolutionary Biology and Ecology of Ostracoda. Hydrobiologia 419: 31–63.

Swanson, K. M., 1989. *Manawa staceyi* n. sp. (Punciidae, Ostracoda): soft anatomy and ontogeny. Courier Forschungsinstitut Senckenberg 113: 235–249.

Tsukagoshi, A. & A. R. Parker, 2000. Trunk segmentation of some podocopine lineages in Ostracoda. In Horne, D. J. & K. Martens (eds), Evolutionary Biology and Ecology of Ostracoda. Hydrobiologia 419: 15–30.

Weitschat, W., 1983. Myodocopid ostracodes with preserved appendages from the Lower Triassic of Spitzbergen. Paläontologische Zeitschrift 57: 309–323.

Wouters, K., 1988. Two interesting new marine interstitial Ostracoda (Crustacea) from the Comoros, with the description of *Danipussella* gen. nov. Bulletin de l'Institut Royal des Sciences Naturelles de Belgique, Biologie 58: 85–93.

Wouters, K., 1997. A new genus of the Pontocyprididae (Crustacea, Ostracoda) from the Indian and Pacific Oceans, with the description of two new species. Bulletin de l'Institut Royal des Sciences Naturelles de Belgique, Biologie 67: 67–76.

Hydrobiologia (2007) 585:201–211
DOI 10.1007/s10750-007-0638-1

OSTRACODA (ISO15)

Ultrastructure of the carapace margin in the Ostracoda (Arthropoda: Crustacea)

Shinnosuke Yamada

Abstract In podocopid ostracods, the "hingement", with teeth/sockets or crenulations, is developed along the attached dorsal margin of the carapace. Dual calcified lamellae, called "duplicatures" are also developed along the free margin. However, the terminology of these marginal areas is often confused by many ostracodologists because of their differing interpretations and viewpoints. In this study, the marginal areas of the carapace were observed in detail, using the transmission (TEM) and scanning (SEM) electron microscopes. Consequently, the terminology of the marginal areas has been revised. The homology between attached and free margins and the structural differences of duplicatures, for some taxa are also discussed.

Keywords Ultrastructure · Marginal infold · Hingement · Ligament · Selvage · Carapace

Introduction

The ostracod carapace, especially the calcified one, covers the whole body of the organism and is a characteristic feature. It has a protective role against external impact, and also serves as a region of metabolism for some physiological functions. In benthonic podocopid ostracods, the "hingement", with teeth/sockets or crenulations, is developed along the attached margin of the carapace, and the dual calcified lamellae along the free margin are known as the "duplicature". Many palaeontologists have taken an interest in these features since their taxonomic viewpoints depend on the calcified carapaces preserved in fossils.

The palaeontologist Sylvester-Bradley (1941) sightly established the terminology for the ostracod carapace which he considered as hard tissue. He divided the dual lamellar structure along the free margin into the outer and inner lamellae, by their relative location to the marginal pore canals. One continuous ridge running along the marginal zone of the outer lamella, and the other developing along the calcified inner lamella, were called "flange" and "selvage" respectively; and an additional continuous minor ridge inside of the selvage was called the "list" (Fig. 1B). These terms were also used by Müller (1898) and by many other palaeontologists quoting Sylvester-Bradley's terminology (Triebel, 1950; Pokorný, 1957; van Morkhoven, 1962); but these structures do not always develop on all ostracod carapaces, and they show differences even between the right and left valves. As a result, there is often confusion in the use of these terms among different researchers, and thus

Guest editors: R. Matzke-Karasz, K. Martens & M. Schudack
Ostracodology – Linking Bio- and Geosciences

S. Yamada (✉)
Department of Earth Sciences, University of Tokyo, Bunkyoku, Hongo 7-3-1, Tokyo 113-0033, Japan
e-mail: r5344004@ipc.shizuoka.ac.jp

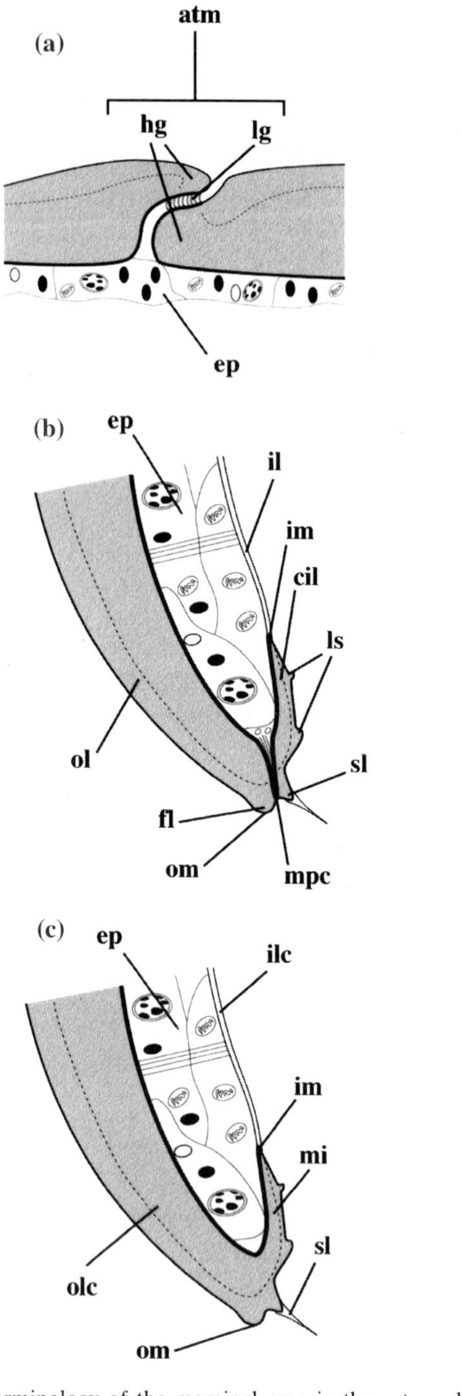

Fig. 1 Terminology of the marginal area in the ostracod carapace. (**a**) terminology of the attached margin; (**b**) terminology of the free margin used in the previous palaeontological studies; (**c**) terminology of the free margin used in this study. Gray areas represent the calcified parts

it is not appropriate to apply these terms universally to the homology of the carapace structure. The conventional terminology, as mentioned above, probably comes from the idea that the ostracod carapace consists of distinct right and left valves, with one valve being formed by concrescence of the outer and inner lamellae.

Arising out of the work of Müller (1898) and Sylvester-Bradley (1941), came some remarkable views on the carapace. Fassbinder (1912), for instance, suggested the homology between the attached and free margins by observation of sections. Harding (1964) established the theory that the ostracod carapace is composed of one continuous sheet, a theory later developed by Kornicker (1969) who divided the continuous sheet into four parts: right and left shells, vestment, and ligament. Kornicker also re-established the terminology of the ostracod carapace from a biologist's viewpoint (see Kornicker, 1969; Fig. 2). These comparative anatomical studies recognised the ligament as an important component at the attached margin, and discussed the homology between ligament and selvage (summarised in Table 1).

In this present study, the marginal areas of the carapace were observed in detail, using the TEM and SEM. Consequently, the terminology of the marginal areas has been revised. The homology between the attached and the free margin, and the structural differences of the duplicature among the some taxa, are also discussed.

Materials and methods

Living specimens used in this study are listed in Table 2, and all of the numbered specimens are deposited in the University Museum, Shizuoka University (SUM-CO-number).

In preparation for observation by SEM (JEOL JSM-5600LV), specimens were fixed in either 5% formaldehyde or 70% ethanol solution. Carapace specimens were then air-dried and ion-coated with gold.

For observations by TEM (HITACHI H-7500), living specimens were initially fixed in 2% glutaraldehyde and 2% paraformaldehyde in 0.1M cacodylate sodium buffer (pH 7.4), with 5% sucrose, for 2 h at 4°C; decalcified in 10% EDTA in the same buffer with 5% sucrose for 3 days at 4°C; post-fixed in 1% osmium tetroxide in the same

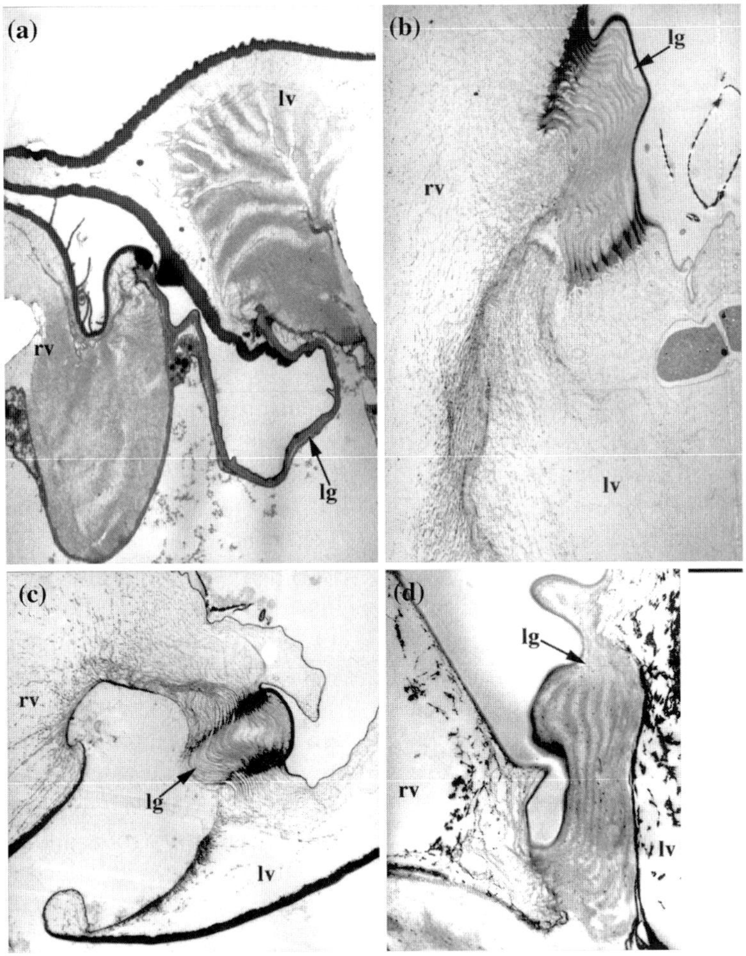

Fig. 2 TEM photographs of the ostracod ligament. (**a**) *Polycope japonica* (SUM-CO-1418); (**b**) *Keijcyoidea inflalittoralis* sp. (SUM-CO-1419); (**c**) *Aurila hataii* (SUM-CO-1420); (**d**) *Paracobanocythere* sp. (SUM-CO-1421). Scale bar is 2.0 μm for **a**; 3.3 μm for **b**; 2.5 μm for **c**; 0.8 μm for **d**

Table 1 Studies on ostracod ligament and selvage

Author name	Homologue of ligament	Nomenclature of ligament	Ultrastructure of ligament	Ultrastructure of selvage
Fassbinder (1912)	Selvage	Ligament	Not figured	Not figured
Harding (1969)	Not mentioned	Soft cuticle	Not figured	Not figured
Kornicker (1969)	Independent	ligament	Not figured	Not figured
Bate & East (1972, 1975)	Not mentioned	Connecting chitin	Layered chitin structure (parabolic pattern)	Layered chitin structure (parabolic pattern)
Jaanusson (1985)	Not mentioned	Intervalvar cuticle	Not figured	Not figured
Keyser (1995)	Not mentioned	Not mentioned	Chitinous bundles	Not figured
This study	Lamella cuticle	Ligament	Chitin bundle of feather-like fibres	Homogeneous with lattice structure

buffer with 5% sucrose for 2 h at 4°C; dehydrated in an ethanol series; and then embedded in Spurr's resin and polymerised. Sections were obtained using the ultramicrotome. Semi-thin sections were stained with 1% toludine-blue in 1% sodium tetraborate solution. Ultra-thin sections were

Table 2 Specimens list

Species name	Locality	Substrate
Neonesidea oligodentata	Shimoda (Shizuoka Pref. Japan)	Algae
Aurila hataii	Shimoda (Shizuoka Pref. Japan)	Algae
Callistocythere setouchiensis	Shimoda (Shizuoka Pref. Japan)	Algae
Paracobanocythere sp.	Shimoda (Shizuoka Pref. Japan)	Algae
Polycope japonica	Shimoda (Shizuoka Pref. Japan)	Algae
Keijcyoidea inflalittoralis sp.	Shimoda (Shizuoka Pref. Japan)	Algae
Loxoconcha pulchra	Obitsu (Chiba Pref. Japan)	Sediments
Vestalenula cornelia sp.	Yaku Island (Kagoshima Pref. Japan)	Sediments

stained with 1% potassium permanganate solution in distilled water for 2 min, and lead citrate (Reynolds, 1963) for 3 min. The following abbreviations were used: atm, attached margin; cil, calcified inner lamella; ep, epidermis; epc, epicuticle; fl, flange; hg, hingement; il, inner lamella; ilc, inner lamella cuticle; im, inner margin; la, lattice structure; lg, ligament; lv, left valve; mf, moulting fluid; mi, marginal infold; nepc, newly formed epicuticle; nsl, newly formed selvage; oilc, old inner lamella cuticle; ol, outer lamella; olc, outer lamella cuticle; om, outer margin; oolc, old outer lamella cuticle; rv, right valve; sl, selvage.

Terminology of the marginal areas of the carapace

In previous studies, the term "attached margin" was identified with "hingement" and "ligament". In this paper, calcified teeth and crenulations are defined as "hingement" and the uncalcified connecting structure is given the name of "ligament" (Fig. 1a).

The ostracod "ligament" is not an elastic structure, like that of bivalves, but an uncalcified cuticular structure connecting both calcified valves with each other as part of a continuous cuticular sheet. Several researchers did not use the term "ligament" (see Table 1), and independently named this connecting flexible cuticular structure as "soft cuticle" (Harding, 1964), "connecting chitin (ligament of hinge)" (Bate & East, 1972) and the "intervalvar cuticle" (Jaanusson, 1985). This study uses the term "ligament" for the structural comparison of previous ostracod studies and for convenience of terminology.

The cuticular dual lamellae structure is called the "duplicature". The inner calcified one along the free margin was formerly called the "calcified inner lamella", but is now identified with the extension of the outer lamella cuticle due to its calcified structure and protective function (Okada, 1982; Yamada et al., 2004). In this study, the term of "marginal infold", after Kornicker (1969) and Hanai & Ikeya (1991), is adopted for this inner structure, and the external margin of this fold is the "outer margin". The boundary between the "marginal infold" and "inner lamella cuticle" (see Yamada et al., 2004) is called the "inner margin". The uncalcified process along the free margin making a seal for the carapace can be observed, and I have adopted name of "selvage" (Fig. 1C). The selvage is often accompanied by a slight ridge at its root, and it is considered that this construction is identified with the selvage as it was defined by Sylvester-Bradley (1941).

Ultrastructure of the marginal areas of the carapace

Ligament

In platycopid and podocopid species bearing strongly calcified carapaces, a thick ligament is recognised without regard to the developmental intensity of the hingement (Fig. 2b–d). In the myodocopid *Polycope japonica*, a thin ligament is connected with each valve (Fig. 2a). The ultrastructure of these ligaments bears certain common features, as listed below:

(1) In transverse section, thick bundles of chitinous fibres are found in the ligament. These fibres show a feather-like structure (Fig. 3a, b).
(2) In longitudinal section, the feather-like fibres are parallel to each other. Curved

parabolic fibres connect to adjacent feather-like fibres (Fig. 3c).

(3) A seamless epicuticular layer (at least for the outer epicuticle) covers the outermost ligament (Fig. 3b).

The bundles mentioned in (1) were found in *Cyprideis torosa*, as illustrated by Keyser (1995). The fibrous structures mentioned in (1) and (2) are similar to the layered structure in the exoskeleton of *Cypridopsis vidua*, shown by Bate & East (1972, 1975) (Table 1). Furthermore, they reported that this structure could be also observed in the chitinous body cuticle and selvage spine. Similar structures have also been reported widely in the arthropod cuticle (e.g., Insecta, Decapoda; Neville (1975)).

Selvage

The selvage is composed only of an epicuticle. Around its root, lattice structures of feather-like fibres develop, and this region seems to be slightly calcified (Fig. 4a–c).

In many species, selvages seem to be homogenous with each other due to the presence of electron-dense material (Fig. 4b, c). Although helicoidal structures exist in the selvage of *Neonesidea oligodentata* (Fig. 4a), these structures are different from the feather-like structure in the ligament (Fig. 3c).

Marginal infold

The broad marginal infold which develops along the free margin of adult podocopids (Fig. 5a) does so to a lesser extent in juveniles (Fig. 5b): the difference being due to the extent of the calcified area (Figs. 6a, b; 7a, b).

In Platycopida and the podocopid Darwinuloidea, which have many primitive characteristics, adults are recognised for the poor development of the marginal infold, like other podocopid juveniles (Fig. 5c, d). But at the marginal areas in both taxa the flexible inner lamella cuticle connects directly to the outer margin (Figs. 6c, d; 7c), and the marginal infold is not fully developed. Therefore, they lack a marginal infold rather than having one which is poorly developed.

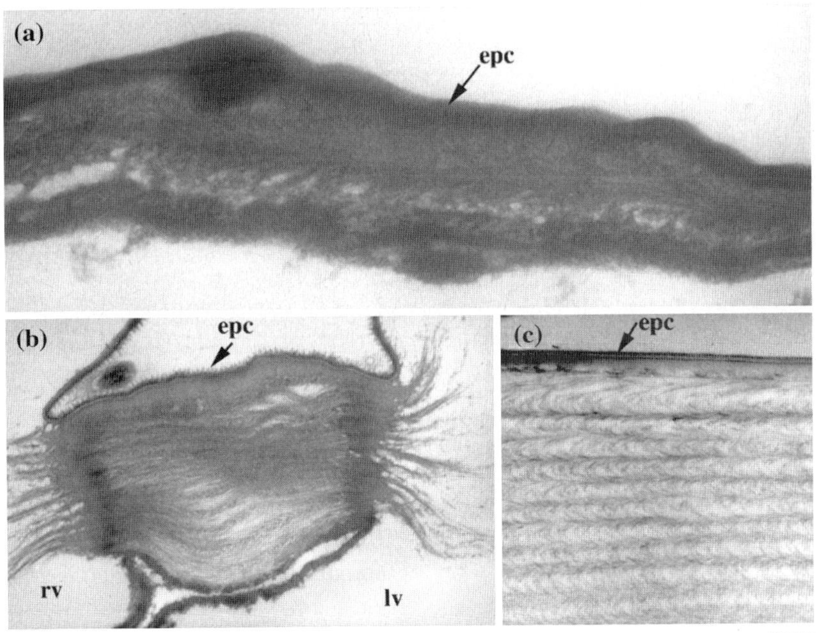

Fig. 3 TEM photographs of the ostracod ligament. (**a**) ligament of *Polycope japonica* in transverse section (SUM-CO-1418); (**b**) ligament of *Loxoconcha pulchra* in transverse section (SUM-CO-1422); (**c**) ligament of *L. pulchra* in longitudinal section (SUM-CO-1423). Scale bar is 0.1 μm for **a**; 0.4 μm for **b**; 0.5 μm for **c**

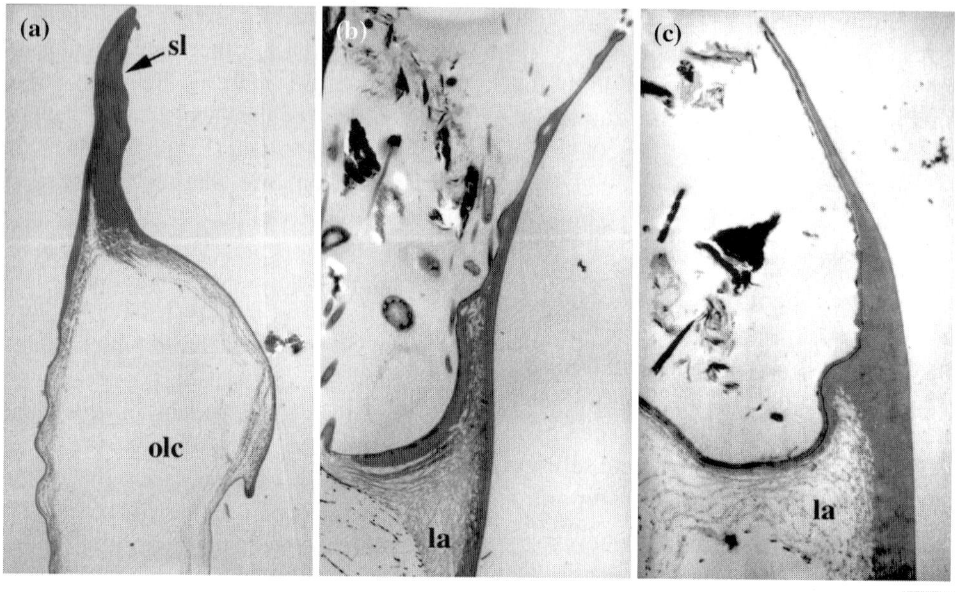

Fig. 4 TEM photographs of the ostracod selvage. (**a**) *Neonesidea oligodentata* (SUM-CO-1424); (**b**) *Aurila hataii* (SUM-CO-1425); (**c**) *Callistocythere setouchiensis* (SUM-CO-1426). Scale bar is 5.0 μm for **a**; 1.3 μm for **b**; 0.7 μm for **c**

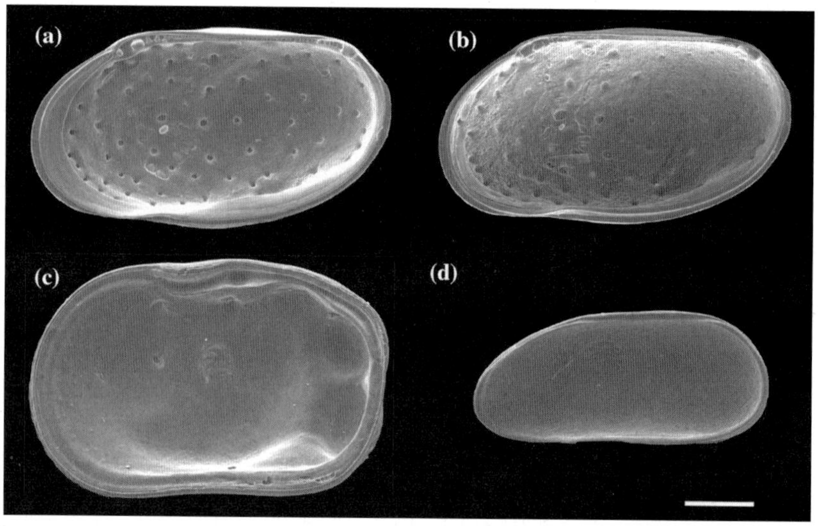

Fig. 5 SEM photographs of the carapace in internal lateral view. (**a**) *Loxoconcha pulchra* (SUM-CO-1427); (**b**) Juvenile (A-1) of *L. pulchra* (SUM-CO-1428); (**c**) *Keijcyoidea inflalittoralis* sp. (SUM-CO-1429); (**d**) *Vestalenula cornelia* sp. (SUM-CO-1430). Scale bar is 100 μm

Discussion

Correlation between "ligament" on the attached margin and "selvage" on the free margin

Only few studies established the homology between the attached and free margins (Table 1). Based on optical observations of sections of *Cypris pubera* (Cypridoidea), Fassbinder (1912) concluded that the selvage is continuous with the ligament, and that both are composed of an epicuticle. Bate & East (1972) reported through TEM observations that the ligament and the selvage consisted of an epicuticle and exocuticle, but they did not mention their homology. On the

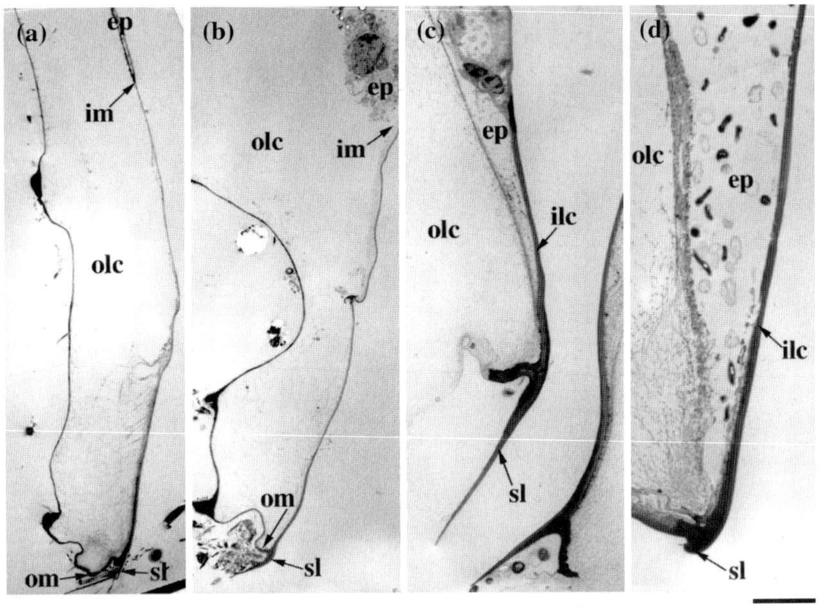

Fig. 6 TEM photographs of the duplicature. (**a**) *Loxoconcha pulchra* (Podocopida, Cytheroidea) (SUM-CO-1422); (**b**) Juvenile (A-1) of *L. pulchra* (Podocopida, Cytheroidea) (SUM-CO-1431); (**c**) *Keijcyoidea inflalittoralis* sp. (Platycopida) (SUM-CO-1432); (**d**) *Vestalenula cornelia* sp. (Podocopida, Darwinuloidea) (SUM-CO-1433). Scale bar is 5.0 μm for **a**; 3.3 μm for **b**, **c**; 2.0 μm for **d**

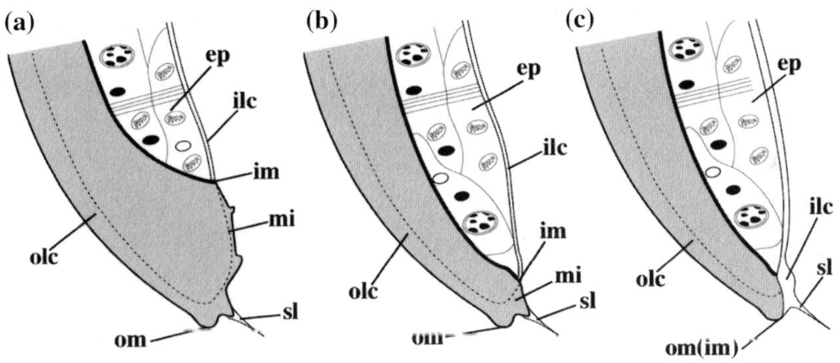

Fig. 7 Schematic drawings of the structure of free margins. (**a**) cytheroid ostracod (adult); (**b**) cytheroid ostracod (juvenile); (**c**) platycopid and darwinuloid ostracods. Gray areas represent the calcified parts

other hand, the homology between the attached and free margins has been discussed by several palaeontologists. Pokorný (1957) and Hanai (1961) noted the correlation between the contact groove, the "selvage" (used in previous palaeontological studies; see Fig. 1), and the "list" in Podocopida (Cytheroidea). Kornicker (1969) divided the myodocopid carapace into four components: "right and left shells", "ligament", and "vestment", and concluded that the ostracod ligament is not continuous with the selvage and that the ligament is independent of other cuticular parts.

The TEM photographs of successive sections around attached and free margins of *Aurila hataii* (Podocopida) and *Keijcyoidea inflalittoralis* sp. (Platycopida) are compared here (Fig. 8). In *Aurila hataii*, the selvage of the right valve is

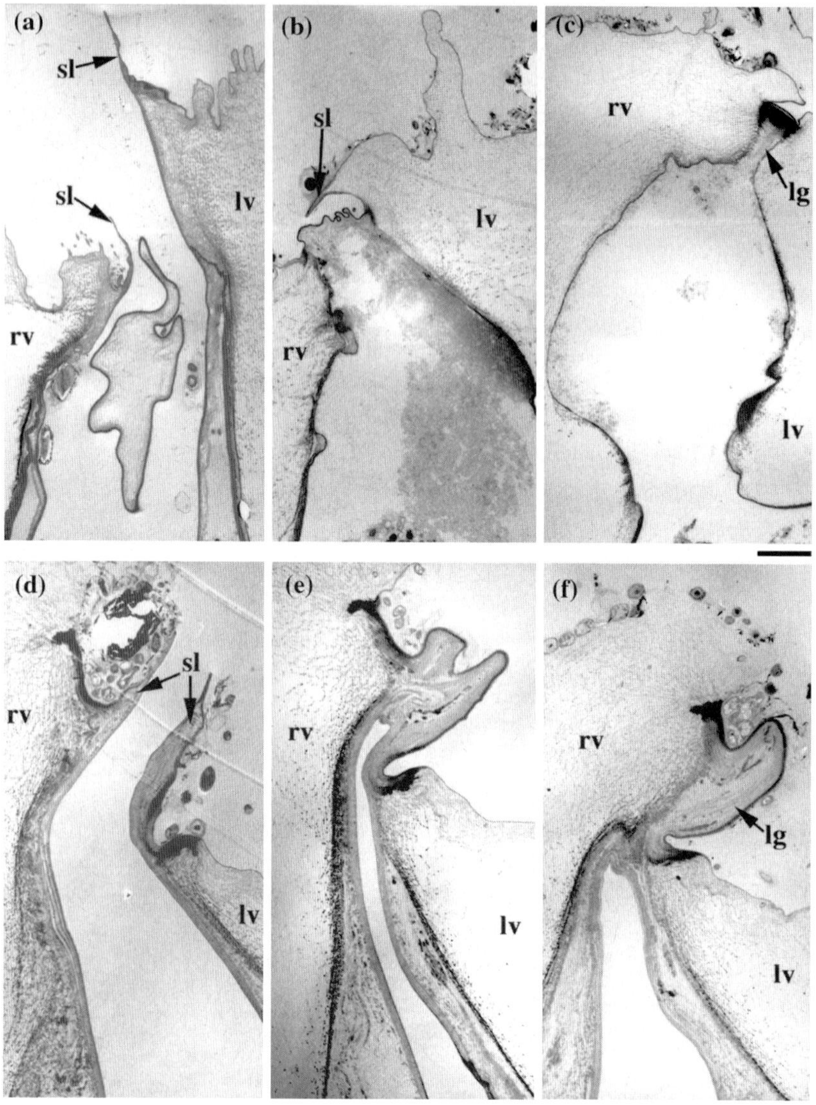

Fig. 8 Successive sections near the boundary between free and attached margins of *Aurila hataii* (**a–c**) (SUM-CO-1434) and *Keijcyoidea inflalittoralis* sp. (**d–f**) (SUM-CO-1435). (**a**, **d**) free margins; (**b**, **e**) transitional areas between free and attached margin; (**c**, **f**) attached margins. Scale bar is 4.0 μm for **a–c**; 2.9 μm for **d–f**

incorporated into the epicuticle of the marginal infold of the left valve. Thus, the selvages of both valves join with the epicuticle of the ligament, and the procuticle of both marginal infolds connect to each other through the procuticle of the ligament, with a chitin-fibrous structure (Figs. 8a–c; 9a–c). In consequence, the lamella cuticle (inside of marginal area) corresponds to the ligament. These facts suggest that the ligament is an uncalcified cuticular structure consisting of the epicuticle and procuticle, and is not continuous with the selvage. This understanding is also supported by the other observations, namely that the ultrastructure of the ligament differs from that of the selvage (Figs. 3, 4), and the formation of the selvage precedes that of the ligament at the pre-moult stage (Fig. 10). In *Keijcyoidea inflalittoralis* sp. which does not have a marginal infold, the selvage of one valve connects to that of the other and both selvages form a continuous epicuticular layer of ligament (Figs. 8d–f; 9d–f). Consequently, the selvages of both valves are

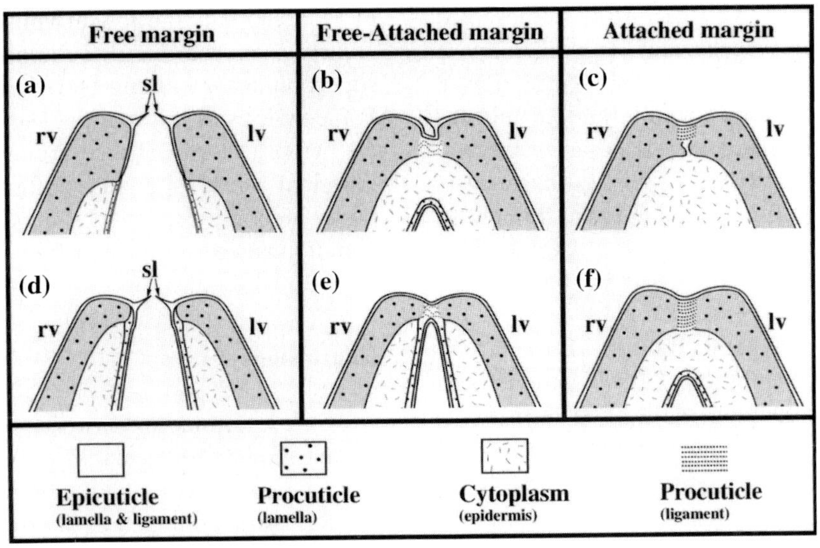

Fig. 9 Schematic drawings of successive sections near the boundary between free and attached margins of *Aurila hataii* (**a–c**) and *Keijcyoidea inflalittoralis* sp. (**d–f**). (**a**, **d**) free margins; (**b**, **e**) transitional areas between free and attached margin; (**c**, **f**) attached margins. Gray areas represent the calcified parts

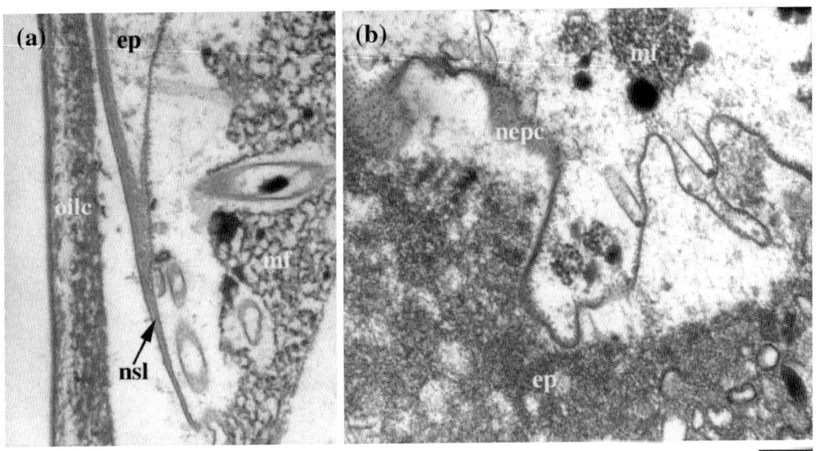

Fig. 10 TEM photographs of the marginal area of *Loxoconcha pulchra* in the pre-moult stage (D2). (**a**) newly formed selvage (SUM-CO-1436); (**b**) newly formed epicuticle in the ligament area (SUM-CO-1436). Scale bars are 0.5 μm

continuous with the epicuticle of the ligament, and the procuticle of the inner lamella cuticle meets the procuticle of the ligament. Also, in this case, the ligament is not completely identified with the selvage.

Elasticity of the ligament

Many previous studies assumed that the ligament of almost all ostracods is not elastic, and that it does not play a role in the opening of valves (cf. Harding, 1964; Smith, 1965; Jaanusson, 1985). Jaanusson (1985), however, reported that dead specimens of *Cypridopsis vidua* had their valves open, and stated the possibility of elasticity of the ligament to open them. According to observations reported here, the feather-like fibres in many ostracod ligaments are thought to correspond to the parabolic pattern made from the helicoidal fibrous arrangement in the arthropod cuticle. This feather-like structure is supposed to contribute only to mechanical strengthening of the cuticle rather than elasticity (Locke, 1964). Besides, the ostracod ligament (and its fibrous structure) occasionally develops along the vertical plane to the animal body (Fig. 2d). In such a case, this arrangement of the ligament is not thought to work effectively for opening the valves. This study also denies that the ligament has enough elasticity to open valves.

Kornicker (1969) and Bate & East (1975) referred to the possibility of the ligament contributing to the opening of the valves in myodocopids with thick ligament, but many ostracod researchers think that myodocopid ostracods open their valves utilising a hydrostatic system (Maddocks, 1992; Vannier et al., 2001).

The elastic protein "resilin" was reported to be distributed in some particular regions of the arthropod cuticle (Neville, 1975). This rubber-like cuticle develops at the root of grasshopper's limb and mechanoreceptive hairs. For elucidation of the elasticity of the ligament, chemical analysis is needed.

Ultrastructure of the marginal infold

Many species of Podocopida develop a marginal infold, although the degree of development depends on the ontogenetic stage (Figs. 6a, b; 7a, b). In *Keijcyoidea inflalittoralis* sp. (Platycopida) and *Vestalenula cornelia* sp. (Podocopida: Darwinuloidea), the "outer lamella cuticle" directly links with the "inner lamella cuticle" along the outer margin (Fig. 6c, d). In these taxa, the outer margin cannot be distinguished from the inner margin, i.e., both margins are identical, because the inner margin is defined as the boundary of calcified cuticle and uncalcified cuticle along the inside of the valves (Fig. 7c).

Fossils of platycopids and darwinuloids without marginal infolds have been reported from Palaeozoic formations and, as is often pointed out, these two taxa retain some primitive characters. On the other hand, extinct Paleocopa and Leiocopa from the Palaeozoic are also lacking a marginal infold (cf. Moore, 1961; Swanson, 1989a, b; Williams & Vannier, 1995). Therefore, the feature of free margin is assumed to be almost equivalent to that of Platycopida and Darwinuloidea. On the contrary, in the fossil carapace of podocopid superfamilies excluding Darwinuloidea, from the Palaeozoic, a broad marginal infold is developed (cf. Moore, 1961). These facts suggest that the marginal infold is a synapomorphic character of these higher taxa, when they derived from the common ancestor in the early Palaeozoic.

Conclusions

1. A new terminology for the ostracod carapace margin has been established, based on homology and ultrastructural observations.
2. The ostracod ligament does not correspond to the selvage, but to the lamella cuticle (inside of marginal area). It is an uncalcified cuticular structure contacting the dorsal margins of both valves. Furthermore, it has specialised fibrous structures to provide for strength, but probably it has no elasticity for opening the valves.
3. Since the Platycopida and podocopid Darwinuloidea do not develop the marginal infold, the structure of their free margin substantially differs from that of podocopid Cytheroidea. Marginal infolds may have appeared in the early Palaeozoic as a synapomorphic character of podocopid superfamilies excluding Darwinuloidea.

Acknowledgements I would like to express my deep gratitude to Dr. Akira Tsukagoshi (Shizuoka University) for critical reading of the manuscript and valuable discussions. I also greatly appreciate Prof. Shigeyasu Tanaka (Shizuoka University) for technical advice. Thanks are also due to Prof. Emeritus Noriyuki Ikeya (Shizuoka University) and members of the Ostracod Research Team of Shizuoka University for their valuable suggestions and continuous encouragements. My deep thanks are extended to Dr. Alicia Moguilevsky (Translations—Group) for checking the English of this text. I also thank Dr. Takahiro Kamiya (Kanazawa University), Dr. Robin Smith (Lake Biwa Museum) and Mr. Tomoatsu Takenaka (Heiden-ji Temple) for kindly helping with the collecting of samples. The staff members of Shimoda Marine Research Center (University of Tsukuba), Misaki Marine Biological Station (University of Tokyo) provided various facilities for this research. This study was partly funded by Research Fellowships of the Japan Society for the Promotion of Science for Young Scientists.

References

Bate, R. H. & B. A. East, 1972. The structure of the ostracode carapace. Lethaia 5: 177–194.

Bate, R. H. & B. A. East, 1975. The ultrastructure of the ostracode (Crustacea) integument. Bulletins of the American Paleontology 65: 529–547.

Fassbinder, K., 1912. Beiträge zur Kenntnis des Süßwasserostracoden. Zoologische Jahrbücher, Abteilung für Anatomie und Ontogenie der Tiere 32: 533–576.

Hanai, T., 1961. Studies on the Ostracoda from Japan: hingement. Journal of the Faculty of Science, Tokyo University section 2: 345–377.

Hanai, T. & N. Ikeya, 1991. Two new genera from Omma-manganji ostracode fauna (Plio-Pleistocene) of Japan—with a discussion of theoretical versus purely descriptive ostracode nomenclature. Transactions and Proceedings of the Paleontological Society of Japan, New Series, 163: 861–878.

Harding, J. P., 1964. Crustacean Cuticle with Reference to the Ostracod Carapace. In Puri, H. (ed.), Ostracods as Ecological and Palaeoecological Indicators. Pubblicazioni della Stazione Zoologica di Napoli 33 (supplemento), 9–31.

Jaanusson, V., 1985. Functional morphology of the shell in platycope ostracods–a study of arrested evolution. Lethaia 18: 73–84.

Keyser, D., 1995. Structural Elements on the Surface of Ostracod Shells. In Ríha, J. (ed.), Ostracoda and Biostratigraphy. Proceedings of the Twelfth International Symposium on Ostracoda, Prague, Czech Republic. Balkema, Rotterdam, The Netherlands, 5–10.

Kornicker, L. S., 1969. Relationship Between the Free and Attached Margin of the Myodocopid Ostracod Shell. In Neale, J. W. (ed.), The Taxonomy, Morphology and ecology of the Recent Ostracoda. Proceedings of the Second International Symposium on Ostracoda. Oliver and Boyd, Edinburgh, 109–135.

Locke, M., 1964. The Structure and Formation of the Integument in Insects. In Rockstein, M. (ed.), The physiology of Insecta 3. Academic Press, London, New York, 379–470.

Maddocks, R. F., 1992. Ostracoda. In Frederick, W. H. (ed.), Microscopic Anatomy of Invertebrates Vol. 9. Wiley-Liss, New York, 415–441.

Moore, R. C. (ed.), 1961. Treatise on Invertebrate Paleontology, (Q) Arthropoda 3. Geological Society of America, University of Kansas Press, Lawrence, Kansas.

Müller, G. W., 1898. Die Ostracoden. Sonderabdruck aus den Abhandlungen der Senckenbergischen naturforschenden Gesellschaft 21: 257–259.

Neville, A. C., 1975. Biology of Arthropod Cuticle. In Hoar W. S., J. Jacobs, H. Langer & M. Lindauer (eds), Zoophysiology and Ecology 4. Springer-Verlag, Berlin, Heidelberg, New York.

Okada, Y., 1982. Ultrastructure and Pattern of the Carapace of *Bicornucythere bisanensis* (Ostracoda, Crustacea). In Hanai, T. (ed.), Studies in Japanese Ostracoda. University Museum, University of Tokyo, Bulletin 20: 229–255.

Pokorný, V., 1957. The phylomorphogeny of the hinge in Podocopida (Ostracoda, Crustacea) and its bearing on the taxonomy. Acta Universitatis Carolinae Geologica 3: 3–22.

Reynolds, E. S., 1963. The use of lead citrate at high pH as an electron opaque stain in electron microscopy. Journal of Cell Biology 17: 208.

Smith, R. N., 1965. Musculature and Muscle Scars of *Chlamydotheca arcuata* (Sars) and *Cypridopsis vidua* (O. F. Müller) (Ostracoda—Cyprididae). In Kesling, R. V., D. G. Darby, R. N. Smith & D. D. Hall (eds), Four Reports of Ostracod Investigations conducted under National Science Foundation. Project GB-26, University of Michigan Museum of Paleontology, Special Publication Report 4, 1–41.

Swanson, K., 1989a. Ostracod phylogeny and evolution—a manawan perspective. Sonderdruck aus CFS-Courier 113 Forschungsinstitut Senckenberg: 11–20.

Swanson, K., 1989b. *Manawa staceyi* n. sp. (Punciidae, Ostracoda): soft anatomy and ontogeny. Sonderdruck aus CFS-Courier 113, Forschungsinstitut Senckenberg: 235–249.

Sylvester-Bradley, P. C., 1941. Shell structure of the Ostracoda. The Annals and Magazine of Natural History 11th series 8: 1–33.

Triebel, E., 1950. Homöomorphe Ostracoden-Gattungen. Senckenbergiana Lethaea 31: 313–330.

Vannier, J., S. Q. Wang & M. Coen, 2001. Leperditicopid arthropods (Ordovician-Late Devonian): functional morphology and ecological range. Journal of Paleontology 75: 75–95.

Van Morkhoven, F. P. C. M., 1962. Post-Palaeozoic Ostracoda Their Morphology, Taxonomy and Economic Use, Vol. 1, General. Elsevier Publications, Amsterdam.

William, M. & J. M. C. Vannier, 1995. Middle Ordovician Aparchitidae and Schmidtellidae: the significance of 'featureless' Ostracods. Journal of Micropalaeontology 14: 7–24.

Yamada, S., A. Tsukagoshi & N. Ikeya, 2004. Ultrastructure of the carapace in some *Semicytherura* species (Ostracoda: Crustacea). Micropaleontology 50: 381–389.

Hydrobiologia (2007) 585:213–223
DOI 10.1007/s10750-007-0639-0

OSTRACODA (ISO15)

Ultrastructure of hepatopancreas and its possible role as a hematopoietic organ in non-marine cypridoidean ostracods (Crustacea)

Radka Symonová

Abstract The arrangement of hepatopancreas and associated cells was examined in freshwater cypridoidean ostracods in the context of comparative microanatomy and cytology using light (LM) and transmission electron microscopy (TEM).
A compact, short, subpyriform hepatopancreas located at the sides of the front part of the fore gut has been observed in the following species of the family Candonidae: *Candona candida*, *C. neglecta, Cyclocypris ovum*, *Cypria ophtalmica*, *Fabaeformiscandona fabaeformis*, *Pseudocandona compressa*. Within the families Cyprididae (in species *Cypridopsis vidua*, *Herpetocypris reptans*, *Heterocypris incongruens*, *Potamocypris unicaudata* and *Psychrodromus olivaceus*), Ilyocyprididae (*Ilyocypris gibba*) and Notodromadidae (*Notodromas monacha*, *N. persica*) a completely different arrangement of the hepatopancreas was observed. In these families the hepatopancreas is tubular and remarkably elongated. Hepatopancreatic cells (sometimes very large) are scattered within a more or less developed layer of other cell types. This organ is also connected to the fore gut. However, it stretches, in comparison with the candonids, much further in the rear part of the body cavity and in the interlamellar space. In transversal and longitudinal sections this organ forms typical lacunae. The lacunae are often longitudinally further subdivided forming a lacunary system. It means that there are often distinct constrictions in some parts of the tubular hepatopancreas. The hepatopancreas verges into multilayer cellular formations (MCFs) in caudal and more peripheral areas. The MCFs can occupy large spaces in the periphery of the body cavity, within the bases of appendages or in the interlamellar space. The morphology of the hepatopancreatic cells in the part close to the fore gut distinctly differs in appearance from the cells in the more caudal areas. The cells in the MCFs strongly resemble "Subdermalzellen" in earlier literature. The development of MCFs often reflects the amount of food in the both parts of the middle gut. The extent and spatial arrangement of the MCF is subject to a high variability. In TEM the hepatopancreatic cells resemble oenocytoid cells described for example in insects. The ultrastructure of some parts of the hepatopancreas viewed in TEM represents typical cecal cells forming a basal labyrinth and microvilli. These specialized cells most probably originate from the hepato-

Guest editors: R. Matzke-Karasz, K. Martens & M. Schudack
Ostracodology – Linking Bio- and Geosciences

R. Symonová (✉)
Department Biologie II - Anthropologie und Humangenetik, Ludwig-Maximilians-Universität München, Grßhaderner Strasse 2, 82152 Planegg-Martinsried, Germany
e-mail: r.symonova@lrz.uni-muenchen.de

pancreatic cells, which are also able to give rise to plasmatocytes.

Keywords Hepatopancreas · Gastric ceca · Hemocytic cells · Hematopoietic organ

Introduction

Very few studies have described the structure and functions of the hepatopancreas (sometimes called the liver or hepatopancreatic gland) in podocopid ostracods. Hitherto it was described only in several species of the family *Cytheridea* as a short globous diverticle of gut epithelium with several large secretory cells (Weygoldt, 1960). Long tenuous ducts reaching rear parts of interlamellar space are briefly mentioned in the families Macrocyprididae and Pontocyprididae (Müller, 1894). There is only a concise description of a long pair of ducti in *Bradleystrandesia fuscata* (Jurine, 1820) of the family Cyprididae (Hartmann, 1968).

Bergold (1910) rejected the hypothesis by Claus (1895) that the hepatopancreas in freshwater ostracods produces lipid secretions and suggested that products of the hepatopancreas in podocopids are most probably the same as products of the mid gut wall cells. The arrangement of the hepatopancreas within the family *Darwinulidae* is only very poorly understood (Hartmann, 1968).

The most detailed work on the structure of hepatopancreas in podocopid ostracods was done by Kesling (1965) in *Candona suburbana*, Hoff, 1942 (Candonidae). The hepatopancreas is here described as a pair of secreting glands connected to the stomach by a short, narrow efferent duct. The hepatopancreas in *C. suburbana* is short, subpyriform to elongate ovate with a strongly lobulate exterior surface. This author provides in his study a series of microphotographs.

The function of the hepatopancreas was studied by McGregor (1967). Stained food particles offered to ostracods were used to visualize food bolus within the guts. Afterwards rhythmic pulsations of the hepatopancreas and the food bolus entering proximal parts of the hepatopancreas were observed.

Bergold (1910) provided drawings of large cells (called by this author "Subdermalzellen") forming layers in the periphery of the body cavity. These cells strongly resemble the cells of the MCF described in this work.

In this study representative species of the following families were studied by light and electron microscopy: Canadonidae, Cyprididae, Ilyocyprididae and Notodromadidae. The goal of the study is to determine in detail the structure and arrangement of the digestive system and its interconnection with other organ systems in podocopid ostracods, since this area of freshwater ostracods biology so far has not been studied (Maddocks, 1992).

Materials and methods

Study samples

The animals used in this study were collected within the framework of survey of ostracod fauna in the Czech Republic in various freshwater habitats. The sampling was carried out using sieves with mesh size 0.2 mm.

The following families and species were examined: Candonidae (*Candona candida* (O.F.Müller, 1776); *C. neglecta* Sars, 1887; *Cyclocypris ovum* (Jurine, 1820); *Cypria ophtalmica* (Jurine, 1820); *Fabaeformiscandona fabaeformis* (Fischer, 1851); *Pseudocandona compressa* (Koch, 1838)), Cyprididae (*Cypridopsis vidua* (O. F. Müller, 1776); *Eucypris virens* (Jurine, 1820); *Herpetocypris reptans* (Baird, 1835); *Heterocypris incongruens* (Ramdohr, 1808); *Potamocypris unicaudata* Schäfer, 1943; *Psychrodromus olivaceus* (Brady & Norman, 1889)), Ilyocyprididae (*Ilyocypris gibba* (Ramdohr, 1808)), Notodromadidae (*Notodromas monacha* (O. F. Müller, 1776); *N. persica* Gurney, 1921).

Histology and microscopy

For the histological analysis and observation in light microscope (LM) the living animals were fixed in modified Bouin-Dubosque-Brasil fluid (Smrž, 1989), embedded in Histoplast (Serva), sectioned by microtome Leica RM 2155 (thickness 5 μm) and stained with Masson's Trichrome (Smrž, 1989) or Domagk's azan staining (Romeis, 1989). For the TEM microscopy the specimens were fixed in 2.5% glutaraldehyde in 0.1 M phosphate-buffer and postfixed in 1% OsO_4. After dehydration in a graded acetone series the specimens were embedded in Spurr's medium epoxy resin (Polysciences Ltd.). Ultrathin sections were cut on an ultratome Leica UCT and stained with lead citrate and a saturated solution of uranylacetate in 50% ethanol. The specimens were observed in TEM microscope JEOL 1011 and the images were acquired by a camera Megaview III (SIS GmbH) equipped with software AnalySIS.

Anatomical terminology

The following terminology of anatomy was used in the present study: the fore gut, sometimes also called anterior intestine (Meisch, 2000) and the rear gut or posterior intestine (Meisch, 2000) here mean the two parts of the middle gut from the embryological point of view. This part of the digestive system is derived from the embryonic endoderm and is divided by the central constriction into the fore gut and the rear gut. The oesophagus and anus, the parts arising from ectoderm, are not taken into account here.

Results

Hepatopancreas in representative species of the families Cyprididae, Ilyocyprididae and Notodromadidae

The hepatopancreatic gland of the following species and families was examined: Cyprididae (*Cypridopsis vidua*, *Eucypris virens*, *Herpetocypris reptans*, *Heterocypris incongruens*, *Potamocypris unicaudata*, *Psychrodromus olivaceus*), Ilyocyprididae (*Ilyocypris gibba*) and Notodromadidae (*Notodromas monacha*, *N. persica*). The hepatopancreas is here developed as a tubular organ varying in its length, but it is always pronouncedly elongate (Figs. 1a, b, 2a). In LM the wall of the hepatopancreas is lined with conspicuously large cells (Fig. 2b), which are detached from each other and invested among other cell types. The hepatopancreatic tube is very often subdivided by distinct constrictions thus forming a typical lacunary system (Fig. 2a). The hepatopancreas is connected with the fore gut, situated along its dorsolateral sides. Posteriorly it stretches into the interlamellar space. In cases, when both parts of the middle gut were filled with a large amount of organic particles, the hepatopancreas also stretched anteriorly from the connection with the fore gut (observed in *H. incongruens* and *N. monacha*). In the part connecting with the fore gut the hepatopancreas was filled up with food. Proximally, the large cells are less numerous and arranged in only one layer. These cells have a conical shape tapering towards the inner cavity of the hepatopancreas. In the tip of these cells there is a distinctly vesicular part. With increasing distance from the connecting point with the fore gut (posteriorly or anteriorly) these cells lose their conical shape, become more tightly arranged to each other and reduce the extent of the vesicular part.

Multilayer cellular formations (MCFs) in the families Cyprididae, Ilyocyprididae and Notodromadidae

Distally, the internal cavity of the hepatopancreas gradually diminishes. In the most distal parts (posterior as well anterior) of the hepatopancreas the cavity does not occur anymore. Here, the hepatopancreas structure is formed only by the cells of the hepatopancreatic wall. The outline of these cells is then determined by the surrounding cells arranged in one- to multilayer cellular formations (MCFs). The cells in the MCFs considerably vary in their shape but always retain a polygonal outline and mostly lack any vesicular parts. Their size is 10–20 μm. The cells of the MCF are stained very intensively by Masson's Trichrome (light violet, Fig. 2c, to dark brown-violet, Fig. 2d inset) as well as by Domagk's Azan (violet to blue, Fig. 2d). With the both staining methods these cells stain very similar as oocytes do (Fig. 2a, c).

The MCFs were hitherto recorded in *E. virens*, *H. incongruens*, *I. gibba*, *N. monacha*

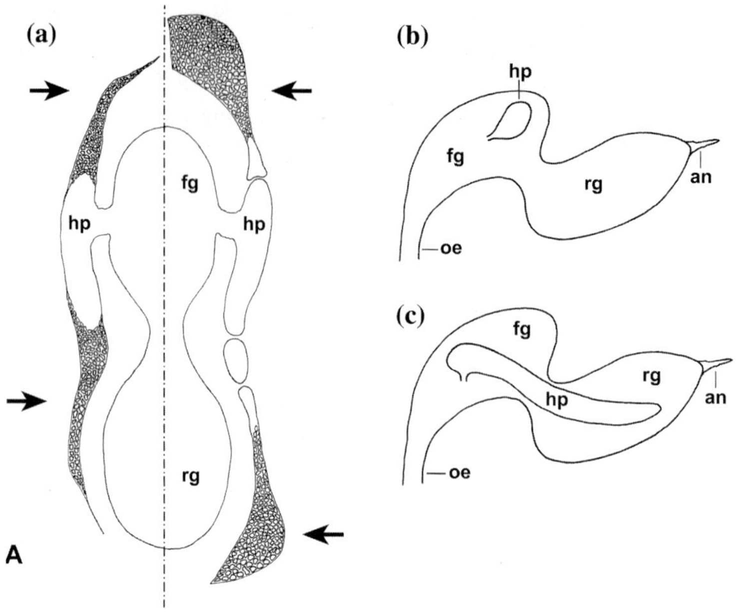

Fig. 1 (**a**–**c**) Schematic drawings showing the different arrangements of the hepatopancreas in podocopid ostracods. (**a**) The hepatopancreas and the MCF (black arrows) when smaller amount of food occurred in both parts of the gut (left) and when the gut was filled with food (right). An, anus; fg, fore gut; hp, hepatopancreas; oe, oesophagus; rg, rear gut. (**b**) Simplified drawing of the hepatopancreas (hp) in the above mentioned species of the family Candonidae. (**c**) The arrangement of the hepatopancreas (hp) observed in the above mentioned species of the families Cyprididae, Ilyocyprididae and Notodromadidae

and *P. olivaceus*. In *H. incongruens* the MCF occurs in most cases in the area around the dorsolateral margin of the rear gut, along the body wall in posterior part, in the interlamellar space, around and between the ovaries and dorsally of eggs (Fig. 2c, d). In *N. monacha* the largest accumulations of cells of the MCF were found in the interlamellar space anteroventrally of the eye in the upper side and anteroventrally the ventral chain of ganglia in the basal part, further in the posterior ventrolateral region, often between testes and Zenker's organ in males and in the surrounding of spermatozoa in the seminal receptacles as well as around the ovaries and eggs in females. In all other species mentioned above the MCFs were observed mostly around the dorsolateral margin of the rear gut in the interlamellar space. In *E. virens, I. gibba* and *P. olivaceus* the MCF occurs as a continuation of the hepatopancreas in the interlamellar space in the posterior parts of the body and in the surround of the rear gut.

The extent of the MCF very often reflects the amount of the food particles in both parts of the gut. The more food particles in the gut the larger MCF with a higher number of layers of cells develop (Fig. 1a).

Hepatopancreas in representative species of the family Candonidae

Within the family Candonidae a uniform arrangement of the hepatopancreas was observed in the following representative species: *Candona candida, C. neglecta, Fabaeformiscandona fabaeformis, Pseudocandona compressa*. The hepatopancreas in the above mentioned species was found to be a paired compact glandular organ adjoining to the dorsolateral part of the fore gut (Figs. 1c, 2e). It usually stretches from the mandibular muscles to the area above the adductors, i.e. the valve closing muscles. The size of the gland reaches in its largest part up to 80 μm in height and 50 μm in

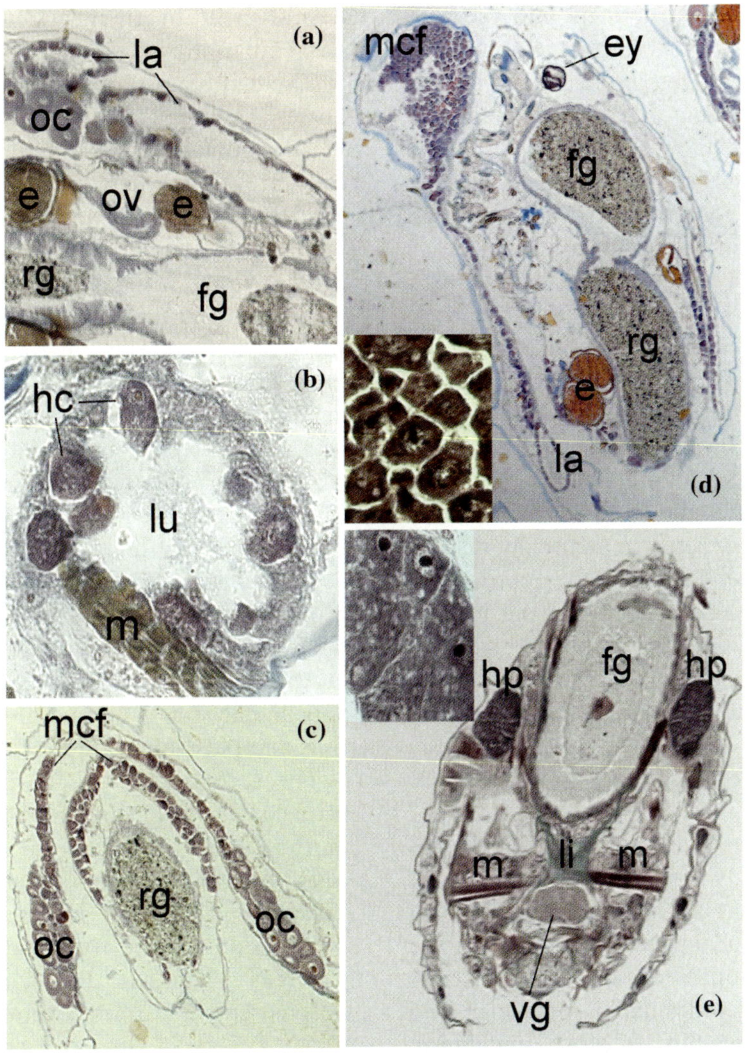

Fig. 2 (**a–e**) Stained by Domagk's Azan unless otherwise indicated. (**a**) *H. incongruens*, longitudinal section in the area of connection between the fore gut (fg) and the rear gut (rg) showing the subdivided lacunae (la) of the hepatopancreas enclosed in the developing oocytes (oc). The mature eggs (e) and the oviduct are also apparent (200x). (**b**) *N. persica*, detailed view of the lumen (lu) of the hepatopancreatic lacuna in transverse section with hepatopancreatic cells (hc) and muscles (m) (200x). (**c**) *H. incongruens*, transverse section through the rear part of the body cavity with the rear gut (rg) and the two prominent MCFs. MCF in the interlamellar space partly overlap the maturing oocytes (100x). (**d**) *H. incongruens*, parasagittal section through the extensive frontal multilayer cellular formation (MCF) connected with another layer of cells ending in lacunae (la). The fore gut (fg), rear gut (rg), eye (ey) and egg (e) are also distinct (100x). Inset shows frontal MCF in *N. persica* stained by Masson's Trichrome (400x). (**e**) *C. candida*, transverse section in the area of the central ligament (li) muscles (m) attachment and the ventral chain of ganglia (vg) showing the paired hepatopancreas (hp) laterally of the fore gut (fg) (100x). Inset represents detailed view of the hepatopancreas closer to the connection with the fore gut (400x). Both stained by Masson's Trichrome

width. The length of the hepatopancreas varies from 80 μm to 150 μm and mostly depends on the amount of food in the fore gut. The hepatopancreas tapers towards both anterior and posterior ends. The gland is composed of large cells (10–20 μm) with numerous vesicular regions (Fig. 1e inset). There is a small cavity situated towards the proximal margin of the hepatopancreas around the place of connection with the fore gut.

The gland releases its products into the fore gut. It is also distinct that the developing and maturing cells are localized in the ventral part of the gland, while the matured strongly vesicular cells are more dorsally.

In the internal cavity of the hepatopancreas, no food particles were observed in the candonids. Within the species of the family Candonidae so far observed neither the MCF nor the lacunary systems occurred. However, another type of conspicuously large cells occurs in the body cavity of the candonids examined. These cells are of variably oval shape measuring approximately 2–4 μm. Typical traits of these cells are the homogenous cytoplasm and the annulated structure observed along the circumference of the nucleus. The cells are more or less symmetrically distributed, in groups or individually, in the anterior and posterior part of the body cavity, in the interlamellar space and were found within the bases of appendages. In the posterior part, these cells are arranged around the rear gut and lying tightly close to the gut wall. The gut wall is conspicuously attenuated in the regions where these cells are enclosed.

Preliminary results in the subfamily Cyclocypridinae (of the family Candonidae) show that a transient arrangement between the short compact hepatopancreas found in the candonids and the long tubular hepatopancreas of the cypridids, ilyocypridids and notodromadids occurs in two species of this subfamily (*Cypria ophtalmica*, *Cyclocypris ovum*). A tubular hepatopancreas, but distinctly shorter and more compact, and with a more prominent cavity, as well as to some extent developed MCF was observed (unpublished observation of the author, not shown).

The hepatopancreas in TEM

In TEM the hepatopancreas was observed in *H. incongruens* and *N. monacha*. The large hepatopancreatic secretory cells already mentioned in the LM level description contain homogenous cytoplasm in most cases only with more or less developed vesicles. Any other membranous organelles are lacking (Fig. 3a). The hepatopancreatic cells and the cells of the MCFs, respectively, were recorded in three forms differing in development of their vesicular structures and basal labyrinth.

The first type is represented by polygonal cells possessing homogenous cytoplasm with numerous vesicles not larger than 200 nm (Fig. 3a, b). These cells prevail in their occurrence in the MCFs. Nuclei of these cells are regularly rounded with scattered heterochromatin areas. Within these cells cavities with developing microvilli (vide infra) were often observed. The intercellular space of these cells is filled up with cells emitting long thin pseudopodia. These cells are morphologically identical with agranular plasmatocytes (Fig. 3a, b) freely moving within the body cavity which were recorded by TEM in *N. persica* (unpublished observations of the author). These cells, which are interconnected to each other possess irregularly shaped nuclei (3–5 μm) with a prominent dark nucleolus. The cytoplasm contains electron-lucent vesicles reaching up to 1 μm in their diameter. For this reason the MCFs appear rather as a mosaic of the oenocytoid-like cells and plasmatocyte-like cells.

The second cell type is represented by cells of a conical shape tapering towards the lumen of the hepatopancreatic cavity (Fig. 3c). These cells typically contain membrane-surrounded vesicles (usually not larger than 500–700 nm) on the side reverse from the lumen of hepatopancreas. Towards the lumen the vesicles become larger to form several vacuoles reaching up to 2.5 μm in diameter and opening by the perforation of the cellular plasma membrane into the lumen.

The third cell type recognized is represented by cells forming a basal labyrinth. These cells can reach up to 40 μm in their length. They emit branching cytoplasmic projections on the side directed to the lumen of the hepatopancreas. Between these projections extensive extracellular space arises – the basal labyrinth (Fig. 3d). These cells also contain lipid droplets and numerous electron-dense as well as electron-lucent vacuoles of unknown composition. Only the part of cytoplasm emitting the projections contains extensive cisternae of rough endoplasmic reticulum (Fig. 4a) and large mitochondria (Fig. 4b). The irregularly shaped nuclei of these are up to 6 μm long.

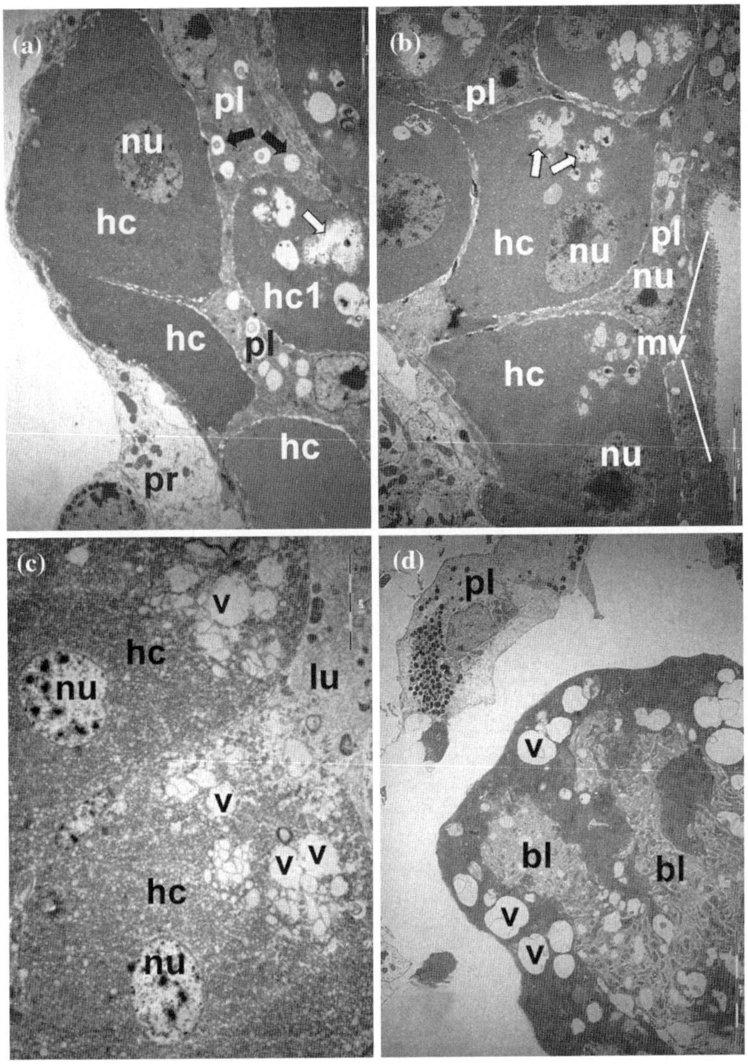

Fig. 3 (**a–d**) (**a**) *N. persica*, peripheral part of hepatopancreas showing hepatopancreatic cells (hc) with homogenous cytoplasm and nucleus (nu) and a cell (hc1) possessing cavity with developing microvilli (white arrow). The space between hepatopancreatic cells is filled up with interconnected plasmatocytes (pl). Rounded vesicles (black arrows) are distinct within the cytoplasm of plasmatocytes. Projection (pr) of freely moving plasmatocyte is apparent of the surface of the hepatopancreas. (scale bar 5 μm). (**b**) *N. persica*, peripheral part of hepatopancreas showing hepatopancreatic cells (hc) surrounding cell emitting microvilli (mv), nu, nucleus; pl, plasmatocyte; white arrows—cavities with developing microvilli. (scale bar 10 μm). (**c**) *N. persica*, central part of hepatopancreas with strongly vesicular (v) hepatopancreatic cells (hc). Vesicles empty into the lumen (lu) of hepatopancres, nu, nucleus. (scale bar 5 μm). (**d**) *N. persica*, part of a hepatopancreatic cell with developed basal labyrinth (bl) toward the centre of the cell and numerous vacuoles in the peripheral parts. (scale bar 10 μm)

Epithelial gastric cecal (= hepatopancreatic) cells possessing microvilli on the margin oriented towards the hepatopancreas lumen (Fig. 4c) were recorded to cover some parts of the inner surface of the hepatopancreatic organ. The microvilli are 80–100 nm wide and not longer than 1 μm, which is in conformity with the situation observed in other crustaceans (Martin, 1992). These gastric cells with numerous smaller (mostly not larger than 1 μm) membrane-delimited electron-lucent and electron-dense vesicles (Fig. 4c) penetrate between already degenerating

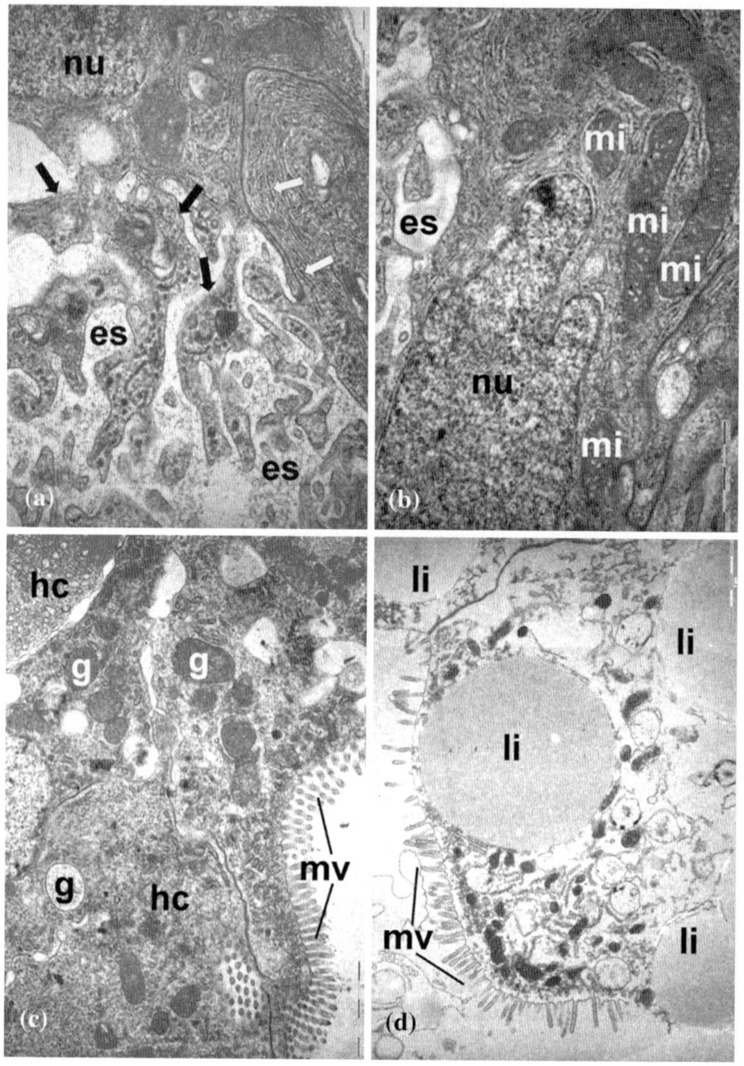

Fig. 4 (**a–d**) (**a**) 10 *N. persica*, detailed view of basal labyrinth showing numerous cytoplasmic projections (black arrows) emitted by hepatopancreatic cell and arising extracellular spaces (es). White arrow—strikingly developed cisternae of rough endoplasmic reticulum, nu, nucleus. (scale bar 1 μm). (**b**) *N. persica*, detailed view of hepatopancreatic cell forming basal labyrinth showing large mitochondria. es, extracellular spaces; nu, nucleus. (scale bar 1 μm). (**c**) *N. persica*, detailed view of gastric cecal cell (hc) developing microvilli (mv) and containing numerous granules. (scale bar 2 μm). (**d**) *N. persica*, degenerating gastric cecal cell with striking lipid droplets (li) and microvilli (mv). (scale bar 2 μm)

gastric cecal cells (Fig. 4d). In the cytoplasm of the degenerating cells there are numerous large (up to 4 μm) lipid droplets of regularly round shape. Further observed traits of these cells are more or less disintegrated peripheral areas, distinctly electron-lucent cytoplasm, numerous small mitochondria and cisternae of rough endoplasmic reticulum. These stages of gastric cecal cells were described by Schrehardt (1978) in *Artemia*.

Discussion

The different arrangement of the hepatopancreas and the associated cells in the candonids examined in comparison with the selected species of cyprid-ids, ilyocypridids and notodromadids indicate also differences in nutrients absorption, their storage and distribution.

Products of secretion of the numerous large secretory cells in the hepatopancreas of *H. incongruens* and *N. monacha* probably contribute to the digestion of the food particles in the guts. The hepatopancreatic cecal cells possessing microvilli and forming lipid droplets within their cytoplasm can play a role in nutrients absorption. As the degenerating cecal cells are engulfed by freely moving plasmatocyte-like cells it is possible that the plasmatocyte-like cells participate in distributing the digested and absorbed nutrients. Since the plasmatocyte-like cells are able to penetrate the spaces among cells of the MCF, they could distribute the nutrients to cells of the MCF and redistribute them among the other cells and tissues.

On the other hand, in the candonids observed it appears that digestion takes place entirely in the fore gut with the contribution of the digestive enzymes coming from the hepatopancreas. The absorption of nutrients according to the LM observation could be accomplished in the rear gut through its wall towards the freely circulating cells. However, this assumption is necessary to be further supported by TEM.

Regarding the absence of any form of vascular system in podocopid ostracods it is possible to consider the two above mentioned arrangements of the hepatopancreas and in particular then the associated cells as bodies ensuring an effective transportation of the nutrients to the other cells.

The basal labyrinth occurs among others in epithelial cells of renal canaliculi and salt glands of many marine bird species (Ude & Koch, 1994). It was also described in gastric cecal cells in *Artemia* (Martin, 1992; Foster & Wolfe, 1986). The basal labyrinth represents a permanent cell surface differentiation resulting in a multiple increase of the cell surface and therefore a higher efficiency of secretion and absorption (Willingham & Pastan, 1984). This function should be taken into account also in freshwater ostracods. Since the carapace of the ostracods is formed by deposition of mineral salts it is also possible that the basal labyrinth participates in retention of the minerals. The minerals could be then transferred either by the cells released from the hepatopancreas or by the MCF to the epidermis forming the carapace.

The larger accumulations of the cells forming the MCF predominantly fill up spaces being not occupied by any other organs. Very well developed MCFs within bases of additional appendages in *Liocypris grandis* are distinct also in Matzke-Karasz & Martens (2005). These additional appendages lack any muscles and provide thus a free space where the cells of the MCF can accumulate. On the other hand smaller groups of these cells and partly likewise the lacunae of the modified hepatopancreas are very often situated strikingly close to other organs or structures (ovaries, testes, receptacle seminis, Zenker's organ, ventral chain of ganglia, eggs, eye). These features, the high potential to be stained, the correlation between the food amount in the guts and the MCF development and the relatively low complexity of the cells (homogenous cytoplasm and/or uniform metabolic vesicles in huge amount) would suggest that these cells serve for storage and distribution of certain metabolites, most probably nutrients absorbed in the hepatopancreas from food. This possibility could be also indicated by the fact, that cells of the MCF are stained in the same colours as the cytoplasm of the oocytes in Masson's Trichrome as well as in Domagk's Azan.

Regarding their internal structural arrangement, the large hepatopancreatic secretory cells resemble oenocytoids described e.g. in the gypsy moth (Lepidoptera, Insecta; Butt & Shields, 1996). That is why these cells are here provisionally called oenocytoid-like cells.

The oenocytoid-like cells are most probably able to give rise to gastric cecal cells possessing microvilli, because such structures were observed to develop inside these cells. Moreover, the plasmatocyte-like cells were observed to divide from large oenocytoid-like cells. Therefore the hepatopancreatic organ could be viewed among others also as a hematopoietic organ sensu lato. The plasmatocyte-like cells can then either interconnect the oenocytoid-like cells among each other or can be released out of the hepatopancreas. In the latter case, the plasmatocyte-like cells can freely move throughout the body cavity and engulf cell debris and/or degenerate gastric cecal cells by means of phagocytosis. The free plasmatocyte-like cells were very often observed close to the hepatopancreas or forming a sheath of this organ.

The hematopoietic organ was described within crustaceans in detail in decapods (Martin & Hose, 1992). Here the hematopoietic organ is a specific structure of high complexity associated with vascular elements or with the stomach (Martin & Hose, 1992). In the group of Branchiopoda much less attention was paid to the blood cells as well as to blood-forming organs in comparison with decapods (Martin, 1992). However, a certain similarity between the MCF and the hepatopancreas of ostracods and the blood-forming organs can be found here. In *Artemia* and *Branchipus* so far poorly defined nodules consisting of loosely accumulated rounded cells differing in size and surrounded by a connective tissue were described (Lochhead & Lochhead, 1941).

Conclusion

Within podocopid ostracods it is possible to distinguish at least two different forms of the hepatopancreas with distinct differential taxonomic distribution. Together with observation in *C. suburbana* (Kesling, 1965) a compact form of hepatopancreas was recorded hitherto in five species of the family Candonidae in detailed LM analysis. This form of hepatopancreas is associated with large free cells displaying a more or less regular pattern of distribution within the body cavity. The second form of hepatopancreas is represented by a very elongate tubular organ associated with large conspicuous cells arranged in the MCFs. This form was observed in several species of the families Cyprididae, Ilyocyprididae and Notodromadidea.

Acknowledgements I thank Jaroslav Smrz (Charles University, Prague) for his help with histological techniques and interpretation of microanatomical structures in ostracods, Renate Matzke-Karasz (Ludwig-Maximilians University, Munich) for her insightful comments on the manuscript and to Dietmar Keyser (Hamburg University) for helpful comments on the microphotographs.This study was funded by the grant from Grant Agency of the Charles University GAUK Nr. 203 135 and by the Ministry of Education of Czech Republic, project Nr. 0021620828.

References

Bergold, A., 1910. Beiträge zu Kenntnis des innern Baues der Süsswasserostracoden. Zoologische Jahrbücher Abteilung für Anatomie und Ontogenie der Tiere 30: 1–42.

Butt, T. M. & K. S. Shields, 1996. The Structure and behavior of Gypsy moth (*Lymantria dispar*) hemocytes. Journal of Invertebrate Pathology 68: 1–14.

Claus, C., 1895. Beiträge zur Kenntnis der Süsswasser-Ostracoden II. Arbeiten aus dem Zoologischen Institute der Universität Wien und der Zoologischen Station in Triest 11: 17–48.

Foster, J. A. & A. F. Wolfe, 1986. Electron microscopic study of the ceca, intestine and associated peritrophic membrane of the brine shrimp, *Artemia*. Proceedings of the Pennsylvania Academy of Sciences 60: 29–32.

Hartmann, G., 1968. Ostracoda (2. Lieferung). In Gruner, H.-E. (ed), Dr. H. G. Bronns Klassen und Ordnungen des Tierreichs. Band **5** (Arthropoda), Abteilung I (Crustacea), Buch 2, Teil IV (Ostracoda) Akademische Verlagsgesellschaft, Leipzig, 217–408.

Kesling, R. V., 1965. Anatomy and dimorphism of adult *Candona suburbana* Hoff. In Kesling, R. V., D. G. Darby, D. D. Hall (eds), Four Reports of Ostracod Investigations (NSF-Project GB-26), 1, University of Michigan Publications, Ann Arbor, Michigan.

Lochhead, J. H. & M. S. Lochhead, 1941. Studies on the blood and related tissues in *Artemia* (Crustacea, Anostraca). Journal of Morphology 68: 593–632.

Maddocks, R. F., 1992. Ostracoda. In Harrison, F. E. & A. G. Humes (eds), Microscopic Anatomy of Invertebrates, Vol. 9. Wiley-Liss: 415–441.

Martin, J. W., 1992. Branchiopoda. In Harrison, F. E. & A. G. Humes (eds), Microscopic Anatomy of Invertebrates, Vol. 9. Wiley-Liss: 25–224.

Martin, J. W. & J. E. Hose, 1992. Vascular Elements and Blood (Hemolymph). In Harrison, F. E. & A.G. Humes (eds), Microscopic Anatomy of Invertebrates, Vol.10. Wiley-Liss: 117–146.

Matzke-Karasz, R. & K. Martens, 2005. The female reproductive organ in podocopid ostracods is homologous to 5 appendages: histological evidence from *Liocypris grandis* (Crustacea, Ostracoda). Hydrobiologia 542: 249–259.

McGregor, D. L., 1967. Rhythmic pulsation of the hepatopancreas in freshwater ostracods. Transaction of American Microscopic Society 86: 166–169.

Meisch, C., 2000. Freshwater Ostracoda of western and central Europe. In Schwoerbel, J., & P. Zwick (eds), Süßwasserfauna von Mitteleuropa 8/3. Spektrum Akademischer Verlag, Heidelberg.

Müller, G. W., 1894. Die Ostracoden des Golfes von Neapel und der angrenzenden Meeres-Abschnitte.

Fauna und Flora des Golfes von Neapel, 21 Friedländer und Sohn, Berlin.

Romeis, B., 1989: Mikroskopische Technik. Neubearbeitet und herausgegeben von Böck, P., 17. Auflage, Urban und Schwarzenberg, München, Wien, Baltimore.

Schrehardt, 1978. Ultrastructural investigations of the filter-feeding apparatus and the alimentary canal of *Artemia*. In Sorgeloos, P., D. A. Bengtson, W. Declair & E. Jaspers (eds), Artemia Research and Applications, Vol. 1. Morphology, Genetics, Strain Characterization, Toxicology. Universa Press, Wetteren, Belgium.

Smrž, J., 1989. Internal anatomy of *Hypochthonius rufulus* (Acari: Oribatida). Journal of Morphology 200: 215–230.

Ude, J. & M. Koch, 1994. Die Zelle, Atlas der Ultrastruktur. Fischer, Jena, Stuttgart.

Weygoldt, P., 1960. Embryologische Untersuchungen an Ostrakoden: Die Entwicklung von *Cyprideis litoralis* (G.S.Brady) (Ostracoda, Podocopa, Cytheridae). Zoologische Jahrbücher Abteilung für Anatomie und Ontogenie der Tiere 78: 369–426.

Willingham, M. C. & I. Pastan, 1984. Endocytosis and exocytosis: current concepts of vesicle traffic in animal cells. International Review of Cytology 92: 51–92.

Hydrobiologia (2007) 585:225–248
DOI 10.1007/s10750-007-0640-7

OSTRACODA (ISO15)

Copulatory behaviour and sexual morphology of three *Fabaeformiscandona* Krstić, 1972 (Candoninae, Ostracoda, Crustacea) species from Japan, including descriptions of two new species

Robin J. Smith · Takahiro Kamiya

Abstract Two new species of *Fabaeformiscandona* and collections of *Fabaeformiscandona japonica* (Okubo, 1990. Bulletin Biogeography Society of Japan 45: 39–50) from Japan have provided an opportunity to study the copulatory behaviour and sexual morphology of these species. The male uses the large fifth limb palps to force open the female carapace for copulation. However, the female has an enlarged overlapping flap over the posterior of the carapace which appears to hinder the male from opening the carapace, thus the female retains some control over potential mates. During copulation, the males' fifth limbs, caudal ramus and *m* process of the hemipenes and the females' protrusion of the female reproductive organ play a crucial role in providing a secure interface between the sexual organs to allow for the transfer of the giant spermatozoa. Development of the sexual organs in both sexes begins during the A-2 instar, and the protrusion of the female genital lobe begins to form during the A-1 instar.

Guest editors: R. Matzke-Karasz, K. Martens & M. Schudack
Ostracodology – Linking Bio- and Geosciences

R. J. Smith (✉)
Lake Biwa Museum, 1091 Oroshimo-cho, Kusatsu-shi, Shiga 525-0001, Japan
e-mail: smith@lbm.go.jp

T. Kamiya
Department of Earth Sciences, Kanazawa University, Kakuma, Kanazawa 920-1192, Japan
e-mail: takamiya@kenroku.kanazawa-u.ac.jp

Keywords Copulation · Sexual behaviour · Ontogeny · Hemipenes · Female reproductive organs

Abbreviations

An1	antennule
An2	antenna
CR	caudal ramus
CRA	caudal ramus attachment
dh	distal hook
FRO	female reproductive organ
Hp	hemipene
L5	fifth limb
L6	sixth limb
L7	seventh limb
LV	left valve
Md	mandible
Mx	maxillula
RV	right valve

Introduction

The genus *Fabaeformiscandona*, belonging to the subfamily Candoninae, has approximately 50 living species (Meisch, 2000) and, like most sexually reproducing groups of ostracods, the taxonomy has relied heavily on sexually dimorphic features, such as the carapace shape, male 5th limbs, hemipenes and the female genital lobes. However, there has been little previous work on the development, function and interaction of such features, despite their taxonomic importance.

Fabaeformiscandona species have giant spermatozoa, a feature that is common and well known in the superfamily (e.g., Zenker, 1854; Stuhlmann, 1886; Müller, 1889, 1894; Retzius, 1909; Lowndes, 1935; Bauer, 1940; Gupta, 1968; Wingstrand, 1988; Matzke-Karasz, 2005). The male sexual apparatus of the Cypridoidea that produces and delivers the giant spermatozoa has been well documented since Zenker's (1850) first detailed description of these features (e.g., Zenker, 1854; Lilljeborg, 1853; Weismann, 1880; Müller, 1880, 1894; Stuhlmann, 1886; Kesling, 1957, 1965; McGregor & Kesling, 1969a, b; Danielopol, 1969, 1978; Matzke-Karasz, 1997) although the function of some components was not well understood in some earlier works (e.g., see the review in Kesling, 1957 of the function of the Zenker organs). The spermatozoa are formed in eight testes that lie in the duplicature of the posterior region of the carapace. From there they travel through a long tube, the vas deferens, to a seminal vesicle before reaching two large pumps called the Zenker organs (or Zenker's organs). The Zenker organs are cylindrical with a central tube supporting rings of spines (or whorls) covered with a muscular sheath. They work by contracting the muscular sheath, thus shortening the cylinder, which pushes the sperm through the central tube. The central tube acts like a spring, so when the muscles relax the Zenker organs regain their original shape. From the Zenker organs the spermatozoa are pumped to the hemipenes. The hemipenes have a winding inner duct that carries the sperm to the tip. From the tip the spermatozoa are pumped into the female, which stores the sperm in two seminal receptacles prior to egg fertilization. Although giant, the whole spermatozoon penetrates the egg, curling itself inside the egg below the shell (Matzke-Karasz, 2005). For general information on reproduction and sexual behaviour in ostracods see the comprehensive reviews by Cohen & Morin (1990) and Horne et al. (1998).

Although a large number of studies have documented the behaviour and position of males and females before, during and after copulation in many different groups (e.g., Sars, 1889; Vávra, 1891; Jensen, 1904; Klie, 1926; Ferguson, 1944; McGregor & Kesling, 1969b; Hart & Hart, 1974; Danielopol, 1976, 1977, 1978; Kamiya, 1988; Danieopol & Horne, 1996; Parker, 1997; Horne et al., 1998) and even possible copulation positions for extinct groups (e.g., Kesling, 1969), there has been little attention given to the interaction between the male and female sexual organs, a result of their tiny size and the often translucent nature of the carapaces which hinders detailed observations. For the giant spermatozoa of the Cypridoidea to be successfully pumped under the forces of the Zenker organs into the females' seminal receptacle the connection between the sex organs must be secure, but how is this achieved?

The most detailed studies of the copulation mechanics of podocopid ostracods were conducted on Candoninae species (Danielopol, 1969, 1978, 1980; McGregor & Kesling, 1969a, b). All these studies included a detailed account of the internal parts of the hemipenes and documented the erect state of the hemipenes. Two of these studies, namely McGregor & Kesling (1969b) and Danielopol (1980), also showed how the male hemipenes interacted with the female reproductive organs. McGregor & Kesling (1969b) studied copulation in *Fabaeformiscandona suburbana* (Hoff, 1942), but they were vague as to how the male and female sexual organs interact; in their drawing of a 'pair in copula' (McGregor & Kesling, 1969b Text-fig. 5, p. 228) there is a gap between the tip of the hemipenes and the vaginas (situated near the base of the seventh limbs; see McGregor & Kesling, 1969a, Text-fig. 4, p. 174). Danielopol (1980) provided a very detailed description of how the hemipenes of *Pseudocandona pratensis* (Hartwig, 1901) interact with the females' reproductive organs.

The discovery of three species of *Fabaeformiscandona* in Kanazawa, western Japan, has provided the opportunity to document the copulatory behaviour to determine the functional morphology of the sexual features and the interaction between the hemipenes and female sexual organs in these species. Furthermore, the ontogeny of the sexual organs is documented to shed light on their origin. Two of the species are new and are described herein. The previous work on candonid copulation, especially Danielopol's (1980) comprehensive study on *P. pratensis*, gives an excellent opportunity to compare and contrast different copulation behaviour and function of sexually dimorphic features within the subfamily. Although many features of *P. pratensis* and the *Fabaeformiscandona* species used in this study can be directly correlated, some features are absent in *P. pratensis*, namely the protrusion from the female's reproductive organ and the proximal lobe of the hemipenes, plus the mating position is different.

Materials, methods and terminology

Material was collected from two sites in Kanazawa, Ishikawa Prefecture, and a third site in Lake Biwa, Shiga Prefecture. The first site at Takao, Kanazawa (N 36°30′15.2″, E 136°38′53.4″, altitude 214 m) was below a small waterfall with an undescribed *Fabaeformiscandona* species living in a small, muddy, shallow pool with lots of leaf litter. The second site at Shinkai pond, Kakuma, Kanazawa (N 36°32′31.6″, E 136°42′12.4″ altitude 94 m) yielded two additional species, *Fabaeformiscandona japonica* (Okubo, 1990) and a second undescribed *Fabaeformiscandona* species in the muddy bottom. Further specimens of *Fabaeformiscandona japonica* were recovered in the littoral zone of Lake Biwa off the Karasuma Peninsular (N 35°04′19.7″, E 135°56′22.2″ altitude 88 m). Two species, *F. japonica* and the species collected at Takao, were chosen to observe copulation in laboratory cultures due to their semi-transparent carapaces (the third species collected at Shinkai pond had a translucent carapace that did not facilitate observations of this kind).

To observe copulation in laboratory cultures, males and females were separated for a period of 24 h. After this time one male was added to the female culture (consisting of approximately 20 individuals) in a glass Petri dish containing a small amount of organic detritus. The male's movements were followed with a video camera attached to an Olympus stereo trinocular microscope while being recorded on Sony DVCAM digital videotape. After a successful copulation the male was removed and the process was repeated with another male. The semi-transparent carapaces of the species used in the study allowed the observation of the sexual organs and appendages through the valves.

Appendages were dissected and mounted in glycerine and drawn with the aid of a camera lucida. Carapaces were stored dry in micropalaeontological cavity slides. Carapaces used for SEM investigation were coated with gold before being photographed with a JEOL SEM. Specimens for soft part investigation using scanning electron microscopy were first either freeze dried using a JEOL freeze drying device, or dried using hexamethyldisilazane (HMDS) (Nation, 1983).

Latitudes and longitudes were recorded with a hand-held Garmin Etrex GPS unit using the Tokyo Datum.

All material is deposited in the Lake Biwa Museum, Shiga Prefecture, Japan (Numbers: LBM 1430000899 to LBM 1430000925).

Terminology of the appendages follows that of Broodbakker and Danielopol (1982), Martens (1987) and Meisch (2000).

Results

Taxonomic descriptions

Superfamily Cypridoidea Baird, 1845
Family Candonidae Kaufmann, 1900
Subfamily Candoninae Kaufmann, 1900
Genus *Fabaeformiscandona* Krstić, 1972
Fabaeformiscandona tora n. sp.

Type locality: Takao, Kanazawa, Ishikawa Prefecture, Japan N 36°30′15.2″, E 136°38′53.4″, elevation 214 m. Area below steep, forested slope along a road. Water running from the slope in a series of springs and waterfalls over saturated leaf litter and moss.

Type material: Holotype, a male (LBM 14300 00899). Allotype, a female (LBM 1430000901). Paratypes, three males (LBM 1430000900, LBM 1430000903 and LBM 1430000904) and two females (LBM 1430000902 and LBM 1430000905).

Derivation of name: Torus is Latin for protuberance or bulge, referring to the rounded, distal bulge of the *m* process of the hemipenis.

Diagnosis: Male carapace with rounded posterior. Hp *a* lobe large and straight sided, *m* process with enlarged, rounded tip and *h* lobe large and hook-shaped. L5 right palp with large rounded tip, left palp with elongate and slender tip with 3 blunt barbs. Female carapace with straight posterio-dorsal margin. Female reproductive organ with small, slender and pointed posterior projection.

Description

Carapace male length 947–997 μm, height 473–514 μm, female length 866–916 μm, height 403–413 μm. Male with rounded posterior, concave ventral margin and curved hinge (Figs. 1a & 2a). In dorsal view laterally compressed. LV larger than RV and overlapping RV at each end of hinge (Fig. 1h). Inner lamella wide anteriorly, narrow posteriorly (Fig. 2a). Female with straight dorsal-posterior margin, tightly curved posterior margin and slightly concave ventral margin (Figs. 1b & 2b). LV overlaps RV with flange on dorsal-posterior and dorsal-anterior margins, the former of which is larger (Fig. 1g). Inner lamella wide anteriorly, narrow posteriorly (Fig. 2b).

An1 with eight segments, first two fused together with two long dorsal setae and two long ventral–apical setae (Fig. 3a). Third segment quadrate with dorsal–apical seta. Fourth segment quadrate with short dorsal–apical seta. Fifth and sixth segments both quadrate and each with two dorsal–apical setae and one ventral–apical seta. Seventh segment with four long apical setae. Eighth segment elongate and slender with four apical setae of differing lengths.

Male An2 with divided fourth segment producing total of six segments (Fig. 3b). *Y* aesthetasc slender almost reaching to end of third segment. Male seta *t2* and *t3* both of equal length and both thickened with rounded ends (Fig. 3c). *t3* curved more tightly than *t2*. *t4* seta tiny. *Gm* slender and short, half length of *GM*. Claw *z1* thicker and slightly longer than *z2*. *G1* short and slender, half length of *G2*.

Md palp, both alpha and beta setae small and slender (Fig. 3c). Terminal segment with two long claw-like setae and two small setae. Coxa well-developed with robust teeth on endite.

Male L5 with large rounded basis, internally with large, well-developed muscles attached to palps (Fig. 4b & c). Basis supporting small endite and large asymmetrical palps. Right palp large and rounded distally, left palp elongate with blunt barbs distally.

L6 slender with elongate segments. First segment with *d1* seta (Fig. 3e).

L7 with 5 segments (Fig. 3f). First segment with small *d1* and long *dp* setae. Final segment quadrate with short *h1* seta and long *h2* and *h3* setae.

CR slightly curved (Fig. 3g). *Sa* half length of *Sp*. *Gp* approximately 63% length of *Ga*.

Hp large and elongate with wide, flattened *a* lobe (Figs. 4a, 5a & b). Surface of *a* lobe with shallow ridges, approximately 0.5–1 μm apart running obliquely to long axis of lobe (Fig. 5c). *m* process S-shaped and with large terminal, rounded tip. *h* lobe relatively large and hook-shaped with rounded terminal tip. Small, curved hook (dh) near inner-proximal edge of *h* lobe. Proximally Hp with a large, elongate, robust, slightly curved projection of the proximal inner shield, pointing in opposite direction to distal lobes (herein termed the proximal projection).

FRO, anterior rounded, with posterior projection small, slender and terminally pointed (Figs. 4d, 5d & e).

Fabaeformiscandona myllaina n. sp.

Type locality: Shinkai Pond, Kakuma, Kanazawa, Ishikawa Prefecture, Japan N 36°32′31.6″, E 136°42′12.4″. Artificial irrigation pond with concrete dam at one end. Substrate muddy with emergent macrophytes.

Type material: Holotype, male (LBM 14300 00913). Allotype, female (LBM 1430000914).

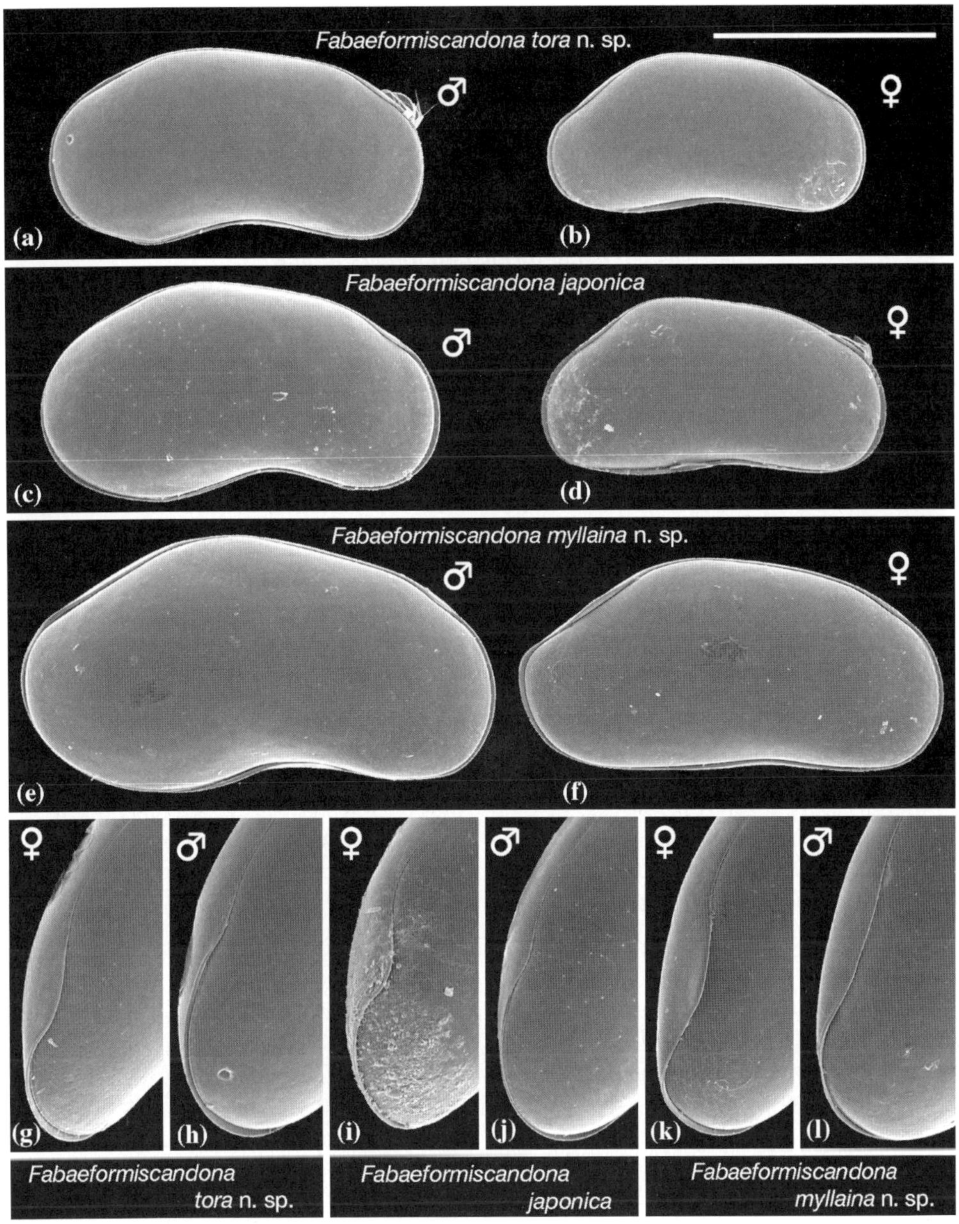

Fig. 1 **a**, **b**, **g** & **h** *Fabaeformiscandona tora* n. sp. (Takao, Kanazawa, Ishikawa). **a**—male, lateral view of whole carapace, anterior to right (LBM 1430000904). **b**—female, lateral view of whole carapace, anterior to right (LBM 1430000905). **g**—female, oblique view of postero-dorsal margin of whole carapace (LBM 1430000905). **h**—male, oblique view of postero-dorsal margin of whole carapace (LBM 1430000904). **c**, **d**, **i** & **j** *Fabaeformiscandona japonica*. **c**—male, lateral view of whole carapace, anterior to right (LBM 1430000922, Shinkai Pond, Kanazawa, Ishikawa). **d**—female, lateral view of whole carapace, anterior to right (LBM 1430000923, Lake Biwa, Shiga). **i**—female, oblique view of postero-dorsal margin of whole carapace (LBM 1430000923, Lake Biwa, Shiga). **j**—male, oblique view of postero-dorsal margin of whole carapace (LBM 1430000922, Shinkai Pond, Kanazawa, Ishikawa). **e**, **f**, **k** & **l** *Fabaeformiscandona myllaina* n. sp. (Shinkai Pond, Kanazawa, Ishikawa). **e**—male, lateral view of whole carapace, anterior to right (LBM 1430000917). **f**—female, lateral view of whole carapace, anterior to right (LBM 1430000918). **k**—female, oblique view of postero-dorsal margin of whole carapace (LBM 1430000918). **l**—male, oblique view of postero-dorsal margin of whole carapace (LBM 1430000917). Scale bar = 600 μm for **a**—**f**, 316 μm for **g** & **h**, 422 μm for **i** & **j**, 456 μm for **k** & **l**

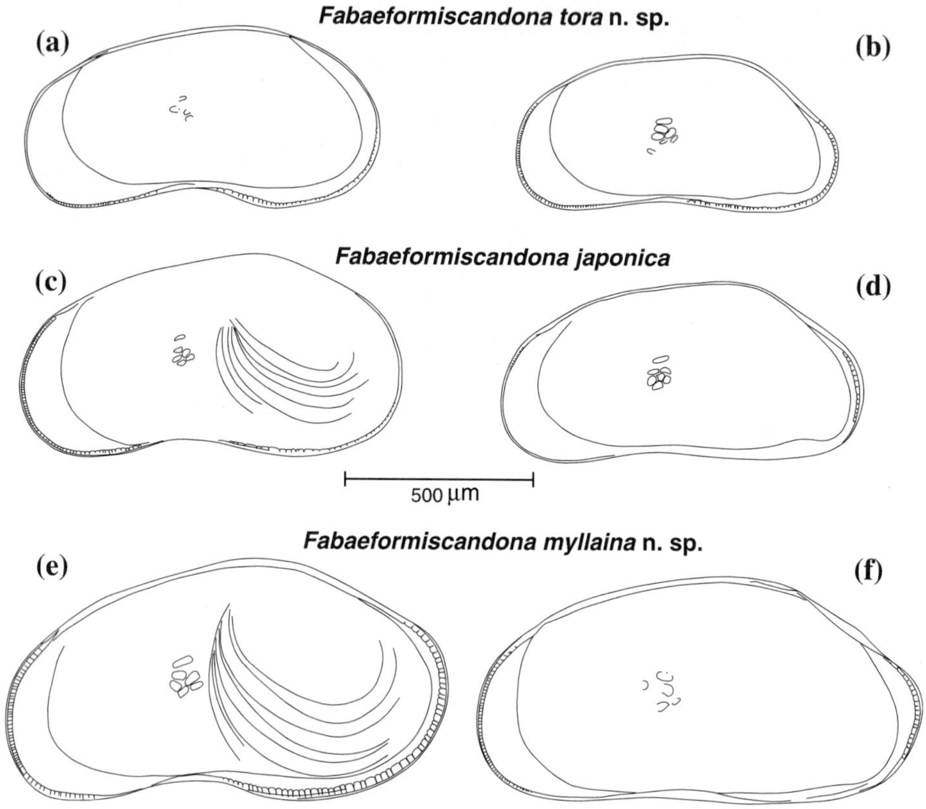

Fig. 2 **a–b** *Fabaeformiscandona tora* n. sp. **a**—male, LV external (LBM 1430000899). **b**—female, LV external (LBM 1430000901). **c–d** *Fabaeformiscandona japonica*. **c**—male, LV external (LBM 1430000924). **d**—female, LV external (LBM 1430000921). **e–f** *Fabaeformiscandona myllaina* n. sp. **e**—male, LV external (LBM 1430000915). **f**—female, LV external (LBM 1430000914)

Paratypes, three males (LBM 1430000915, LBM 1430000916 and LBM 1430000917) and one female (LBM 1430000918).

Derivation of name: Myllos is Greek for crooked or bent, referring to the shape of the *a* lobe of the Hp.

Diagnosis: Carapace large, female with truncated posterior. Male left L5 palp elongate with curved distal end. Right male L5 palp rounded distally, drawn out to elongate protrusion anteriorly. Hp with S-shaped *a* lobe, *m* process with slightly broadened distal end, *h* lobe curved and striated. FRO protrusion inflated and rounded, FRO angular in lateral view, pointed anteriorly.

Description

Carapace male length 1,269–1,329 µm, height 645–705 µm, female length 1,138–1,168 µm, height 534–554 µm. Male—Dorsal margin with prominent hump posterior of mid-length (Figs. 1e & 2e). Posterior of hump postero-dorsal margin straight sloping down to rounded posterior margin. Anterior of hump dorsal margin straight, sloping down to rounded posterior margin. Ventral margin strongly concave. Dorsal view very compressed. LV overlaps RV with flanges on postero-dorsal and antero-dorsal margins (Fig. 1l). Inner lamella wide anteriorly, very narrow posteriorly (Fig. 2e)

Female more elongate than male with prominent angle between postero-dorsal margin and

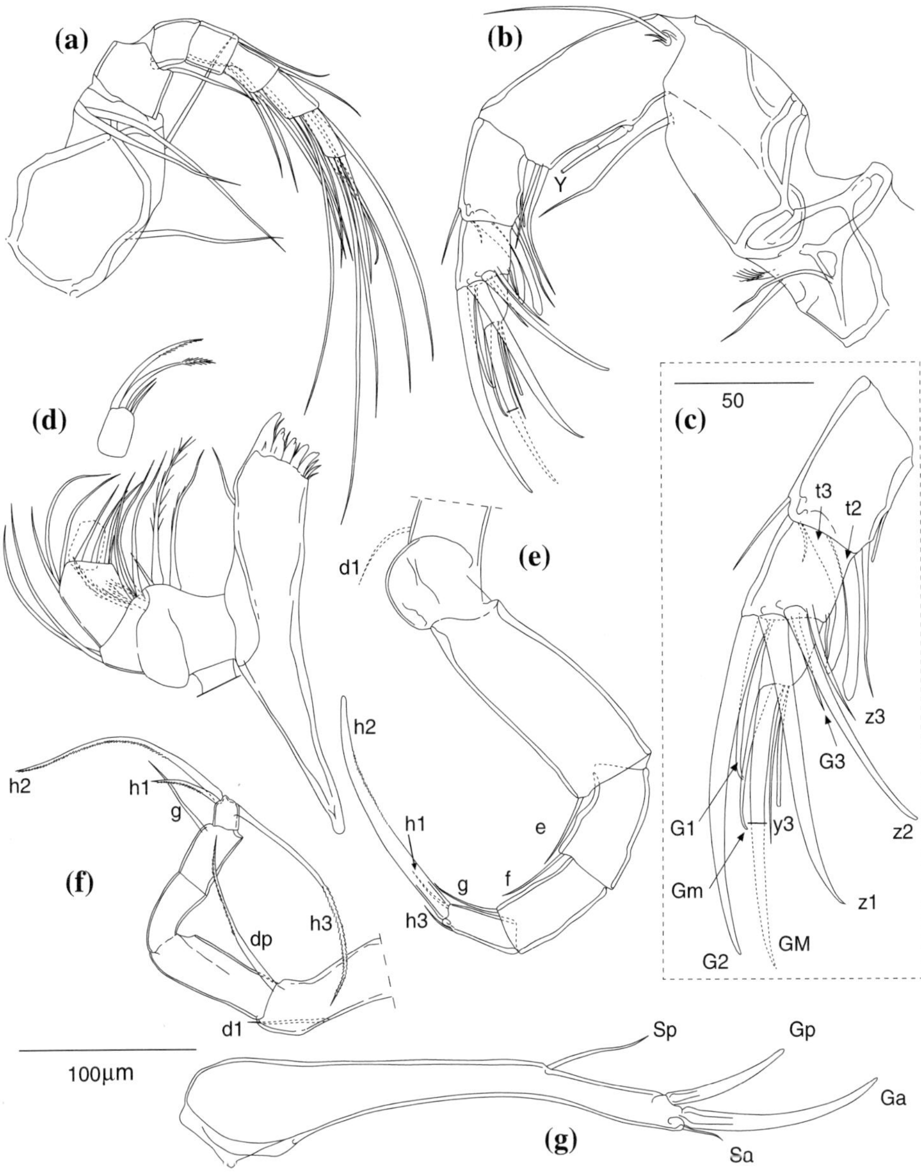

Fig. 3 *Fabaeformiscandona tora* n. sp. **a**—An1 (LBM 1430000900); **b**—An2 (LBM 1430000900); **c**—detail of distal part of male An2 (LBM 1430000900); **d**—Md (LBM 1430000900); **e**—L6 (LBM 1430000900); **f**—L7 (LBM 1430000901); **g**—CR (LBM 1430000900)

rounded posterior margin (Figs. 1f & 2f). Ventral margin only slightly concave. LV overlaps RV with flanges on postero-dorsal and antero-dorsal margins, the former of which is larger and more strongly developed than in male (Fig. 1k). Inner lamella wide anteriorly, narrow posteriorly (Fig. 2f).

An1 with eight segments, first two fused together forming large base (Fig. 6a). First segment with robust, setulous seta on dorsal–apical corner. Second segment with robust setulous apical–dorsal seta and two long apical–ventral setae. Third segment quadrate with one apical–dorsal seta. Fourth segment quadrate with long apical–dorsal seta. Fifth segment slightly elongate with two long apical–dorsal setae and one shorter apical–ventral seta. Sixth segment slightly elongate with two long apical–dorsal setae and one

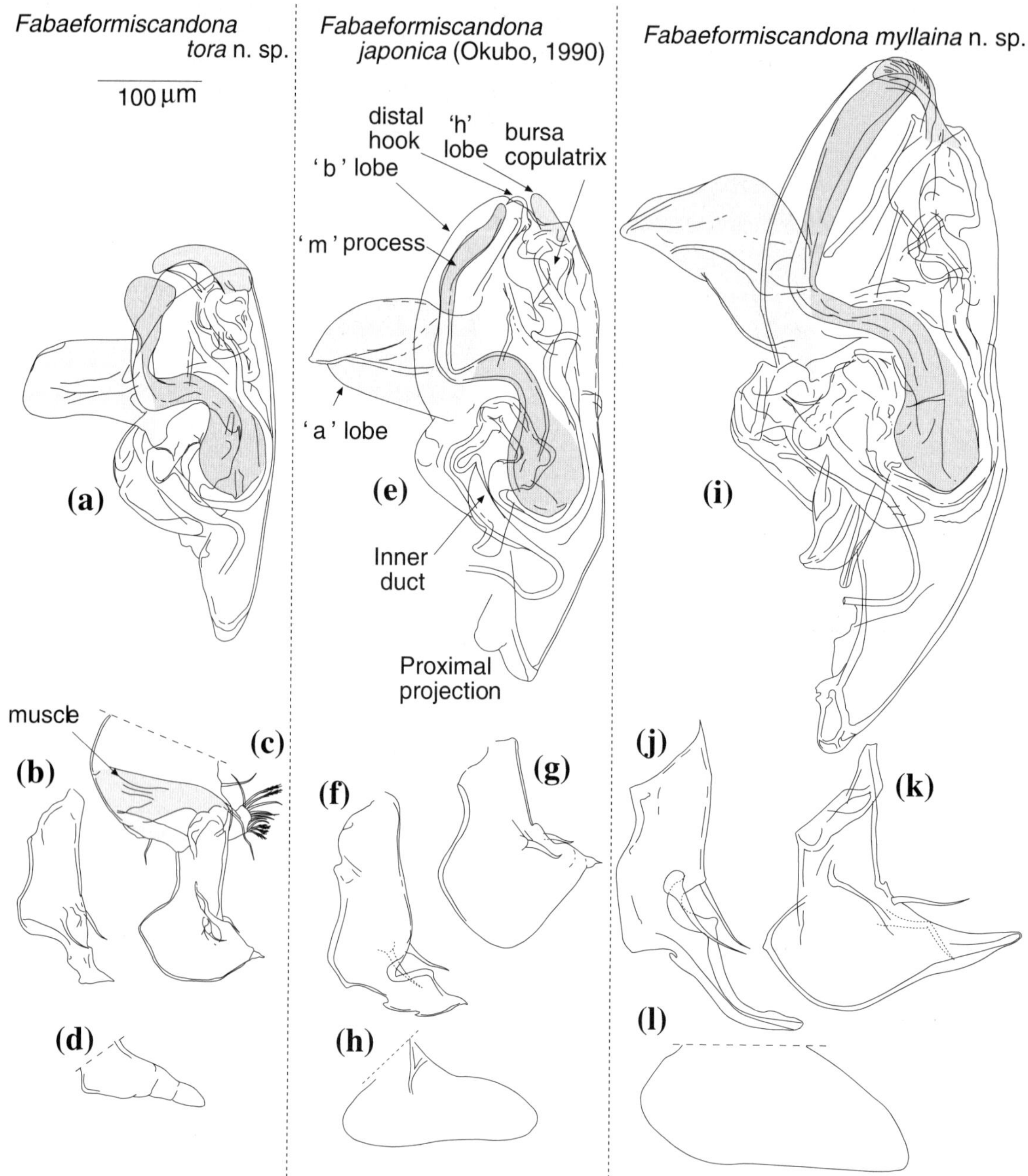

Fig. 4 **a–d** *Fabaeformiscandona tora* n. sp. (Takao, Kanazawa, Ishikawa). **a**—Hp (LBM 1430000900); **b**—L5 left palp (LBM 1430000900); **c**—right L5 (LBM 1430000900); **d**—FRO projection (LBM 1430000901). **e–h** *Fabaeformiscandona japonica* (Shinkai Pond, Kanazawa, Ishikawa). **e**—Hp (LBM 1430000920); **f**—L5 left palp (LBM 1430000920); **g**—right L5 (LBM 1430000920); **h**—FRO projection (LBM 1430000921). **i–l** *Fabaeformiscandona myllaina* n. sp. (Shinkai Pond, Kanazawa, Ishikawa) **i**—Hp; **j**—L5 left palp; **k**—right L5; **l**—FRO projection

shorter apical–ventral seta. Seventh segment with three long and one very short apical setae. Terminal segment with three setae of unequal length and one aesthetasc (*ya*).

An2 male with six segments, female with five (Fig. 6b). Male *t2* and *t3* setae large and inflated, *t4* very small (Fig. 6c). *z3* seta-like, *z1* and *z2* well developed claws, *z2* shorter than *z1*. *G3* reduced

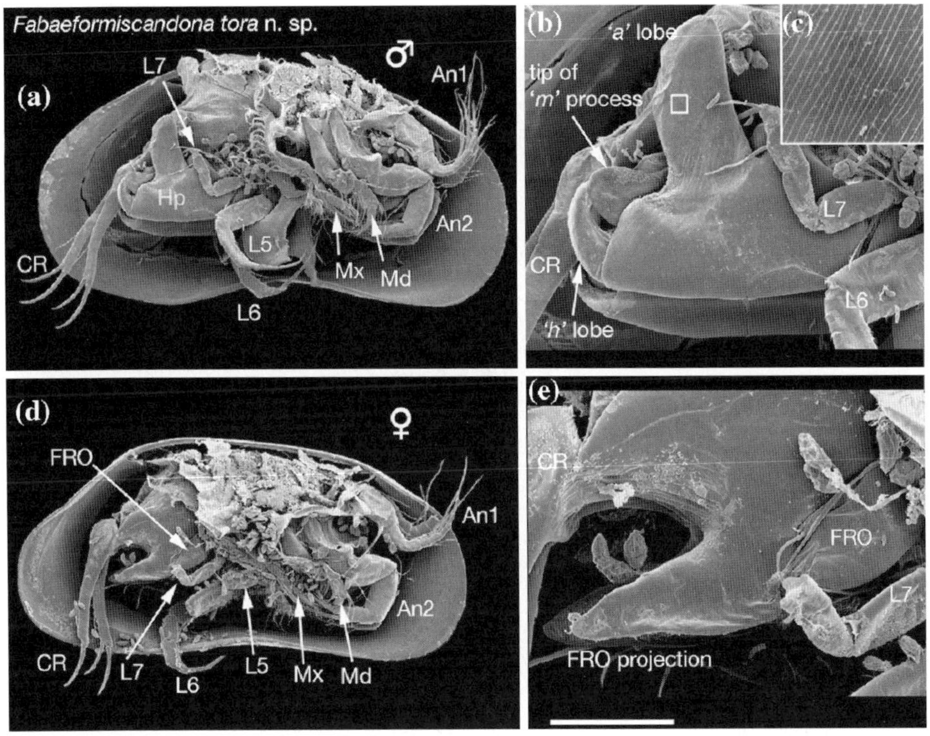

Fig. 5 *Fabaeformiscandona tora* n. sp. (Takao, Kanazawa, Ishikawa) **a**—male, freeze-dried specimen with right valve removed (LBM 1430000906). **b**—male, detail of Hp (LBM 1430000906). **c**—detail of *a* lobe of Hp, position of which marked by square on b (LBM 1430000906). **d**—female, freeze-dried specimen with right valve removed (LBM 1430000907). **e**—detail of posterior of female body (LBM 1430000907). Scale bar = 247 μm for a & d, 92 μm for b, 12 μm for c, and 66 μm for **e**

to small setae. *Gm* approximately half length of *GM*.

Md *s2* reduced and setulous (Fig. 6d). Alpha and beta setae very small, slim and smooth. Gamma seta slender and smooth. Md coxa well developed with long seta on outer edge and short setules on inner edge of endite (Fig. 6e).

L5 males with very large, asymmetrical, fused palps (Fig. 4j & k). Right palp with rounded, beak-shaped distal end. Left palp elongate with long projection distally, slightly hook-shaped.

L6 elongate with long *d1* seta and short *e* and *f* setae (Fig. 6f). Seta *g* relatively long. Distal quarter of claw *h2* smooth.

L7 with 5 segments, *d1* relatively short, *dp* long (Fig. 6g). Seta *g* long, *h3* and *h2* sub-equal in length. Seta *h1* relatively long.

CR slightly curved with inflated proximal base (Fig. 6h). *Sp* relatively long, *Sa* short. *Gp* approximately 75% length of *Ga*.

Hp elongate and distally acutely rounded (Fig. 4i). *a* lobe curved roughly S-shaped and pointed distally. *m* process with slightly inflated, rounded distal end. *h* lobe relatively small and curved, widening and with curved striations distally. Proximally Hp with a large, elongate, robust, slightly curved projection of the proximal inner shield, pointing in opposite direction to distal lobes (the proximal projection).

FRO in lateral view with angular shape and pointed anteriorly and distally. FRO protrusion large and stubby (Fig. 7).

Remarks

All female specimens of *F. myllaina* n. sp. had a polygonal pattern on the posterior part of the carapace (Fig. 8). Danielopol (1978) figured a similar pattern on the carapace of *Candona angulata* Müller, 1900 and attributed it to problems of calcification.

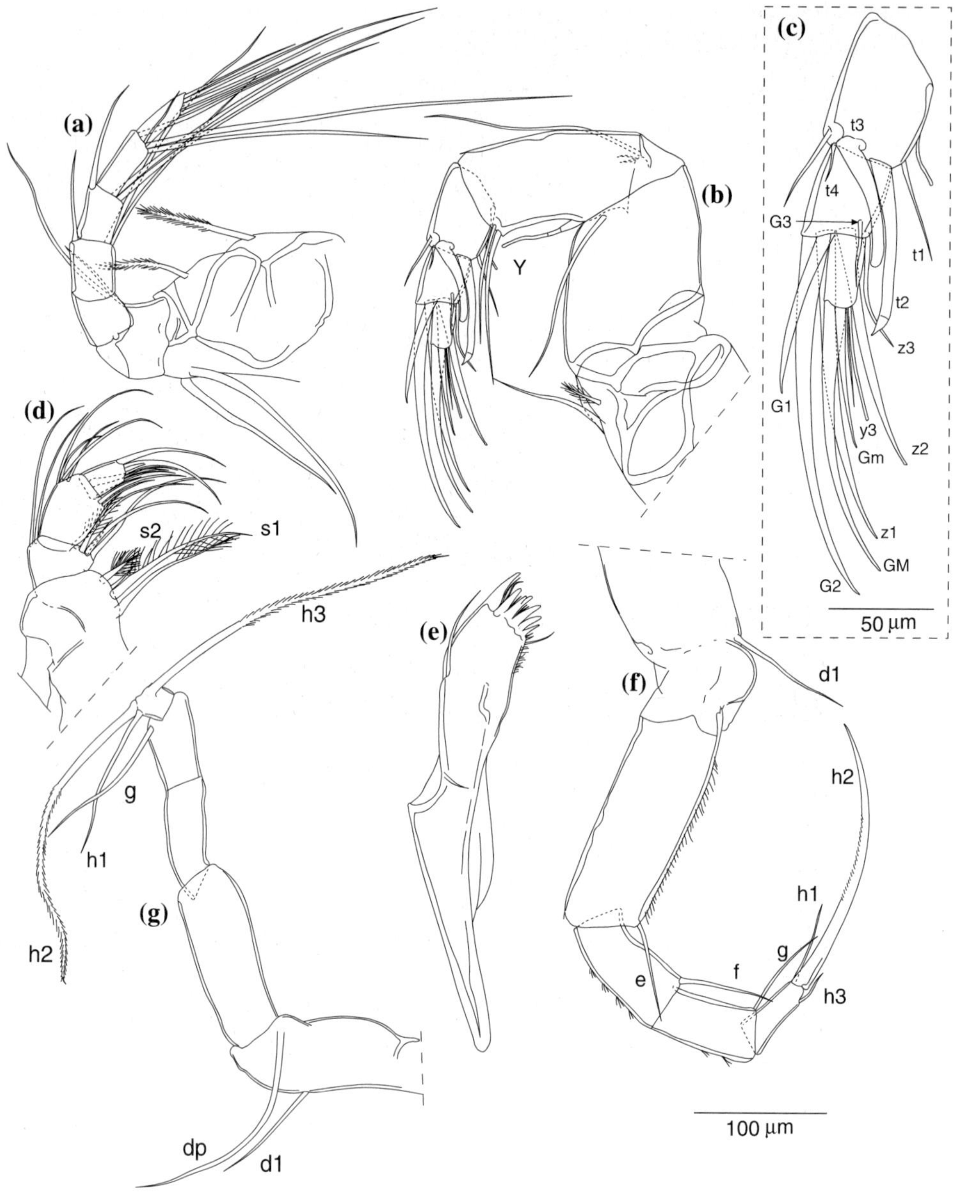

Fig. 6 *Fabaeformiscandona myllaina* n. sp. (Shinkai, Kanazawa, Ishikawa) **a**—An1 (LBM 1430000913); **b**—An2 (LBM 1430000913); **c**—detail of distal part of male An2 (LBM 1430000913); **d**—Md palp (LBM 1430000913); **e**—Md coax (LBM 1430000913); **f**—L6 (LBM 1430000913); **g**—L7 (LBM 1430000913)

Of the three species encountered in the study, *F. myllaina* n. sp. is the largest and is easily distinguished from the other two species by the lateral outline of the carapace (Fig. 1 & 2). The shape of the hemipenes, L5 palps and female genital lobes are also different between the species (Fig. 4). The Hp large hooked *h* lobe of *F. tora* n. sp. is much larger than the one found in *F. japonica* (Fig. 4e), which is straighter; the *h* lobe of *F. myllaina* n. sp. is also hooked but is striated. The Hp *a* lobe of *F. japonica* has a characteristic indentation at its distal end, in contrast to the rounded, blunt end of this lobe in *F. tora* n. sp. The *a* lobe of *F. myllaina* n. sp. is pointed distally and S-shaped. The male L5 palps vary in shape between species, with the left palp broadest in *F. japonica* (Fig. 4f) and most elongate in *F. myllaina* n. sp. The right L5 palp is most

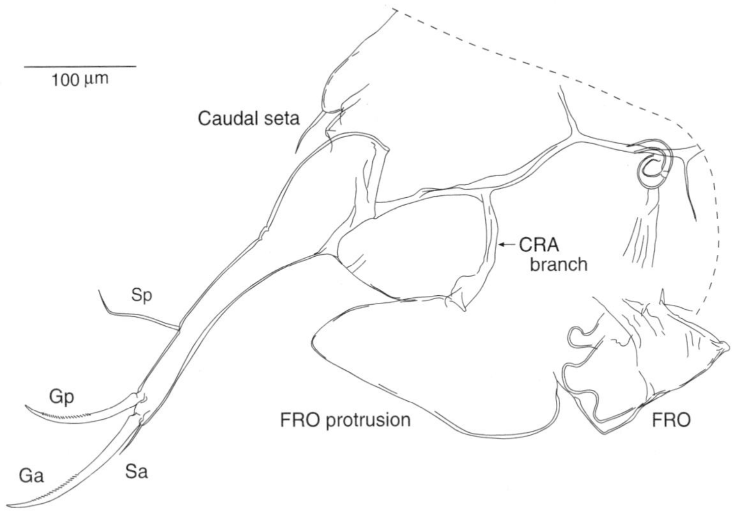

Fig. 7 *Fabaeformiscandona myllaina* n. sp. Posterior part of female body (LBM 1430000914)

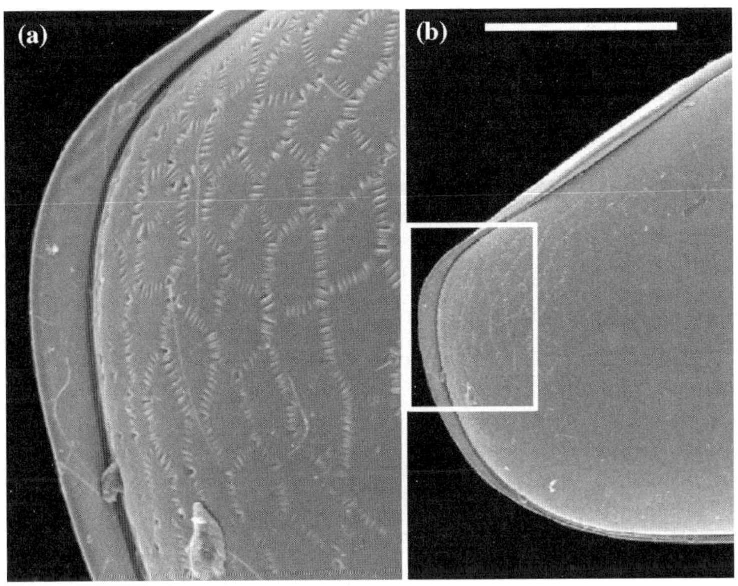

Fig. 8 *Fabaeformiscandona myllaina* n. sp. (LBM 1430000919). **a**—detail of posterior of female carapace. **b**—Posterior third of female carapace. White rectangle indicates position of **a**. Scale bar = 62 μm for A, 208 μm for **b**

inflated in *F. japonica* (Fig. 4g), while that of *F. myllaina* n. sp. is drawn out anteriorly into a beak-shaped protrusion. The protrusion of the female genital lobe is most slender in *F. tora* n. sp., and most rounded in *F. myllaina* n. sp., while that of *F. japonica* is of intermediate shape between the other two species.

Within the genus, three groups can be recognized by the absence and presence of the posterior flaps of the female carapace: one group with no flaps on either valve, one group with a flap on the left valve and a third group with a flap on both valves (Absolon, 1978). The species studied belong to the second group, which includes both *Fabaeformiscandona holzkampfi* (Hartwig, 1900) and *Fabaeformiscandona fabaeformis* (Fischer, 1851). *Fabaeformiscandona tora* n. sp. is similar to *F. holzkampfi* in carapace shape but differs in

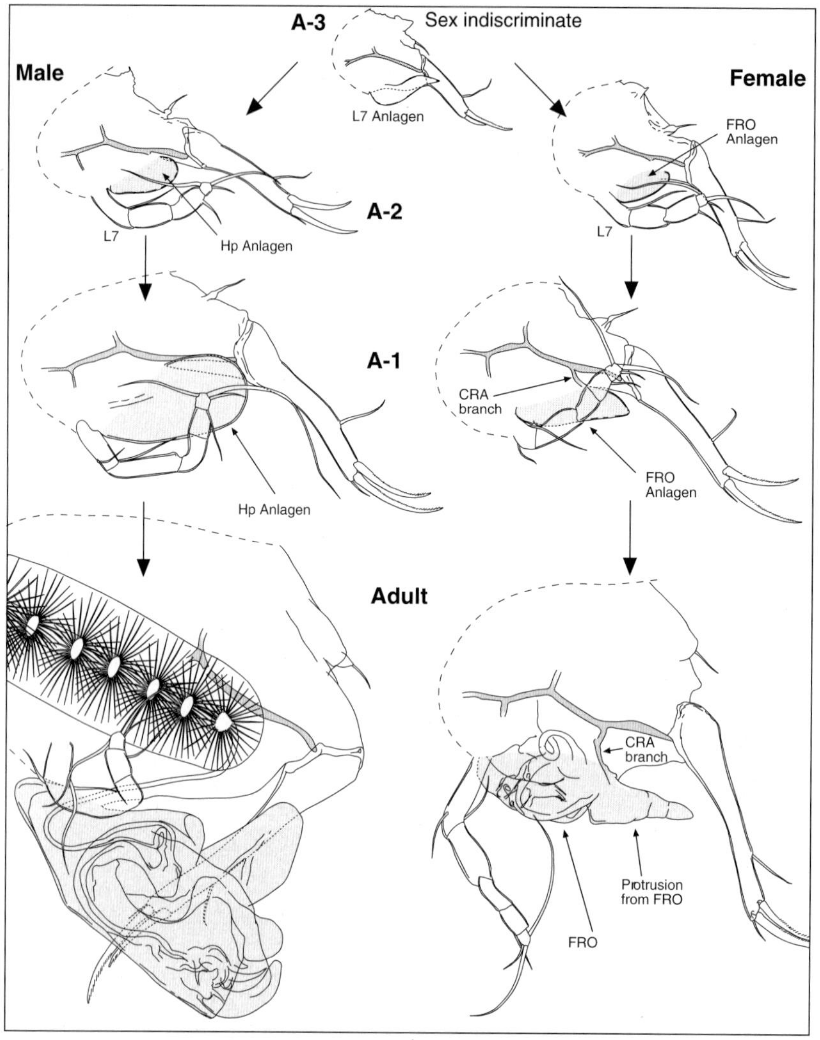

Fig. 9 Ontogeny of the sexual organs of *Fabaeformiscandona tora* n. sp. (Takao, Kanazawa, Ishikawa). A-3 instar (LBM 1430000911); A-2 instar, male (LBM 1430000908); A-2 instar, female (LBM 1430000912); A-1 instar, male (LBM 1430000909); A-1 instar, female (LBM 1430000910); Adult, male (LBM 1430000903); Adult, female (LBM 1430000902)

the shape of the male L5 and Hp and by the lack of the prominent posterior collar of the female carapace that typifies *F. holzkampfi*. *Fabaeformiscandona fabaeformis* is also similar to *F. tora* n. sp. in carapace shape, but the former is larger with different shaped male L5 and Hp, plus the segments of the antennule of *F. fabaeformis* are significantly more elongate than those of *F. tora* n. sp. *Fabaeformiscandona myllaina* n. sp. is easily distinguished from other species of *Fabaeformiscandona* by carapace shape, Hp and male L5 morphology.

Ontogeny of the sexual organs

A lobe posterior to the L7 first appears in the A-2 instars of *Fabaeformiscandona tora* n. sp. (Fig. 9). At this stage discrimination between males and females can be made by the shape of the L5 palps; by this instar stage the males have an enlarged L5 palp compared with that of the females. The shape of the posterior lobe in both sexes is of similar shape, although slightly smaller in females. At the A-1 instar stage the L5 palps can again be used to discriminate between males and

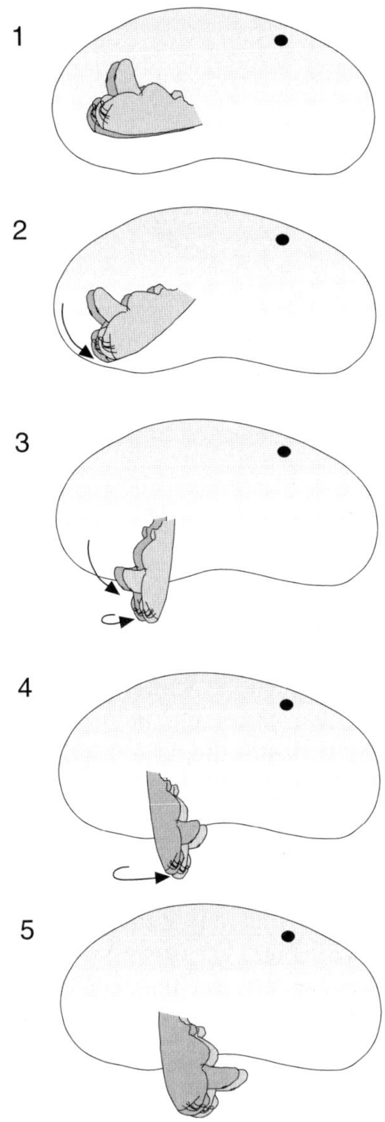

Fig. 10 Movement of the Hp from storage position to pre-copulative position

females, but additionally the males have incipient testes visible below the posterior region of the carapace. The lobe behind the L7 is also a different shape between males and females in this instar stage; large and rounded in the males and small and pointed in the females. Additionally, the females have a small branch of the CR attachment that curls down to the dorsal, posterior corner of the lobe (Fig. 9); this is missing in the male A-1 instars. In the adult females the lobe consists of two parts, a rounded lobe anteriorly containing the sexual organs (the FRO), and a posteriorly pointing lobe with no internal structure (the female genital lobe of other authors). The branch of the CR attachment linked to the posterior lobe in the A-1 instars is also present in adult females, but absent in the adult males.

The Zenker organs and the internal structure of the Hp are not fully developed immediately after moulting from the A-1 to the adult stage. These structures develop after the moult as indicated by dissections of freshly moulted adult males that reveal incipient Zenker organs and internal parts of the Hp (see Wouters, 1998 for a review of immature Zenker organs in other species). The length of time it takes for an adult male to fully develop its sexual organs and become sexually active is currently unknown. Danielopol (1980) reported that it took some days after the final moult for the Zenker organs and internal parts of the Hp to fully form in the species *Candona angulata*, *Candona sanociensis* (Sywula, 1971) (as *C. dancaui*) and *Pseudocandona pratensis*.

Copulatory behaviour (observations of living material)

The copulatory behaviours of two species, *Fabaeformiscandona japonica* and *Fabaeformiscandona tora* n. sp., were investigated and were found to be similar.

The males of both species often tried unsuccessfully to copulate with females; one male of *F. tora* n. sp. attempted copulation six times (each time with a different female) within the space of 30 min before a successful copulation was achieved. As a male approached, females usually closed their carapaces. The male then orientated the stationary female using the antennae and sixth limbs so that the ventral part of the male carapace was in contact with the posterior margin of the female carapace (Figs. 11 & 14) and clamped the female carapace with the anterior-ventral part of their carapace and the antennae. The male then repeatedly tried to insert the left L5 into the female's carapace, the female remaining stationary during this time with the carapaces closed. The male either let the female go after 5–10 s, or waited longer (up to a couple of minutes), stationary, until the female opened the carapace to move away. The male was then sometimes

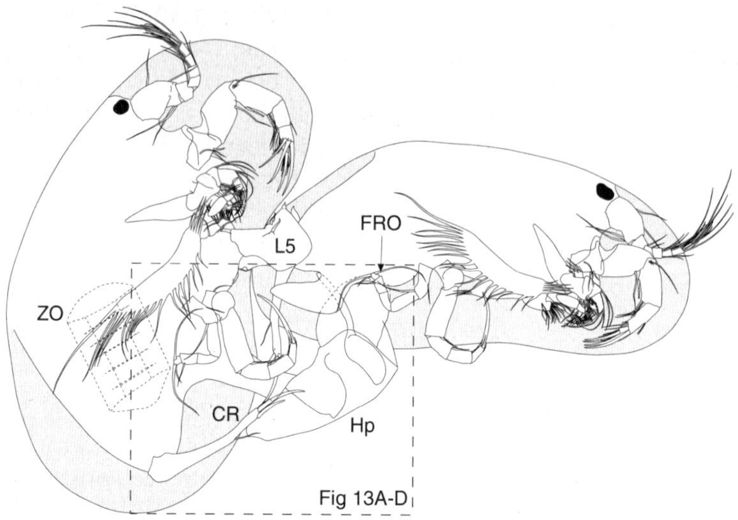

Fig. 11 Copulation position of *Fabaeformiscandona japonica*, right valve removed

successful in inserting the L5 palps into the female's carapace. On achieving insertion of the left L5 the right L5 was also inserted and the female valves spread apart. The male swung down the Hp from the posterior of the carapace (Fig. 10). As the Hp swung down they rotated as one unit through 180° (counter-clockwise as viewed from the dorsal), so that the Hp on the right side in the stored position was rotated to the left side of the body, and vice versa with the left hand side Hp. To achieve erection and insertion of the Hp into the female carapace three movements happen simultaneously: 1. In the pre-copulatory position (Fig. 13a), the male pushed its CR down onto the proximal projection of the Hp so that the tip pushed up into the female carapace, the Hp being pivoted at the base (Fig. 13b). 2. The *a* lobe of the Hp flipped outwards and down towards the main body of the Hp. 3. The large tip, including the bursa copulatrix, of the Hp was thrust outwards at an angle of approximately 45° to the long axis of the main Hp body. At erection the distal hook (*dh*) of the Hp became extruded at the tip of the Hp (Fig. 12). As soon as the Hp was inserted into the female carapace the Zenker organs started pumping (Fig. 13c). This had the effect of straightening the inner duct of the Hp, which in turn pushed up against the *m* process. This movement pushed the *m* process up and around, so that it hooked over the FRO protrusion (Fig. 13d). During some copulations the *m* process was repeatedly rotated back and forth before a secure connection with the female was achieved. Other times the *m* process was successfully rotated into the correct position on the first try. With this achieved, the male was mostly stationary with the only movement observed being the beating of the Mx branchial plates and the pumping of the Zenker organs. The male antennae during this time were withdrawn inside the carapace (Fig. 11), occasionally being extruded to rest gently on the female's carapace. The female was often more active than the male, moving the CR and An2 in addition to the Mx branchial plate movement. During an individual copulation event the angle and position of the male and female varied very little once the sexual organs were in contact. However, the position adopted by various copulating couples varied somewhat with the angle between the males' and females' ventral margins of the carapaces ranging from 48° to 78° (Fig. 14).

During the first seconds of copulation the rate of contraction of the Zenker organs was at the highest and varied from approximately two times per second (Fig. 14) to once a second. After the initial contractions the rate reduced and towards the end of the period of copulation, stopped altogether. The rate and number of contractions varied considerably from individual to individual, with some males contracting the Zenker organs

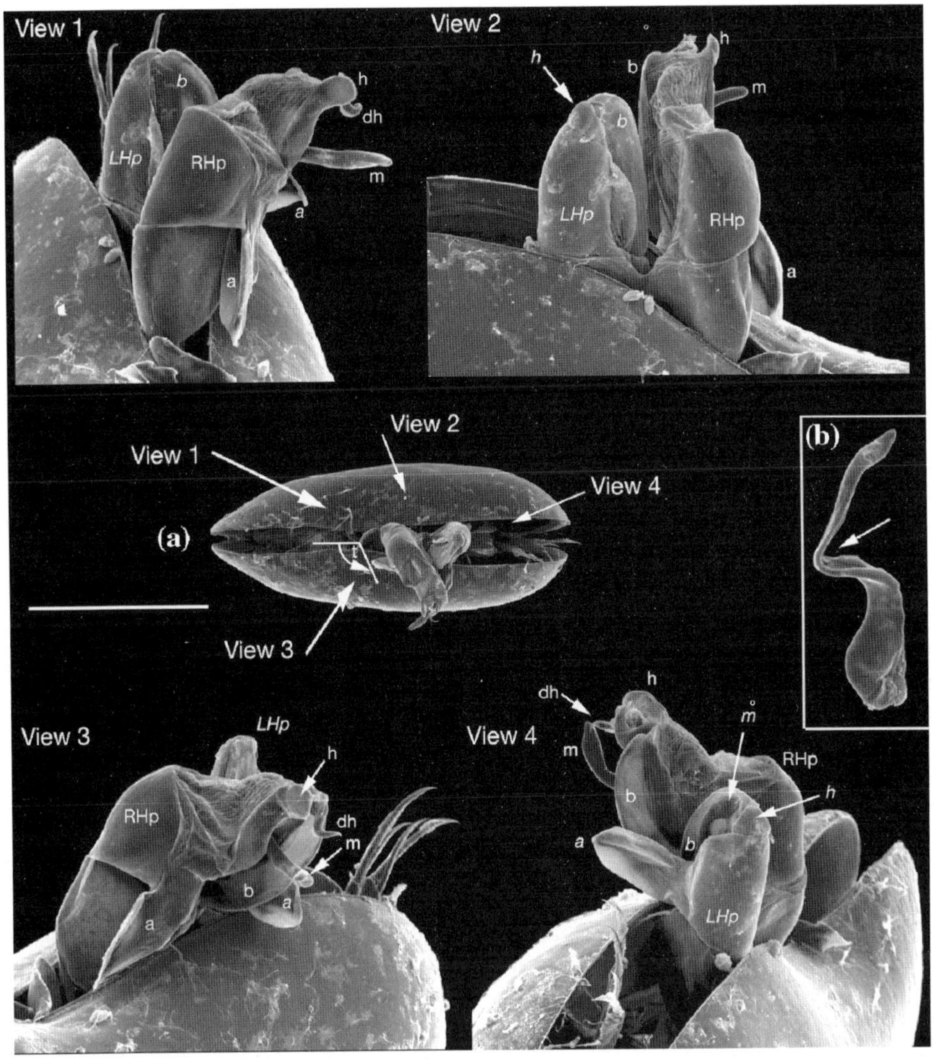

Fig. 12 *Fabaeformiscandona japonica*, Lake Biwa, male, **a**—Ventral view of male dried with HMDS (LBM 1430000925). Specimen was killed using ethanol during copulating with female. Upon death the male and female became separated and Hp of male rotated about 100° from copulatory position (marked by angle t on a). The left Hp, LHp (on the left side of the body when in storage position, but on the right side during copulation, see Fig. 9) reverted back to the non-erect position. The right Hp, RHp (on the right side of the body when in storage position, but on the left side during copulation) remained in the erect position with *m* process rotated. Views 1–4 correspond to those marked on a. **b**—*m* process removed from Hp. Arrow marks point of flexibility. Scale bar = 500 μm for a, 193 μm for views 1–4 and 121 μm for b

frequently throughout copulation, while others contracted them just a few times. During copulation the Zenker organs were usually not fully relaxed, but rather alternated between being fully contracted and partially contracted. Prior to the Hp being withdrawn the Zenker organ became fully relaxed, presumably allowing the inner duct to spring back into its original shape and the *m* process to be rotated and moved back down into the Hp body. The L5 palps were removed from the female carapace and the female released. The Hp was then returned to its non-erect state (i.e., the tip withdrew within the main body of the Hp and the *a* lobe flipped back to its original position) and rotated back into the storage position. The process of returning the Hp to the storage position was much slower than the initial unfolding of the Hp prior to copulation, and some

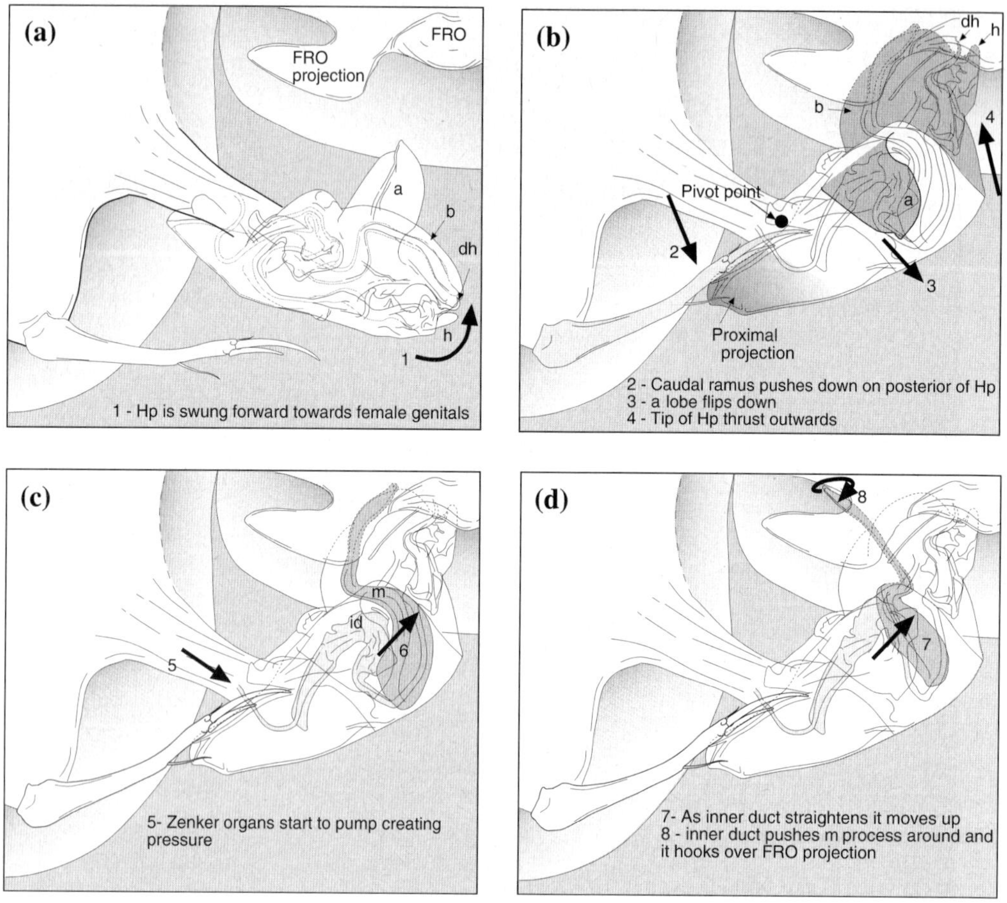

Fig. 13 Interaction of Hp and FRO during copulation of *F. japonica*. L6, L7 and female CR not drawn for clarity. **a**—Pre-copulative position. **b**—Erection of Hp. **c**—Movement of inner duct. **d**—Rotation of *m* process of Hp

males took several attempts (rotating the Hp back and forth) until successfully returning the Hp to the original storage position.

The length of time copulation lasted varied considerably between specimens (Fig. 14). After initial contact between males and females further contact, prior to copulation, lasted from between 10 s to over three minutes. Copulation lasted from seven seconds to nearly 17 min.

Discussion

The ontogeny and origin of the sexual organs

The total number of segments represented between the L7 and the CR is not directly known in *Fabaeformiscandona* as no limbs or segmentation are expressed in this region. However, Martens (2003) and Matzke-Karasz & Martens (2005) demonstrated that the giant cyprid *Liocypris grandis* Sars, 1924 has five features with various morphologies associated with the female genitals. They numbered these five structures R1 to R5 and considered them to be remanent limbs. They concluded that the female genitals in all podocopid ostracods are formed of five segments (i.e., one pair of limbs per segment), but the limbs of these segments are only seen in one species of modern ostracod, *Liocypris grandis.* The R5 of *L. grandis* is somewhat similar in morphology and position to that of the FRO protrusion of some *Fabaeformiscandona* species (e.g., *F. japonica* and *F. tora* n. sp.) (Fig. 5e). However, in other *Fabaeformiscandona* species the FRO protrusion is of varied shape (e.g., *Fabaeformiscandona siliquosa* (Brady, 1910) and *Fabaeformiscandona brisiaca* (Klie, 1938)), reduced (e.g.,

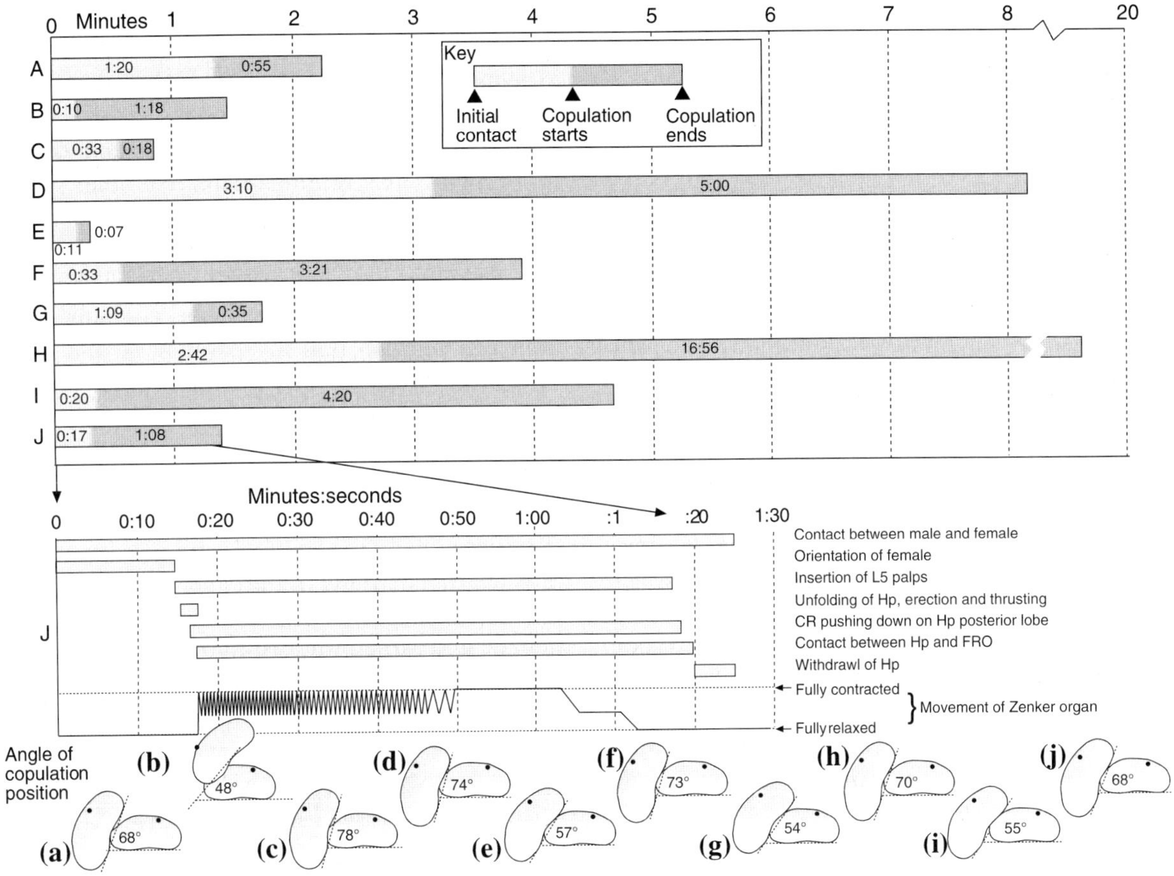

Fig. 14 Duration and positions of copulations in *F. japonica* (Lake Biwa, Shiga). A to J are copulation events. Copulation J is also displayed as individual events that occurred during copulation. Angles of copulation positions are measured between the ventral margins of the males' and females' carapaces when the sexual organs were in contact

Fabaeformiscandona breuili (Paris, 1920) and *Fabaeformiscandona latens* (Klie, 1940)) or totally missing (e.g., *Fabaeformiscandona protzi* (Hartwig, 1898) and *Fabaeformiscandona brevicornis* (Klie, 1925)). It is therefore not clear at present if the FRO protrusion of *Fabaeformiscandona* species is homologous with the R5 of *L. grandis* and further work is required to document its morphology and position in different groups. There is no evidence of any structures homologous to R1-4 in any *Fabaeformiscandona* species so far investigated by the authors.

The FRO protrusion of *Fabaeformiscandona* is associated with a curling branch of the CR attachment (Figs. 8 and 9) and this may indicate that it is of abdominal origin, perhaps a remanent pair of abdominal appendages. The *Anlagen* of the FRO protrusion in *F. tora* n. sp. appears in the A-1 instar as evidenced by the appearance of the branch of the CR attachment. Following Martens (2003) hypothesis that the FRO of ostracods is formed by five segments and assuming that segments form sequentially from the anterior to the posterior, it would mean that these five segments are already formed by the A-1 stage. This would suggest a rapid and compressed ontogeny of the posterior of the body of ostracods. All major podocopid groups of ostracods add one pair of limbs at each instar from the A-7 to A-2 stages, with the exception of the A-6 instar (e.g., Claus, 1868; Kesling, 1951; Weygoldt, 1960; Fox, 1964; Roessler, 1983; Smith & Martens, 2000; Smith & Kamiya, 2002, 2003; Horne et al., 2004). This gap in the ontogeny at the A-6 instar is attributed to the loss of the maxilla (e.g., Smith & Martens, 2000). Speculating that the ancestral ostracod also added one segment/pair of limbs per instar stage would indicate a loss of three instar stages during the evolution of the Podocopida.

The male An2

The An2 of the Candonidae is sexually dimorphic with the males having six segments, one more than the females, the result of the fourth segment in the A-1 stage dividing into two (this segment doesn't divide in females). From the extra division four setae protrude, two of which are enlarged, curved and have a rounded or pointed tip (either called the *t2* and *t3* setae or the male setae) (Figs. 3c & 7c). The *t2* and *t3* setae are presumed to be connected with copulation; McGregor & Kesling (1969b) suggested that they assist the male in holding on to the smooth dorsal margin of the female carapace, while Meisch (2000) noted that although the exact function is unknown, "they are no doubt connected with the important sensorial and prehensile role of the male antennae before and during copulation" (p. 12). Horne et al. (1998) assumed mechano- and chemo-sensorial roles for the male setae, based on observations of the morphology. With *Fabaeformiscandona japonica* and *F. tora* n. sp., the male An2 were used extensively prior to copulation in order to orientate the female into the correct copulatory position. During such handling the *t2* and *t3* setae (on the inner side of the appendage) frequently came into contact with the female carapace. However, during copulation the An2 were mostly tucked out of the way inside the carapace (Fig. 11), only occasionally being extruded so that the tips of the claws rested gently on the female carapace; the *t2* and *t3* setae did not come into contact with the females *during* copulation. During mating in a reversed ventral to ventral position, the antennae of males of *P. pratensis* are also tucked out of the way (Danielopol, 1980). Contrary to this, other authors (e.g., Kaufmann, 1892; Rioja, 1940; Ferguson, 1944) stressed the role of the An2 in attaching the male to the dorso-posterior of the female carapace during copulation, indicating that the precise method of attachment varies between different groups, even though the copulation position may be similar.

The male L5 and the female carapace

The male L5s each have a large rounded base that supports the palps (Fig. 4). The palps are enlarged and asymmetrical in shape, and do not have a moveable terminal hook like other groups (e.g., the Cypridinae). During copulation the blade-shaped left palp is inserted first into the female carapace and the large, inflated right palp is inserted second. The basis contains large muscles that can pull the palps back and forth. The muscles are the largest in any appendage and it must be assumed that the male can move the palps back and forth with considerable force. Prior to copulation the male's large L5s are used to force the female's carapace open. McGregor & Kesling (1969b) also noted similar behaviour in *Fabaeformiscandona suburbana* with the males inserting their L5 palps "inside the valves of the female and turned to assist in forcing and holding the valves apart for the intromission of the voluminous hemipenes" (p. 230).

The carapaces of both sexes in the three *Fabaeformiscandona* species studied each have two flaps at each end of the hinge where the left valve overlaps the right valve, the purpose of which is to produce a tight closure between the valves (Absolon, 1978). In the females, the posterior flap on the left valve is significantly enlarged compared with that of the males (Fig. 1g–l) and covers the area where the males try to insert the L5 palps. We hypothesize that the female posterior flaps have a secondary function of hindering the males from inserting the L5 palps when the female's carapace is tightly closed. Observations of laboratory cultures included frequent failed attempts of the males to open the females' carapaces, which supports this view.

One consequence of sexual reproduction is a conflict of interests between males and females, which can be reflected in sexually selected features. Males want to mate with as many females as possible as sperm is 'cheap' to produce and reproductive success is mainly a result of the number of copulations achieved. Females, on the other hand, may only need to mate once to obtain enough sperm to fertilize their eggs, and hence want to choose the best 'quality' male rather than just mating lots of times with many males (Butlin & Menozzi, 2000). This conflict of interests can result in an 'arms race' between males and females with males evolving traits that promote frequent copulation, while females evolve traits that assist with choosing the best mate. With the *Fabaeformiscandona* species studied, we hypothesize that such a conflict type of sexual selection exists between the sexes as evidenced by the males' L5s and the females' carapace. The males' very large and powerful L5s have evolved to increase the chances of males being able to force females to copulate. The females have countered this by evolving a defensive posterior flap on the carapace to foil unwanted male attention and thus are able to retain some control with whom they mate with.

On one hand, an 'arms race' between the sexes appears to be counterproductive as energy that could be used to produce more offspring is instead used to develop, maintain and improve features of the 'arms race'. It also means that in addition to the species generally evolving to external factors (e.g., habitat) that improve overall survivability, both sexes are also co-evolving intraspecifically just to stay in an equal position to each other, and with no overall benefit to either sex. On the other hand, such a conflict may have a beneficial outcome for the species; the strongest (fittest) males may have a better chance of opening a female's carapace compared with weaker males and thus may have a better copulation success rate (essentially an indirect contest between the males). However, if copulation success is a result of the males' strength alone, the weaker males may also be able to enjoy a certain rate of success, albeit lower than the strongest males. These weaker males can have their success rate lowered still if the females are also able to select appropriate males for copulation (i.e., select only the strongest males). This double safeguard against weaker males passing on their genes will therefore maintain or improve the genetic 'strength' of the species.

The function of the male CR during copulation

The process of the CR pushing down of the proximal projection of the Hp during copulation to push the Hp into the female carapace in *F. japonica* and *F. tora* n. sp. (Figs. 11 & 13) has not been previously reported for any other species. However, other species with a well-developed proximal projection on the Hp (e.g., *F. myllaina* n. sp.) perhaps copulate in a similar way. The distal lobes (*a*, *b* and *h*) and the *m* process of the Hp have generally been the parts that are considered taxonomically important and little attention is usually paid to the proximal part. Therefore the proximal part of the Hp is poorly represented in the literature (i.e., often it is not figured) and it is difficult to assess how widespread a Hp proximal projection is from the literature alone. However, *Fabaeformiscandona hyaline* (Brady & Robertson, 1870) clearly does have a well-developed Hp proximal projection (as figured in Plate 3 of Martens & Dumont, 1984). Other species that, from the literature, appear to have similar structures include *Fabaeformiscandona fabaeformis* (Fischer, 1851) (as figured in Fig. 33 of Sywula, 1974), *Fabaeformiscandona holzkampfi* (Hartwig, 1900) (as figured in Fig. 35 of Sywula, 1974), *Fabaeformiscandona fragilis* (Hartwig, 1898) (as figured in Fig. 34 of Sywula, 1974), *Candona candida* (O.F. Müller, 1776) (as figured in Fig. 19 of Sywula, 1974), *Candona weltneri* Hartwig, 1899 (as figured in Fig. 26 of Sywula, 1974), *Candona neglecta* Sars, 1887 (as figured in Fig. 27 of Meisch, 2000), *Candona lindneri* Petkovski, 1969 (as figured in Fig. 28 of Meisch, 2000), *Pseudocandona lobipes* (Hartwig, 1900) (as figured in Danielopol, 1969) and *Schellencandona triquetra* (Klie, 1936) (as figured in Marmonier, 1988; see Meisch 2000, Fig. 64). It is not possible to determine the presence of a Hp proximal projection in other species from the literature alone, but this feature could be widespread in the subfamily Candoninae.

The Hp lobes a, b *and* h *and the distal hook*

The *a* lobe of the Hp is large, flattened and different in shape in all three species in this study (Fig. 4). Internally they have no structure apart from a strengthening chitinous support. The outer surface of the *a* lobes in all three species have shallow striations approximately spaced at 0.5–1 μm intervals, running obliquely to the long axis of the lobes (e.g., Fig. 5c). The shallow nature of the striations indicates that their function is unlikely to be structural (i.e., to strengthen the lobe). Prior to copulation the tip of the *a* lobe forms the front edge of the Hp and comes into contact with the ventral margin of the female's carapace (Fig. 13a), but when the Hp achieves erection the *a* lobe is flipped down and outwards and no further contact with the female is made (Figs. 12 & 13b). The exact function of the *a* lobe is not clear at present but we speculate that it is perhaps sensory and used prior to copulation to indicate to the male that the female's carapace has successfully been forced open. The rapid downward movement of the *a* lobe as the Hp becomes erect may also help to push the female's valves further apart as the extruded tip of the Hp is thrust upwards towards the FRO. The flipped down position of the *a* lobe during erection was also noted by McGregor & Kesling (1969a) and Danielopol (1969, 1978, 1980). Danielopol (1969, 1978, 1980) did not venture a guess as to its function, but McGregor & Kesling (1969a), although unsure of its function did speculate that it may have a sensory role.

The *h* lobe is much smaller than the *a* lobe and is situated at the tip of the extruded part of the Hp during erection (Figs. 4 & 12). It comes into contact with the FRO during copulation, but due to the inability to clearly observe the *h* lobe during copulation its function is currently unknown. Its hook shape in both *F. tora* n. sp and *F. myllaina* n. sp. may indicate that it hooks into, or onto, the FRO. The *b* lobe is larger than the *h* lobe but is thinner, forming a curtain-like flap on the inner edge of the Hp during erection (Fig. 12). Again its role in copulation is unclear, but it possibly covers the FRO. Danielopol (1980) noted that the *b* and *h* lobes of *P. pratensis* only marginally come into contact with the FRO, but it should be noted that the *h* lobe of the *Fabaeformiscandona* species is significantly smaller and different in shape to that of *P. pratensis*.

The distal hook (*dh*) is extruded during erection so that it is positioned at the tip of the Hp (Fig. 12). Its position, size and shape would suggest that it hooks into the FRO opening, although this is currently speculative.

The Hp m *process*

The *m* process is a bent, chitinous rod that becomes extruded from the Hp body during erection (Fig. 12). It has a rounded distal section that the inner duct comes into contact with. Dissections of the Hp revealed that at the point of maximum curvature (marked with arrow on Fig. 12b), the *m* process has a limited amount of flexibility and it is partly elastic. Immediately after erection and positioning of the Hp to the FRO, the *m* process is pushed up and rotated through 90° and clipped on to the projection of the FRO, thus providing a secure contact between male and female. McGregor & Kesling (1969a) stated that in *Fabaeformiscandona suburbana* during erection the *m* process "seems to serve as a prop to keep the distal shield extended" and did not note that it rotated. It is possible that they saw the Hp in an erect state but before the *m* process had rotated. Danielopol (1969) did not see the interaction between the *m* process and the FRO of *Candona* aff. *C. compressa* but speculated that the *m* process either holds the female genital lobe (=FRO protrusion of this paper) or has a sensorial role; his first conjecture is similar to that seen in this study, although he did not surmise it may rotate. In *Pseudocandona pratensis* however, Danielopol (1980) demonstrated that the *m* process hooks into the internal edge of the genital opening. It should be noted that females of *P. pratensis* do not have a FRO protrusion for the *m* process to clip on to. The function of the *m* process in the *Fabaeformiscandona* species and *P. pratensis*, is essentially the same, i.e. to securely connect the male and female genitals. However, the actual way this is achieved is different; the *m* process hooks on the FRO protrusion in *F. tora* n. sp. and *F. japonica*, and hooks into the genital opening in *P. pratensis*.

The difference in the type of *m* process connection may play a part in the copulation position available to the male and female. The *Fabaeformiscandona* species copulate with the males attached to the posterior region of the females' carapace (Figs. 11 & 14). While *P. pratensis* starts copulation in a similar position to that of the *Fabaeformiscandona* species, the male later slips down and rotates, so that the male and female are in a reversed ventral to ventral position (Danielopol, 1980). A reserved ventral to ventral copulation position for *Fabaeformiscandona* species studied would mean that the *m* process would no longer reach the FRO projection. With *P. pratensis* there is no FRO protrusion and so the male instead clips the *m* process to the inner edge of the FRO genital opening. The results of this method is that *P. pratensis* is not restricted to a postero-dorsal copulation position.

Hemipenes/female genital lobe interaction

During copulation the Hp are held in place with the CR pushing down on the proximal projection and the *m* process hooked over the FRO protrusion (Fig. 13d). This provides a secure interface between the tip of the Hp and the FRO, the functions of which could be twofold. Firstly it would aid in the transfer of giant spermatozoa typical of the family. The spermatozoa are moved from the testes, through the Hp and into the narrow duct of the vagina and finally into the seminal receptacle. To do this, the males have two large pumps, the Zenker organs that move the spermatozoa. Presumably, the sperm must therefore be exchanged under some pressure and the corresponding interface between the male and female needs to be tight to prevent the male and female sexual organs from moving apart as the Zenker organs pump.

A second advantage of a secure interface is that the shape of the hemipenes and the female genital lobe are species specific, highlighted by the fact that these features are considered taxonomically important (e.g., Meisch, 2000). Different species of *Fabaeformiscandona* are often found together in the same habitat and such a difference in the shape of the sexual organs could act as a method to stop wasted interspecific copulation (i.e., only a male with the appropriately shaped sexual organs can mate with a particular female). A similar 'lock & key' hypothesis was considered by Horne et al. (1998) for ostracods, but they noted that even though male sexual organs can be of diverse shape, the corresponding female sexual organs are often similar in shape and do not appear to be correlated with the shape of the male Hp. The female sexual organs in *F. tora* n. sp. and *F. japonica* are of similar shape thus supporting Horne et al. (1998) view, but a lock & key' scenario does not just have to play on the FRO opening and tip of the Hp. The shape of the males' *m* processes and the FRO projections of the females are different and possibly do correlate, thus acting as a displaced 'lock & key'. Additionally, a similar displaced 'lock & key' role could be played by the carapace of the females and the L5 of the males; only males with the appropriately shaped L5s will succeed in forcing open a female's carapace. In this sense, the *Fabaeformiscandona* species could have a double combination, displaced 'lock & key' barrier against interspecific copulation.

Copulation in other species/groups

The copulation observed in the *Fabaeformiscandona* species in this study relied heavily on the Hp posterior lobe, *m* process, movement of the inner duct and the females FRO protrusion. However, not all species of *Fabaeformiscandona* have similar structures, for example the FRO protrusion is reduced in *F. breuili* and *F. latens* and totally missing in *F. protzi* and *F. brevicornis* so copulation must be different in these species. As noted above, the *m* process in *P. pratensis* serves a similar role as the *m* process in *Fabaeformiscandona tora* n. sp and *F. japonica*, namely to provide a secure connection, but the actual way this is achieved is different. The *m* process hooks over the FRO protrusion in the two *Fabaeformiscandona* species studied and hooks into the genital opening in *P. pratensis*. As *P. pratensis* has no FRO protrusion, it is possible that other species without FRO protrusions also copulate in a similar way to *P. pratensis*, such as *F. protzi* and *F. brevicornis*.

Species of the Cyclocypridinae, another candonid subfamily, lack both the *m* process and proximal projection on the Hp, and the FRO protrusion is very reduced, but species of this group have spermatozoa that are often longer than the carapace. Therefore the mechanism of producing a secure interface in this subfamily must also be different to that observed in this study for *Fabaeformiscandona*. The authors were unable to find Hp proximal projections in any species of the families Cyprididae, Ilyocyprididae and Notodromadidae and thus presumably the CR is not used in copulation in these families (indeed the subfamilies Cyprettinae and Cypridopsinae of the Cyprididae only have a reduced CR). Additionally, these groups all appear to lack both the FRO protrusion and the *m* process. This high variation of morphology of sexual features within the superfamily indicates that the mechanisms of copulation within the Cypridoidea must vary considerably between groups and species, even between species of the same genus and subfamily. Study of the variation of copulation mechanics within the podocopes may help in the understanding of the phylogeny and evolution of the group.

Acknowledgements The authors thank Renate Matzke-Karasz (Ludwig-Maximilians University, Munich, Germany) for her many useful comments and discussions, Dan Danielopol (Institute of Limnology, Austria) for his encouragement, advice and help in obtaining literature and David Horne (Natural History Museum, London, UK) for comments made on the manuscript.

References

Absolon, A., 1978. Die Gattung *Candona* (Ostracoda) im Quartär von Europa. Rozpravy Cheskoslovenske Akademie ved, Rada Matematickych a Prirodnich ved 88: 1–76.

Bauer, H., 1940. Über die Chromosomen der bisexuellen und der parthenogenetischen Rasse des Ostracoden *Heterocypris incongruens* Ramd. Chromosoma 1: 621–637.

Broodbakker, N. W. & D. L. Danielopol, 1982. The chaetotaxy of Cypridacea (Crustacea, Ostracoda) limbs: proposals for a descriptive model. Bijdragen tot de Dierkunde 52: 103–120.

Butlin, R. K. & P. Menozzi, 2000. Open questions in evolutionary ecology: do ostracods have the answers? Hydrobiologia 419: 1–14.

Claus, C., 1868. Beiträge zur Kenntnis der Ostrakoden. I. Entwicklungsgeschichte von Cypris. Schriften der Gesellschaft zur Beförderung der gesamten Naturwissenschaften zu Marburg 9: 151–166.

Cohen, A. C. & J. G. Morin, 1990. Patterns of reproduction in ostracodes: a review. Journal of Crustacean Biology 10: 184–211.

Danielopol, D. L., 1969. Recherches sur la morphologie de l'organe copulateur mâle chez quelques Ostracodes du genre *Candona* Baird (Fam. Cyprididae Baird). In Neale, J. (ed.), The taxonomy, morphology and ecology of Recent Ostracoda, Edinburgh, 136–153.

Danielopol, D. L., 1976. Supplementary data on *Pussella botosaneanui* Danielopol, 1973 (Ostracoda, Bairdiidae). Vie Milieu 26: 261–273.

Danielopol, D. L., 1977. Recherches sur les Ostracodes Entocytheridae. Données sur *Sphaeromicola cebennica juberthiei* nov. ssp. et *Sphaeromicola cirolanae* Rioja. International Journal of Speleology 9: 21–41.

Danielopol, D. L., 1978. Über Herkunft und Morphologie der Süßwasser-hypogäischen Candoninae (Crustacea, Ostracoda). Sitzungsberichte der Österreichischen Akademie der Wissenschaften, Mathematisch-Naturwissenschaftliche Klasse, Abteilung 1 187: 1–162.

Danielopol, D. L., 1980. Sur la biologie de quelques ostracods Candoninae épigés et hypogés d'Europe. Bulletin du Musee National d'Histoire Naturelle de Paris, Séries 4 2: 471–506.

Danielopol, D. L. & D. J. Horne, 1996. Some observations on sexual behaviour and the morphology of the male copulatory appendages in *Metacypris cordata* (Ostracoda, Timiriaseviinae). In Keen, M. C. (ed.), Proceedings of the 2nd European Ostracodologists Meeting. British Micropalaeontological Society, 171–174.

Ferguson, E., 1944. Studies on the seasonal life history of three species of freshwater Ostracoda. American Midland Naturalist 32: 713–727.

Fox, M. H., 1964. On the larval stages of cyprids and on *Siphlocandona* (Crustacea, Ostracoda). Proceedings of the Zoological Society of London 142: 165–176.

Gupta, B., 1968. Aspects of the motility in the non-flagellate spermatozoa of freshwater Ostracods. In Miller, P. L. (ed.). Aspects of motility. Symposia of the Society of Experimental Biology 22:117–129.

Hart, D. G. & C. W. Hart Jr., 1974. The ostracod family Entocytheridae. The Academy of Natural Sciences of Philadelphia, Monograph 18: 1–239.

Horne, D. J., D. L. Danielopol & K. Martens, 1998. Reproductive behaviour. In Martens, K. (ed.), Sex and parthenogenesis: evolutionary ecology of reproductive modes in non-marine ostracods. Backhuys Publishers, Leiden, The Netherlands: 157–195.

Horne, D. J., R. J. Smith J. E. Whittaker & J. W. Murray, 2004. The first British record and a new species of the superfamily Terrestricytheroidea (Crustacea, Ostracoda): morphology, ontogeny, lifestyle and phylogeny. Zoological Journal of the Linnean Society 142: 253–288.

Jensen, S., 1904. Biologiske og systematiske Undersøgelser over Ferskvands-Ostracoder. Videnskabelige Meddelelser fra Dansk naturhistorisk Forening i København 7: 1–78.

Kamiya, T., 1988. Morphological and ethological adaptations of Ostracoda to microhabitats in *Zostera* beds. In Hanai, T. et al. (eds), Evolutionary biology of Ostracoda. Kodansha Scientific, Tokyo, 303–318.

Kaufmann, A., 1892. Über die Gattung *Acanthopus* Vernet und eine neue Süßwassercytheride. Zoologischer Anzeiger 15: 393–395.

Kesling, R. V., 1951. The morphology of ostracod molt stages. Illinois Biological Monograph 21: 1–126.

Kesling, R. V., 1957. Notes on Zenker's Organs in the ostracod *Candona*. American Midland Naturalist 57: 175–182.

Kesling, R. V., 1965. Anatomy and dimorphism of adult *Candona suburbana* Hoff. 56pp. In Kesling, R. V., D. G. Darby, R. N. Smith, & D. D. Halls (eds), Four Reports of Ostracod Investigations, conducted under National Science Foundation Project GB – 26, The University of Michigan.

Kesling, R. V., 1969. Copulatory adaptations in ostracods Part III. Adaptations in some extinct ostracods. Contributions from the Museum of Paleontology, The University of Michigan 22: 273–312.

Klie, W., 1926. Ostracoda, Muschelkrebse. In Paul Schulze, (ed.), Biologie der Tiere Deutschlands 22: 1–55.

Krstić, N., 1972. Rod *Candona* (Ostracoda) iz Kongerijskikh Slojeva Juzhnog dela Panonskog Basena. Monographs of the Serbian Academy of Sciences and Arts, section of Natural and Mathematical Sciences 39: 1–145.

Lilljeborg, W., 1853. De Crustaceis ex Ordinibus Tribus: Cladocera, Ostracoda et Copepoda, in Scania Occurentibus. Om de inom Skåne förekommande crustaceer af ordningarne Cladocera, Ostracoda och Copepoda. Lund, Berlingska Boktryckeriet.

Lowndes, A. G., 1935. The sperms of freshwater ostracods. Proceedings of the Zoological Society of London 1935: 35–48.

McGregor, D. L. & R. V. Kesling, 1969a. Copulatory adaptations in ostracods. Part I. Hemipenes of *Candona*. Contributions from the Museum of Paleontology, The University of Michigan 22: 169–191.

McGregor, D. L. & R. V. Kesling, 1969b. Copulatory adaptations in ostracods. Part I. Adaptations in living ostracods. Contributions from the Museum of Paleontology, The University of Michigan 22: 221–239.

Marmonier, P., 1988. Biocénoses interstitielles et circulation des eaux dans le sous-écoulement d'un chenal aménagé du Haut-Rhône français. Thèse de doctorat de l'Université Claude Bernard, Lyon 1, part 1: 1–160, part 2: 1–108.

Martens, K., 1987. Homology and functional morphology of the sexual dimorphism in the antenna of *Sclerocypris* Sars, 1924 (Crustacea, Ostracoda, Megalocypridinae). Bijdragen tot de Dierkunde 57: 183–190.

Martens, K., 2003. On a remarkable South African giant ostracod (Crustacea, Ostracoda, Cyprididae) from temporary pools, with additional appendages. Hydrobiologia 500: 115–130.

Martens, K. & H. J. Dumont, 1984. The ostracod fauna (Crustacea, Ostracoda) of Lake Donk (Flanders): a comparison between two surveys 20 years apart. Biologisch Jaarbock Dodonaea 52: 95–111.

Matzke-Karasz, R., 1997. Descriptive nomenclature and external morphology of the Zenker's organs of Cypridoidea (Crustacea, Ostracoda). Sonderveröffentlichungen, Geologisches Institut der Universität zu Köln 114: 295–315.

Matzke-Karasz R., 2005. Giant spermatozoon coiled in small egg: fertilization mechanisms and their implications for evolutionary studies on Ostracoda (Crustacea). Journal of Experimental Zoology 304B: 129–149.

Matzke-Karasz, R. & K. Martens, 2005. The female reproductive organ in podocopid ostracods is homologous to five appendages: histological evidence from *Liocypris grandis* (Crustacea, Ostracoda). Hydrobiologia 542: 249–259.

Meisch, C., 2000. Freshwater Ostracoda of Western and Central Europe. In Schwoerbel, J. & P. Zwick (eds), Süßwasserfauna von Mitteleuropa 8/3. Spektrum Akademischer Verlag, Heidelberg, Berlin.

Müller, G. W., 1880. Beitrag zur Kenntnis der Fortpflanzung und der Geschlechtsverhältnisse der Ostracoden. Zeitschriften für die gesammten Naturwissenschaften 53: 221–246.

Müller, G. W., 1889. Die Spermatogenese der Ostracoden. Zoologische Jahrbücher, Abteilung für Anatomie und Ontogenie der Tiere 3: 677–726.

Müller, G. W., 1894. Die Ostracoden des Golfes von Neapel und der angrenzenden Meeres-Abschnitte. Fauna und Flora des Golfes von Neapel und der angrenzenden Meeres-Abschnitte 21(I–VIII): 1–404.

Nation, J. L., 1983. A new method using hexamethyldisilazane for preparation of soft insect tissue for scanning electron microscopy. Stain Technology 58: 347–351.

Okubo, I., 1990. Sixteen species of freshwater ostracoda from Japan. Bulletin Biogeography Society of Japan 45: 39–50.

Parker, A., 1997. Mating in Myodocopina (Crustacea, Ostracoda): results from video recordings of a highly iridescent cypridinid. Journal of the Marine Biological Association, UK, 77: 1223–1226.

Retzius, G., 1909. Die Spermien der Crustaceen. Biologische Untersuchungen von Prof. Dr. Gustaf Retzius. Leipzig: Biologische Untersuchungen, Neue Folge. XIV: 1–54.

Rioja, E., 1940. Estudios Carcinologicos 5: Morfologia de un ostracod epizoario observado sobre Cambarus (Cambarellus) montezumae Sauss. De Mexico, *Entocythere heterodonta* n. sp., y descripcion de algunos de sus estados larvarios. Anales del Instituto de Biologia de la Universidad Nacional de Mexico. Ciencia, Revista Hispano-Americana de Ciencias puras y aplicadas 3: 201–204.

Roessler, E. W., 1983. Estudios taxonomicos, ontogeneticos, ecologicos y etologicos sobre los ostracodos de agua dulce en Colombia. IV. Desarrollo postembrionario de *Heterocypris bogotensis* Roessler (Ostracoda, Podocopa, Cyprididae). Caldasia 13: 755–776.

Sars, G. O., 1889. On some freshwater Ostracoda and Copepoda, raised from dried Australian mud. Forhandlinger I Christiania Videnskabs-Selskabet 1889: 1–79.

Smith, R. J. & T. Kamiya, 2002. The ontogeny of *Neonesidea oligodentata* (Bairdioidea, Ostracoda, Crustacea). Hydrobiologia 489: 245–275.

Smith, R. J. & T. Kamiya, 2003 The ontogeny of *Loxoconcha japonica* Ishizaki, 1968 (Cytheroidea, Ostracoda, Crustacea). Hydrobiologia 490: 31–52.

Smith, R. & K. Martens, 2000. The ontogeny of the cypridid ostracod *Eucypris virens* (Jurine, 1820) (Crustacea, Ostracoda). Hydrobiologia 419: 31–63.
Stuhlmann, F., 1886. Beiträge zur Anatomie der inneren männlichen Geschlechtsorgane und zur Spermatogenese der Cypriden. Zeitschrift für wissenschaftliche Zoologie Abt A 44: 536–569.
Sywula, T., 1974. Malzoraczki (Ostracoda). Fauna Slodkowodna Polski 24: 1–315.
Vávra, W., 1891. Monographie der Ostracoden Boehmens. Archiv der Naturwissenschaftlichen Landesdurchforschung von Boehmen 8: 1–116.
Weismann, A., 1880. Parthenogenese bei den Ostracoden. Zoologischer Anzeiger 3: 82–84.
Weygoldt, P., 1960. Embryologische Untersuchungen an Ostrakoden: Die Entwicklung von *Cyprideis litoralis* (G. S. Brady), (Ostracoda, Podocopa, Cytheridae). Zoologische Jahrbücher, Abteilung für Anatomie und Ontogenie der Tiere 78: 369–426.
Wingstrand, K. G., 1988. Comparative spermatology of the Crustacea Entomostraca. 2. Subclass Ostracoda. Biologiske Skrifter. The Royal Danish Academy of Sciences and Letters, Copenhagen 32: 1–149.
Wouters, K., 1998. A new thalassocypridine genus (Crustacea, Ostracoda) from brackish waters in the Indian and Pacific Oceans, with the description of a new species. Bulletin de l'Institut Royal des Sciences Naturelles de Belgique, Biologie 68: 111–122.
Zenker, G. F. W., 1850. Über die Geschlechtsverhältnisse der Gattung *Cypris*. Archiv für Anatomie und Physiologie 1850: 193–202.
Zenker, G. F. W., 1854. Monographie der Ostracoden. Archiv für Naturgeschichte 20: 1–87.

Hydrobiologia (2007) 585:249–253
DOI 10.1007/s10750-007-0641-6

OSTRACODA (ISO15)

Early release of eggs and embryos in a brooding ancient asexual ostracod: brood selection or a gambling strategy to increase fecundity?

Ricardo L. Pinto · Carlos E. F. Rocha · Koen Martens

Abstract Asexual lineages lack the means to purge their genomes of (deleterious) mutations through recombination. Evolutionary theory thus predicts that such lineages will be prone to early extinction. In brooding animals, brood selection might provide a mechanism to counter the accumulation of mutations. Of the three putative ancient asexual animal groups, only the darwinulid ostracods are brooders. Here, we test the incidence of egg and juvenile abortion in a darwinulid species, *Penthesilenula brasiliensis*, under two temperature treatments. Part of the offspring is released without brooding (close to 30% in one treatment). The majority of these aborted eggs hatches and develops. As it is unlikely that females are such bad judges of offspring quality, either the surviving animals will present deficiencies later on in development (brood selection) or early egg release can be a (bet-hedging) strategy to increase fecundity in favourable conditions.

Keywords Brooding · Ancient asexual · Ostracoda · Darwinulidae · Reproduction · Evolution · Bet-hedging

Introduction

Evolutionary theory holds that fully asexual lineages are short-lived (Maynard Smith, 1990). Reproduction without meiosis (including recombination) and karyogamy is indeed assumed to have many deleterious side effects. Amongst others, asexual reproduction prevents genomes from purging deleterious mutations, thus causing mutational meltdown through Muller's Ratchet and Kondrachov's Hatchet. Nevertheless, three animal groups claim the status of ancient asexuals: bdelloid rotifers, some lineages within the oribatid mites and the darwinulid ostracods.

The Darwinulidae (c. 30 recent species) are thought to have been asexual for at least 200 million years (Martens et al., 2003). Absence of recombination and the presence of a limited

Guest editors: R. Matzke-Karasz, K. Martens & M. Schudack
Ostracodology – Linking Bio- and Geosciences

R. L. Pinto (✉) · C. E. F. Rocha
Departamento de Zoologia, Instituto de Biociências, Universidade de São Paulo, Rua do Matão, trav.14, n.101, São Paulo, SP CEP 05508-900, Brazil
e-mail: rlpinto@ib.usp.br

K. Martens
Freshwater Biology, Royal Belgian Institute of Natural Sciences, Vautierstraat 29, Brussels 1000, Belgium

K. Martens
Deparment of Biology, University of Ghent, K. L. Ledeganckstraat 35, Ghent 9000, Belgium

amount of gene conversion have been demonstrated for *Darwinula stevensoni* (Brady & Robertson, 1870) (Schön & Martens, 2003). All darwinulids are brooders and the number of embryos and eggs that can be retained at the same time in the brood chamber varies among species. *Darwinula stevensoni*, for example, can hold up to 10 or 15 juveniles at once (McGregor, 1969; Ranta, 1979; Horne et al., 1998; Van Doninck et al., 2003a).

It has been suggested (Lively & Johnson, 1994) that brooding asexuals have an additional advantage over non-brooders to counter, or at least slow down, the accumulation of deleterious mutations by actively removing defective embryos from the brood pouch. Such brood selection could indeed purge genomes with a high incidence of deleterious mutations from the gene pool, provided that females have the means to detect such deficient embryos. Horne et al. (1998) showed that *Darwinula stevensoni* has the ability to detect, select and eject unwanted material from the brood pouch. But whether or not the observed ejection of eggs truly constitutes brood selection in the sense of Lively & Johnson (1994) remained equivocal.

Here, we test the brood selection hypothesis in the darwinulid *Penthesilenula brasiliensis* (Pinto & Kotzian, 1961) by checking for the incidence of early ejection of eggs and first instar juveniles from the brood pouch (here termed *abortion*) and for the viability of the ejected eggs. Also, effects of temperature on the incidence of abortion and on the rate of early development are checked.

Materials and methods

Our model organism *Penthesilenula brasiliensis* was collected from forest leaf litter at the Boracéia Biological Station, Municipality of Salesópolis, São Paulo State, Brazil (GPS co-ordinates: 23°38′16.9″ S; 45°50′24.5″ W).

Two temperature treatments were applied, namely 17 and 22°C, the former being the temperature (at the moment of sampling) of the leaf litter from which the animals were retrieved. The experiment was conducted for a period of 90 days and animals were checked daily. For each temperature, 17 females were individually placed in small glass vials with natural mineral water and with a piece of leaf from the original sample to serve as substrate and food. Vial, water and leaf were replaced three times weekly. Eggs and first and second instar juveniles ejected from the brood pouch were removed to separate vials and were incubated at similar conditions and their early development (until first moult, i.e. beginning of second stage) was monitored.

Mann–Whitney *U* tests, with STATISTICA® (version 6.0) software, were used for all analyses, as the results did not fit to normal distributions (even after several transformations), thus excluding the use of parametric tests. We tested for: (A) the effect of temperature on the mean number of eggs produced per female; (B) the effect of temperature on the abortion ratio (per cent); (C) differences in mortality of the aborted specimens between the two temperatures; (D) the effect of temperature on the development speed, i.e. time spent from egg laying, through hatching, to first moult, for individuals brooded up to second stage; and differences in development speed between brooded and precociously ejected animals, at (E) 17°C and (F) 22°C.

As these animals take a long time to mature (more than one year from egg to adult, Pinto, unpubl.), it was not possible to acclimatise animals to lab conditions over several generations in order to exclude maternal effects. This is inherent to experimental work with these ostracods (Van Doninck et al., 2002, 2003b).

Results

Thirty-one out of the initial 34 mother individuals survived throughout the 90 days: 16 at 17° and 15 at 22°C. During this period, the fate of 171 eggs was followed. In each of the treatments, the mean number of eggs produced per female varied between 5 and 6 (no significant differences between treatments were observed—Table 1A).

From our observations, *Penthesilenula brasiliensis* can brood its juveniles up to second stage, but part of the offspring was released precociously, mainly as eggs. At the higher temperature, there was a significant decrease in the

Table 1 Results of the Mann–Whitney *U* tests checking: (A) the effect of temperature on the mean number of eggs produced per female; (B) the effect of temperature on the percentage of abortion; (C) the mortality before second stage of aborted specimens between the two temperature treatments; (D) the effect of temperature on the rate of development inside the pouch; (E) the development rates at 17°C between aborted and brooded specimens; and (F) the development rates at 22°C between aborted and brooded specimens

	Rank Sum Group1	Rank Sum Group 2	U	Z	*P*-level	Z adjusted	*P*-level	Valid *n* Group 1	Valid *n* Group 2	2*1sided exact *P*
A	286.500	209.500	89.500	1.2056	0.2279	1.2275	0.2196	16	15	0.231 6
B	320.000	176.000	56.000	2.5298	0.0114	2.6312	0.0085	16	15	0.0105*
C	126.000	45.000	30.000	0.2464	0.8053	0.2729	0.7849	13	5	0.8489
D	287.000	91.000	0.00	4.4158	0.0000	4.4348	0.0000	14	13	0.0000*
E	78.000	58.000	30.000	0.1587	0.8738	0.1606	0.8723	9	7	0.9181
F	33.500	71.500	16.500	0.4949	0.6206	0.5214	0.6020	4	10	0.6353

(*) = Significant *P* value, $\alpha = 0.01$

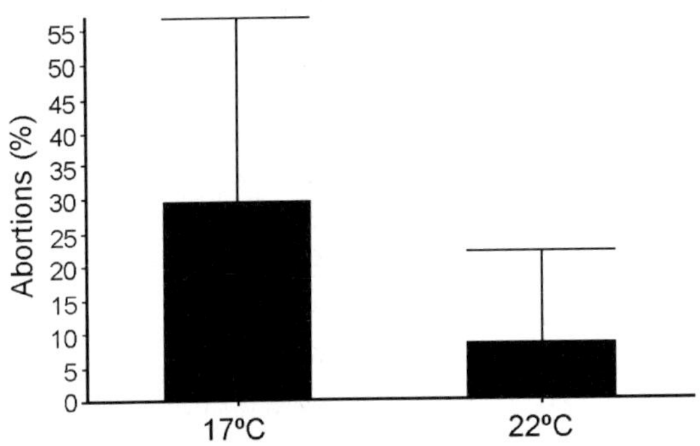

Fig. 1 Graphic showing the proportion of abortions (average + standard deviation) at 17°C ($n = 16$) and at 22°C ($n = 15$). The difference is significant ($P = 0.0105$)

percentage of abortion ($P = 0.0105$) (Fig. 1, Table 1B). The proportion of abortion averaged 29.7% at 17°C and 8.4% at 22°C. The mortality before first moult was similar and not significantly different (Table 1C) in both temperatures: at 17°C the mean mortality was 35.9% and at 22°C it was 30.0%. The rate of development of brooded animals increased strongly and significantly at the higher temperature ($P < 0.0001$) (Fig. 2, Table 1D). Inside the mother, animals take approximately 32 and 15 days to reach the second stage at 17 and 22°C, respectively. Comparisons of development speed between aborted and brooded animals showed no significant difference either at 17°C (Table 1E) or at 22°C (Table 1F). However, we stress the occurrence of two conspicuous cases of slow developing aborted eggs: at 17°C, a juvenile took 38 days to reach second stage, that is 6 days longer than the average at the same temperature and 4 days longer than the second slowest developing individual; at 22°C, a juvenile took 22 days to reach second stage, which is 7 days longer than the average at the same temperature and 5 days longer than the second slowest developing individual at the same temperature. None of the offspring (either brooded or aborted) reached adulthood, the majority (80%) having died within a year.

Discussion

Brooding behaviour of *P. brasiliensis* differs from that of *D. stevensoni* (the only other species in the whole Family that had its life cycle studied so far) in that juveniles are brooded up to the second stage, while *D. stevensoni* releases its juveniles in the third stage (Van Doninck et al., 2003a).

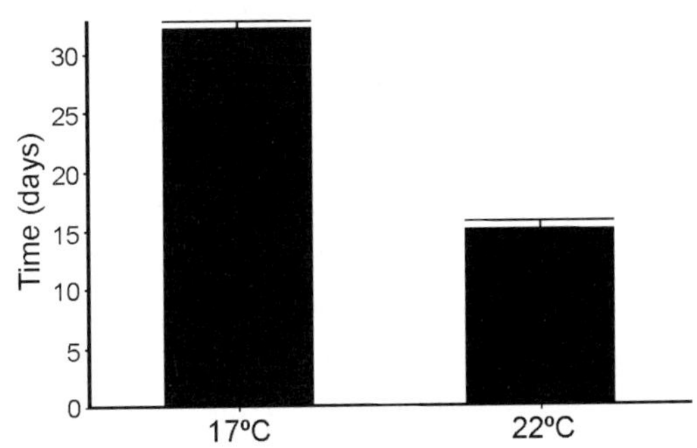

Fig. 2 Graphic showing time span from egg laying until first moult (average + standard deviation) at 17°C ($n = 14$) and at 22°C ($n = 13$). The difference is significant ($P < 0.0001$)

Besides this, we observed a maximum of six juveniles inside the brood chamber of *P. brasiliensis* at one time, while *D. stevensoni* is known to carry up to 15 juveniles.

The number of aborted eggs during the present experiment was surprisingly high, with an average of almost 30% being aborted at 17°C. Additionally, a remarkable proportion of these aborted eggs hatched and moulted at least once (65–70% in different treatments). There can be three causes for this unexpectedly high incidence of abortion:

(1) Females are bad in judging quality of their eggs, and so eject a majority of viable eggs. It is difficult to imagine that females would allow abortion of such a significant part of their offspring, the majority of which appears to be at least viable during early development.
(2) Deficient offspring cannot only be measured by their success in early development, but will become apparent later on in the development. Rossetti & Martens (1996, 1998) and Pinto et al. (2004) have remarked upon the high incidence of teratological specimens in the ancient asexual Darwinulidae. Even though malformations can have different origins (e.g. diseases, traumas, environment), some of these malformed animals could have resulted from defective embryos that were detected and aborted by their mother but still managed to reach adulthood. Initially, we intended to screen for malformations in the offspring obtained during the present experiment, once they attained adulthood. Unfortunately, this was not possible as most offspring died and none reached the adult stage.
(3) The early release of eggs is not, or at least not only, related to brood selection. The high success in the early development of aborted eggs at 17°C (here considered the optimal temperature, as this was the ambient temperature in the native habitat) might indicate that females have adopted an optional strategy to risk release of eggs without brooding. This double strategy, similar to bet-hedging (i.e. to reduce the possible losses on a bet, risk, or speculation by betting or speculating on both sides), could lead to an increase in fecundity, as there is limited space in the brood pouch and early evacuation of eggs allows faster production of new eggs. The development rate conforms well to this idea. When development of juveniles was faster (at 22°), there was a relatively reduced number of abortions, but when they developed slowly (at 17°), the number of abortions was high. As the mean number of eggs per female was similar in both temperatures, we can conclude that slow development could cause overcrowding within the brood cavity. It is possible that problems with the experimental set-up might have biased direct release of eggs to a higher incidence. Females might release fewer eggs in natural conditions. Nevertheless, the effect of temperature (and of the development rate, indirectly) on the proportion of abortions seems clear.

The fact that almost always eggs were aborted in favour of juveniles supports the idea that overcrowding is a problem. In this situation, one could expect that preferably a juvenile would be retained instead of an egg as the female has already invested time and energy in rearing that juvenile. There is less energy invested in an egg so, if overcrowding occurs, the excess eggs go first. On the other hand, abortions did not occur exclusively when the brood chamber was full. If eggs are aborted when the chamber is still half empty, that may indicate brood selection.

If the direct release of eggs is indeed an optional strategy in favourable conditions, it could skew fecundity estimates from field populations in this and possibly more brooding species. There would be less eggs and juveniles in the brood pouch than the amount actually produced, while the aborted eggs and juveniles in the sediment would not be counted (only eggs and juveniles in the pouch are generally counted in these animals—Van Doninck et al., 2003a).

The hypotheses of brood selection and of fecundity enhancing (bet-hedging) are not mutually exclusive, both could be simultaneously causing the abortions. Indeed, the two cases mentioned in the results, of exceedingly slow developing aborted individuals, could be the indication of defective animals aborted as a result of brood selection. The aborted individuals that hatched and moulted normally could be a result of the fecundity increase strategy.

Speculation aside, the high incidence of early egg release is most unexpected. There is as yet no definitive verdict on whether all of these observed events are abortions as a result of brood selection, or if a thus-far unknown strategy for increase of fecundity in favourable circumstance is here demonstrated, or if this behaviour is a mixture of both.

Acknowledgements Fábio Lang da Silveira (São Paulo, Brazil) is acknowledged for allowing the use of equipment in his laboratory. Roberto Munehiza Shimizu (São Paulo, Brazil) provided advice with statistics. FAPESP allocated a Ph.D. grant (Processo: 03/13510-5) to RLP and provided financial assistance to KM (Processo: 98/05073-4). The study was also supported with a scholarship from 'Pós-Graduação, Área Zoologia, IB-USP' and 'CAPES/PROAP/IBUSP-2003'. Robin Smith (Kusatsu, Shiga, Japan), Isa Schön (Brussels, Belgium) and several anonymous referees read the manuscript and suggested improvements.

References

Horne, D. J., K. Martens & F. Mösslacher, 1998. A short note: is there brood selection in *Darwinula stevensoni*? In Crasquin-Soleau, S., E. Braccini & F. Lethiers (eds), What about Ostracoda! Actes du 3e Congrès Européen des Ostracodologistes, Paris-Bierville, France 8–12 juillet 1996. Bulletin des Centres de Recherches Exploration-Production Elf-Aquitaine 20: 33–35.

Lively, C. M. & S. G. Johnson, 1994. Brooding and the evolution of parthenogenesis: strategy models and evidence from aquatic invertebrates. Proceedings of the Royal Society of London, Series B 256: 89–95.

Martens, K., G. Rossetti & D. J. Horne, 2003. How ancient are ancient asexuals? Proceedings of the Royal Society of London, Series B 270: 723–729.

Maynard-Smith, J., 1990. Evolution of Sex. Cambridge University Press, Cambridge.

McGregor, D. L., 1969. The reproduction potential, life history and parasitism of the freshwater ostracod *Darwinula stevensoni* (Brady and Robertson). In Neale, J. W. (ed.), The taxonomy, morphology and ecology of recent Ostracoda. Oliver & Boyd, Edinburgh, 194–221.

Pinto, R. L., C. E. F. Rocha & K. Martens, 2004. On the genus Penthesilenula Rossetti & Martens, 1998 (Crustacea, Ostracoda, Darwinulidae) from (semi-) terrestrial habitats in Sao Paulo State (Brazil), with the description of a new species. Journal of Natural History 38: 2567–2589.

Ranta, E., 1979. Population biology of *Darwinula stevensoni* (Crustacea, Ostracoda) in an oligotrophic lake. Annales Zoologici Fennici 16: 28–35.

Rossetti, G. & K. Martens, 1996. Redescription and morphological variability of *Darwinula stevensoni* (BRADY & ROBERTSON, 1870) (Crustacea, Ostracoda). Bulletin de l'Institut Royal des Sciences Naturelles de Belgique, Biologie 66: 73–92.

Rossetti, G. & K. Martens, 1998. Taxonomic Revision of the Recent and Holocene representatives of the Family Darwinulidae (Crustacea, Ostracoda), with a description of three new genera. Bulletin de l'Institut Royal des Sciences Naturelles de Belgique, Biologie 68: 55–110.

Schön, I. & K. Martens, 2003. No slave to sex. Proceedings of the Royal Society of London, Series B 270: 827–833.

Van Doninck, K., I. Schön, L. De Bruyn & K. Martens, 2002. A general purpose genotype in an ancient asexual. Oecologia 132: 205–212.

Van Doninck, K., I. Schön, K. Martens & B. Goddeeris, 2003a. The life-cycle of the ancient asexual ostracod *Darwinula stevensoni* (Brady & Robertson, 1870) (Crustacea, Ostracoda) in a temporate pond. Hydrobiologia 500: 331–340.

Van Doninck, K., I. Schön, F. Maes, L. De Bruyn & K. Martens, 2003b. Ecological strategies in an ancient asexual animal group. Freshwater Biology 48: 1285–1294.

Hydrobiologia (2007) 585:255–272
DOI 10.1007/s10750-007-0642-5

OSTRACODA (ISO15)

The ontogeny of appendages of *Heterocypris salina* (Brady, 1868) Ostracoda (Crustacea)

Nerdin Kubanç · Oya Özuluğ · Cüneyt Kubanç

Abstract The post-embryonic development of the appendages of the Cyprididae ostracod *Heterocypris salina* (Brady, 1868) are described in detail and compared with those of other podocope species documented in previous studies. Generally, the appearence of limbs during onotgeny of *H. salina* is similar to that of other species, but small differences in limb morphologies were identified between *H. salina* and other Cyprididae species, including other *Heterocypris* species. Some features appear either earlier or later in the development of *H. salina* compared with other species, even species of the same genus. These features may be useful characters for phylogenetic analyses at the genus and family levels.

Keywords Ontogeny · Appendages · Ostracoda · Cyprididae

Introduction

Ostracods grow by moulting (ecdysis) and in the Podocopida there are usually seven or eight moult stages, or instars, between the egg and adult stages. The adult stage is usually termed 'A', with the previous instars designated as 'A-1' (one stage before the adult stage), 'A-2' (two stages before the adult stage) etc. The first instar (A-8) of the Cypridoidea exhibits three pairs of appendages, namely the antennules, antennae and mandibles, and a poorly calcified carapace. A properly calcified carapace appears at the A-7 stage, and each successive moult adds soft-part and carapace features, the larger instars showing an increasing likeness to the adult of the species. Only the last stage (the adult) is fully formed and sexually mature (Athersuch et al., 1989).

The majority of studies of the ontogenetic development of podocopid ostracod appendages and carapaces have concerned the superfamily Cypridoidea, e.g., various instars of *Cyclocypris ovum* (as *Cypris ovum*), *Cypridopsis vidua* (as *Cypris vidua*) and *Dolerocypris fasciata* (as *Cypris fasciata*) by Claus (1865, 1868); *Heterocypris incongruens* (as *Cyprinotus incongruens*) by Schreiber (1922); *Cypridopsis vidua* by Kesling (1951); *Herpetocypris agilis, Heterocypris incongruens, Cypridopsis vidua* and *Cypria ophthalmica* by Fox (1964); undetermined species of the genera *Stenocypris*, *Potamocypris* and *Cypridopsis* by Ghetti (1970); some features of the latter stages of *Pseudocandona serbani* by Broodbakker & Danielopol (1982); *Heterocypris bogotensis* by Roessler (1983); the antenna of the last two instars

Guest editors: R. Matzke-Karasz, K. Martens & M. Schudack
Ostracodology – Linking Bio- and Geosciences

N. Kubanç (✉) · O. Özuluğ · C. Kubanç
Faculty of Science, Department of Biology, Istanbul University, 34118 Vezneciler, Istanbul, Turkey
e-mail: nerdin.kubanc@gmail.com

of *Sclerocypris* species by Martens (1987); *Eucypris virens* by Smith & Martens (2000); the carapace shape of *Eucypris virens* by Baltanás et al. (2000); carapace shape and ornamentation of most stages of *Potamocypris humilis* by Horne & Smith (2004); and the development of the antennule of *H. incongruens* by Smith & Tsukagoshi (2005).

Additionally, the ontogenetic development of species belonging to the superfamily Cytheroidea, e.g. *Limnocythere inopinata* by Scheerer-Ostermeyer (1940), *Cyprideis torosa* (as *C. litoralis*) by Weygoldt (1960), *Loxoconcha japonica* by Smith & Kamiya (2003), *Uncinocythere occidentalis* by Hart et al., 1985 and Smith & Kamiya (2005); the superfamily Bairdioidea, e.g. *Neonesidea oligodentata* by Smith & Kamiya (2002); the superfamily Terrestricytheroidea, e.g. *Terrestricythere elisabethae* by Horne et al. (2004); and the superfamily Darwinuloidea e.g. *Darwinula stevensoni* by Scheerer-Ostermeyer, 1940 have also been studied.

Heterocypris salina prefers both small and slightly salty coastal and inland waters. Additionally, it also occurs in pure freshwater habitats, including springs. Only female populations are known; males have never been found. This species can be successfully cultured in pure fresh water in the laboratory (Meisch, 2000).

This is the first detailed ontogenetic study of the appendages of this species. As most work on the ontogeny of ostracods have concerned the Cyprididae, this provides and opportunity to document the differences in development of species within the same family. In particular, it aims to identify possible developmental variations within the Cyprididae that could be used for future phylogenetic analyses of the family.

Materials and methods

Heterocypris salina was collected alive from Büyükçekmece Lake (41°06′16″ N, 28°31′53″ E) Istanbul. Live adults were kept in 24-well tissue culture plates (17 mm diameter, 20 mm deep) and were fed periodically (once in 3 days) with a mixture of *Chroococcus* (Cyanobacteria) and *Nitzschia* (Diatoms). During the study, the water temperature was constant at 23°C. The appendages were dissected and mounted in lactophenol + orange G on glass slides. The appendages were drawn with the aid of a camera lucida.

Terminology

The species was identified according to Meisch (2000). Chaetotaxy of limbs follows the model proposed by Broodbakker and Danielopol (1982), revised for the antennae by Martens (1987) and Smith & Martens (2000).

Results

The following section does not offer a full description of the chaetotaxy of each instar as this can be determined from the illustrations.

Instar A-8 (Fig. 1)

Carapace length 160–170 μm; maximum height 120 μm. Antennule consisting of four podomeres, first podomere long and rectangular. Second podomere with one apical seta; third podomere with two long apical setae; last podomere with two long, one medium length setae and an aesthetasc (*ya*). Antenna consists of a protopodite, endopodite and exopodite. Protopodite consists of two podomeres. Exopodite is located on the apical outer side of a small base with two very short and one long seta. Endopodite is made up of three podomeres; first podomere with one long seta and anlage of aesthetasc *Y*; second podomere with one short seta and one large dorso-apical claw (*G3*) and *Anlage* of *G1*; terminal podomere with two large claws (*GM* and *g*), an aesthetasc *Y3* and a medium length seta. Mandibular palp is a very simple structure. First podomere with one seta; second podomere with one seta and a strong, curved apical claw.

Instar A-7 (Fig. 1)

Length of carapace 190–210 μm, maximum height 150 μm. Antennule similar to A-8. Antenna, one seta appear on the protopodite; Exopodite similar to instar A-8 as first endopodial podomere. Second podomere of endopodite with claw *G3*

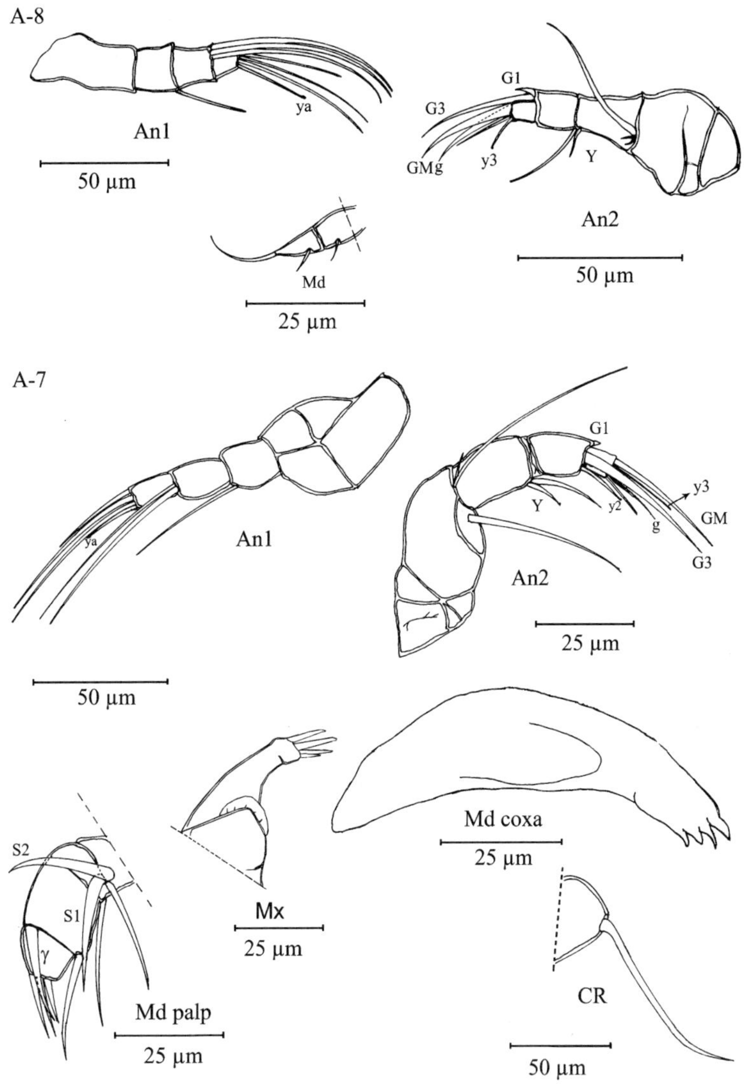

Fig. 1 *Heterocypris salina* instar A-8; An1, An2, Md and A-7; An1, An2, Md palp, Md coxa, Mx, CR

and the *Anlage* of *G1* as in the previous instar, but first time aesthetasc *y2* appears. Claw *g* on the apical podomere now transformed into a seta different from the previous instar, thus with only two claws on this appendage. Mandible developed into a feeding appendage with coxa and palp. Coxal endite bears four teeth. First palp segment is large, with *S1*, *S2* and one seta on the ventral edge. Although not clearly visible, endopodite three-segmented, first endopodial segment with smooth seta on the ventro-apical edge. Second endopodal segment with a stout smooth gamma seta on the anterior side and a stout seta on the opposite; terminal segment with three short, stout apical claws. *Anlage* of maxillula consists of an elongated and curved palp, just behind mandibles. *Anlage* of caudal ramus appears the first time on the posterior region of the animal. Caudal ramus consisting of basal segment with a long terminal claw.

Instar A-6 (Fig. 2)

Length of carapace 230–290 µm; height 160–200 µm. Antennule similar to that of instar A-7, only one short additional seta formed at the end of the third podomere. Antenna *G1* developed into a large claw. Chaetotaxy of other podomeres unchanged. Mandibular coxa now with five teeth. Three podomeres of the endopodite

Fig. 2 *Heterocypris salina* instar A-6; An1, An2, Md palp, Md coxa, Mx, CR

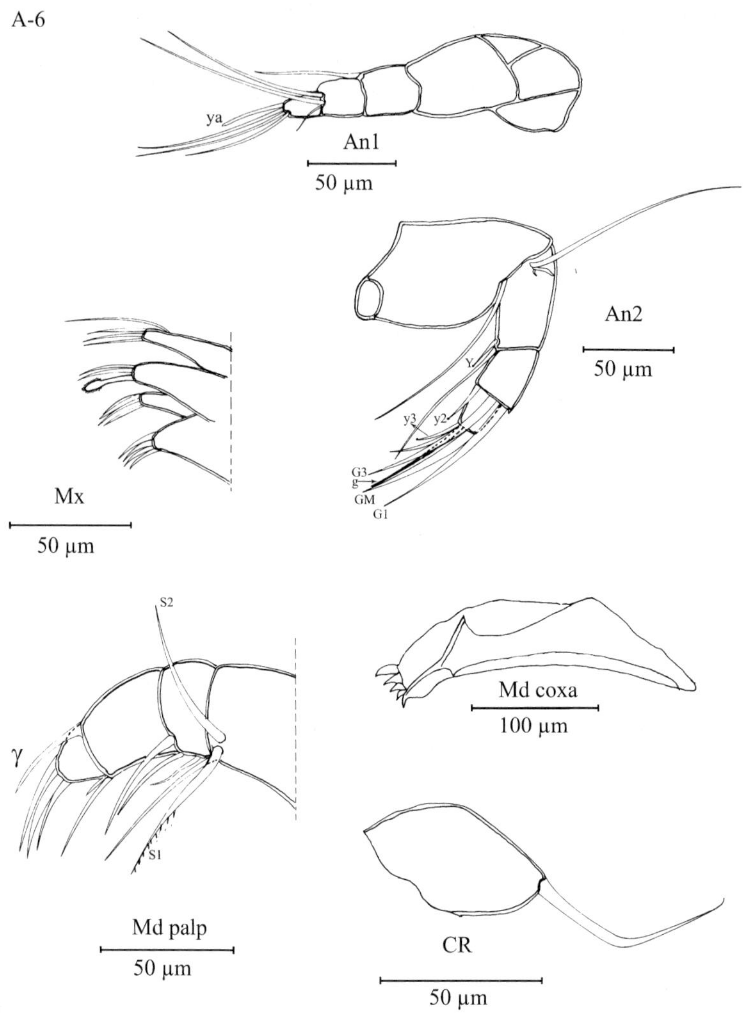

now clearly visible. Endopodite with an additional medio-ventral seta and a ventro-apical seta on the first endopodal podomere and small seta on the ventro-apical edge of the second endopodal segment. Maxillula with endopodite and three endites, all elongate. Endopodite with three claw-like terminal setae and one medio-dorsal seta. Third endite with one pinnated and curved *Zahnborste*, accompanied by two medium-length setae. Second and first endite with four terminal setae. Caudal ramus slightly longer than in instar A-7 but is still an *Anlage*.

Instar A-5 (Fig. 3)

Carapace length 300–310 μm; maximum height 240 μm. Protopodite of antennule large and subrectangular, bearing one small seta. Second podomere more quadrate than in instar A-6. One additional medium seta on the second and third podomere; terminal podomere unchanged. Protopodite of antenna longer than in instar A-6 with a small seta. First natatory seta appearing on the inner edge of the first podomere of the endopodite. Position of aesthetasc *Y* changed, now located in ventro-median position. Second and terminal endopodal podomere unchanged. Mandibular coxa similar to instar A-6. First endopodal podomere with three ventro-apical setae (including beta seta); second endopodal podomere with an additional seta on the ventral edge in sub-apical position. Third endopodal podomere unchanged. Endopodite of maxillula two-segmented; first podomere large, elongate and

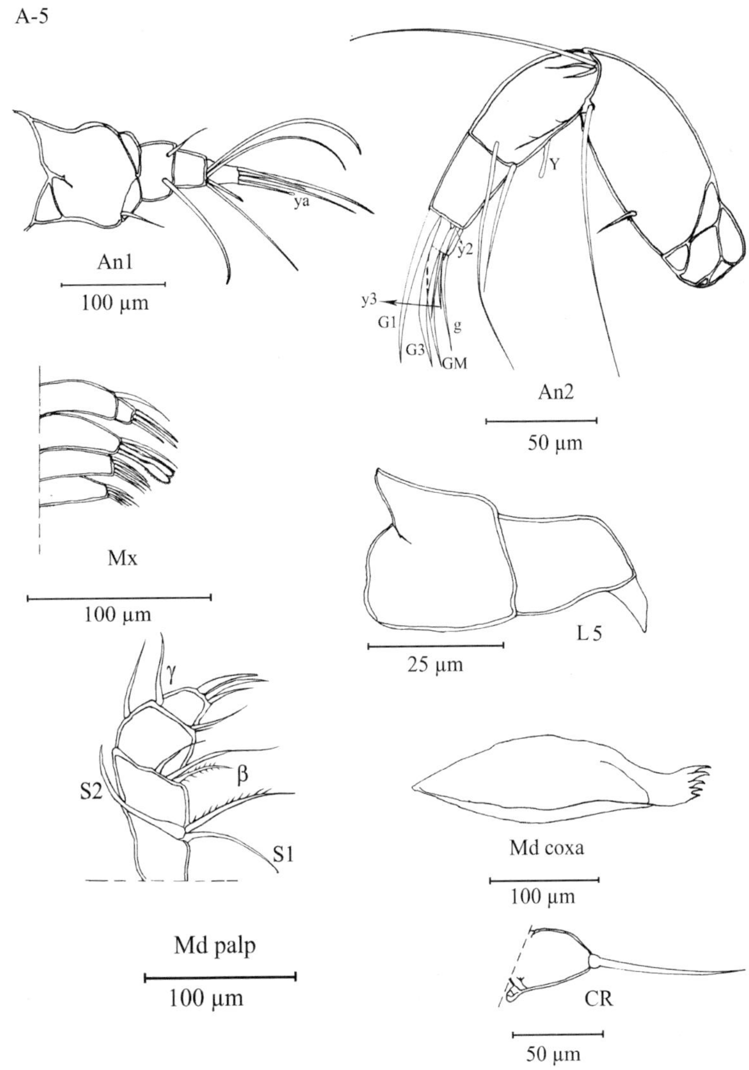

Fig. 3 *Heterocypris salina* instar A-5; An1, An2, Md palp, Md coxa, Mx, L5, CR

subrectangular with one dorso-apical seta; second podomere smaller, subquadrate and bearing three apical setae, one of these shorter than the others. Third endite with two pinnated and curved *Zahnborsten*, accompanied by two setae. Second and first endite with five terminal setae. *Anlage* of the fifth limbs consisting of an elongate podomere, capped by a beak like podomere, appearing near the base of the caudal ramus. Caudal ramus is still an *Anlage* and unchanged from instar A-6.

Instar A-4 (Fig. 4)

Carapace length of 350–390 μm; height 240–260 μm. Antennule protopodite divided into two podomeres now consisting of five podomeres, two setae on the first podomere; second podomere with one small seta on dorsal edge; next podomere with a small medium-length seta, opposite to a long apical seta. Fourth and terminal podomere unchanged. Antenna two natatory setae in the mid apical position present on the inner face. Second endopodal podomere bears *Anlage* of *G2*; terminal podomere unchanged. Mandibular coxa is still similar in shape but with the addition of an extra tooth on the endite. Endopodite with additional seta in beta group at apex of first podomere and another additional seta on the opposite side; second and third podomere unchanged but terminal podomere bearing four stout apical claws.

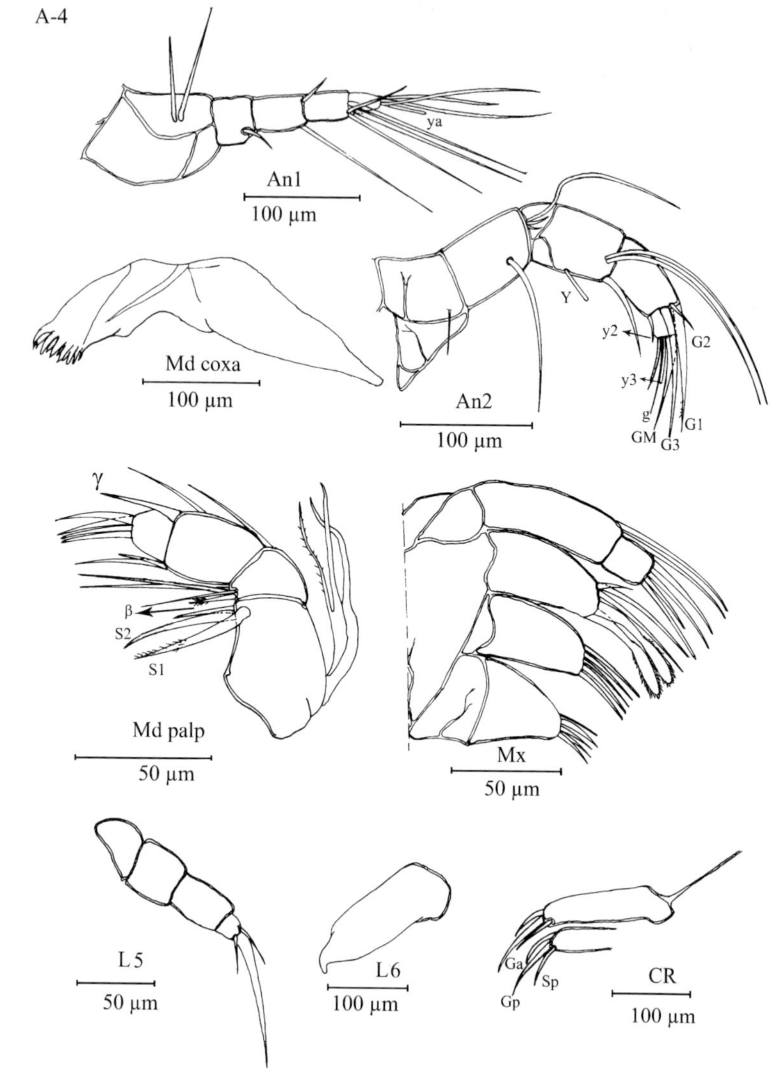

Fig. 4 *Heterocypris salina* instar A-4; An1, An2, Md palp, Md coxa, Mx, L5, L6, CR

First podomere of endopodite of maxillula with two medium-length setae on its outer apical edge; second podomere bearing three equal setae of medium length and one short seta. Third endite with two long, one short setae accompanying the two *Zahnborsten*. First and second endites remaining unchanged. Fifth limbs developed into walking legs, consisting of four podomeres. First podomere rounded. Second podomere subquadrate. Third podomere rectangular. Terminal podomere trapezium-shaped with two short setae and one long stout seta. *Anlage* of sixth limbs appearing in this instar for first time. Its shape is similar to the *Anlage* of the fifth limb in previous instar. Caudal ramus altered from its *Anlage* state to a form similar to that in adults. Caudal ramus has a rounded end, bearing two claws (*Ga, Gp*) and a seta (*Sp*) of medium length on the ventral edge; distal claw (*Ga*) short; adjacent claw (*Gp*) approximately two times longer.

Instar A-3 (Figs. 5, 6)

Carapace length 470–530 μm; maximum height 310 μm. Antennule consisting of six podomeres, resulting from the fourth podomere dividing into two smaller quadrate podomeres. First and second podomeres remaining unchanged. Third podomere longer than others, being approximately twice as long as wide. Fourth podomere

Fig. 5 *Heterocypris salina* instar A-3; An1, An2, Md palp, Md coxa, Mx

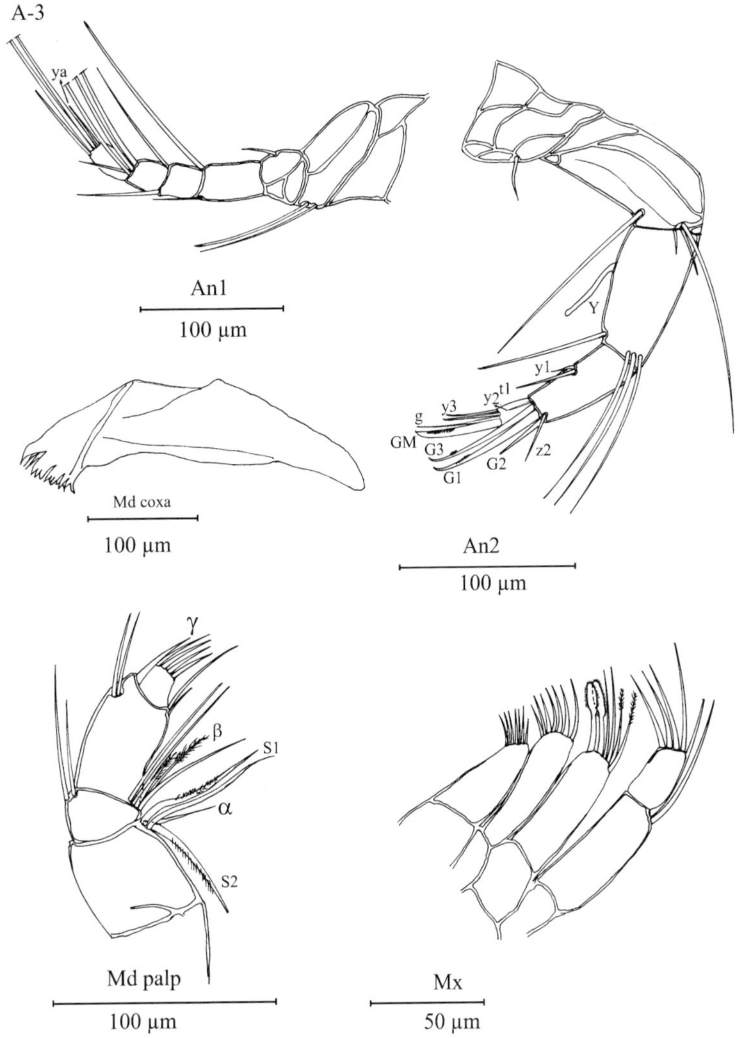

subquadrate, bearing one long and one short setae positioned similar to third segment. Fifth and terminal podomeres unchanged; first endopodal segment of antenna now with three natatory setae; second endopodal podomere with seta (*t1*) on the ventral side, accompanied by a small seta (*y1*); this segment apically with the large claw *G1*, seta *G2*, and the other claw *G3* and the first appearance of seta *z2*. Terminal podomere unchanged. Mandibular coxa is still similar in shape. First palp-segment of the mandibula with an additional alpha seta accompanying *S1* and *S2*. First endopodal podomere (second palp-segment) with additional more hirsute seta near the beta seta and with one additional seta on the apical dorsal edge. Second endopodal podomere with an additional seta and terminal podomere and coxa unchanged. Number of setae of endopodite of maxillula is similar to instar A-4 but all setae are now equal in length. Third endite with two long smooth setae, two medium hirsute setae and two *Zahnborsten*. First and second endites with seven apical setae. Fifth limbs changed in form and now recognizable as a maxilliped, consisting of an endopodite pointing posteriorly and a feeding process (endite) pointing anteriorly. Elongated endopodite three-segmented, terminating with one long and two short setae. Endite bearing seven setae on the apical edge and one medium seta on the on the protopodite side a position about ½ way along the margin. Sixth limbs recognizable as walking legs consisting of four

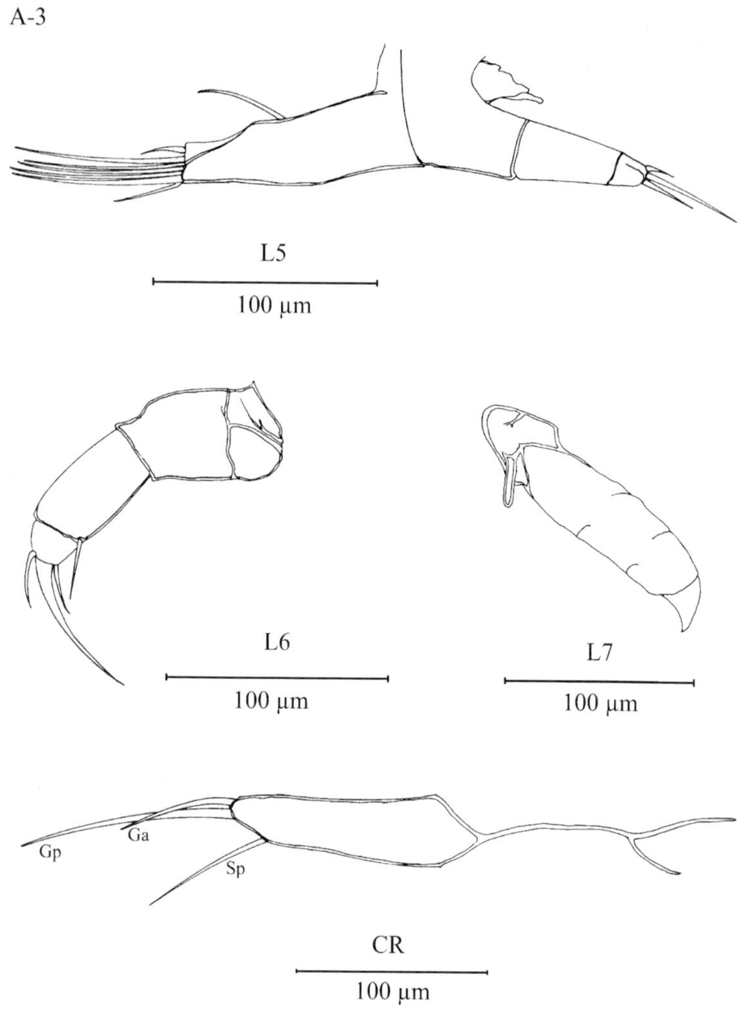

Fig. 6 *Heterocypris salina* instar A-3; L5, L6, CR

podomeres. Third podomere elongated and with a short seta on its ventro-apical corner. Fourth podomere small and subquadrate with one subapical seta, a large, terminally serrated claw and a ventro-apical seta. Seventh limbs are in *Anlage* state, similar in shape to that of the sixth limbs in instar A-4. Caudal ramus more elongate and *Ga*, *Gp* and seta *Sp* longer than in instar A-4.

Instar A-2 (Figs. 7, 8)

Carapace length of 610–700 μm; height 330–400 μm. Antennule consisting of seven podomeres, as a result of the third podomere dividing to form two quadrate podomeres. First podomere with one additional seta on the dorsal side, second podomere unchanged, fourth podomere (third in instar A-3) with two small, one long seta; fifth and sixth podomeres (fourth and fifth in instar A-3) each with one additional seta. Terminal podomere (sixth in instar A-3) unchanged. Antenna with one additional long seta on the ventral margin of the protopodite, first podomere of endopodite with four long natatory setae. Two new setae *t2* and *z1* (both of total of two now), appearing on second endopodal segment. Chaetotaxy of third endopodal podomere unchanged. Mandibular coxa similar to that in instar A-3. First mandibular endopodal podomere with one additional seta in beta-group; second endopodal podomere with two additional setae, one of these near dorso-apical corner (total of three) and the other in an interio-apical position. Terminal podomere with two claw-like and three normal

Fig. 7 *Heterocypris salina* instar A-2; An1, An2, Md palp, Md coxa, Mx

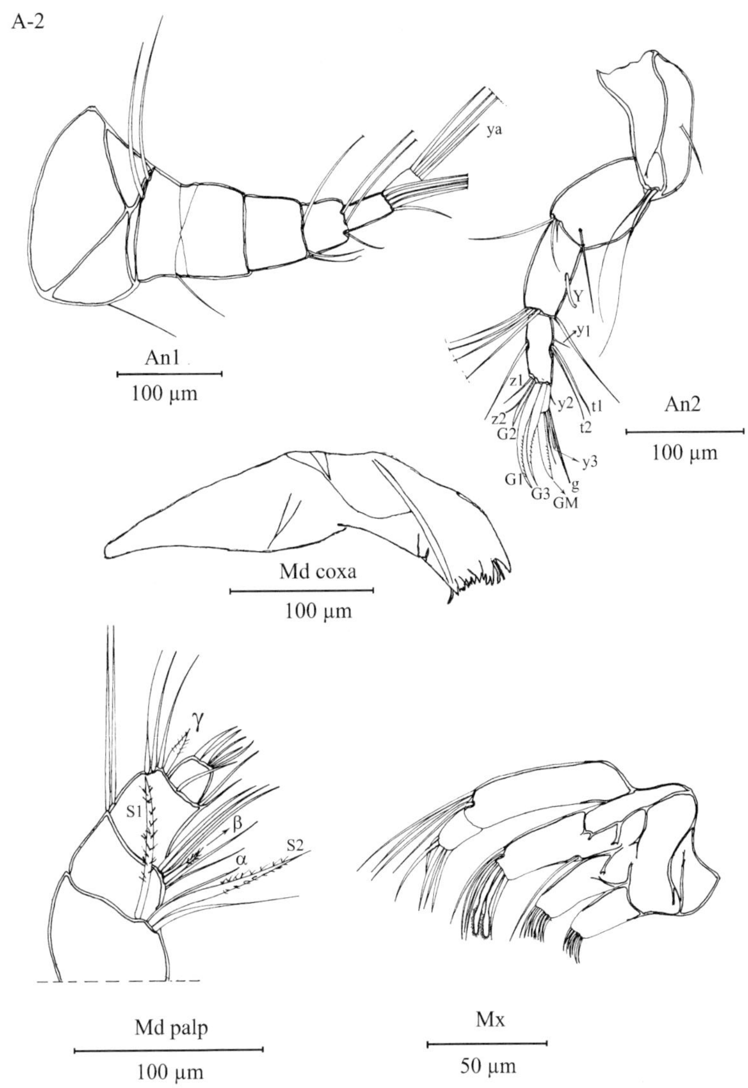

setae. Maxillula endopodite with two additional short setae on top of the first podomere; second podomere with additional one shorth seta on its outer apical edge. Third endite with an additional seta on the inner edge, close to the two, stout pinnated *Zahnborsten* and additional long seta on the inner side. First and second endites now with approximately eight terminal setae. Endopodite of fifth limbs now two segmented. Protopodite with two small setae *a* and one long seta *b*; endite with four additional long setae, forming a group of eleven. Proximal to this group one isolated seta (*d*). Sixth limbs consisting of five podomeres (third podomere in instar A-3 divided into two segments). Second and third podomere with two additional setae on the ventral apical corner. Fourth podomere with a small seta on the ventral apical corner; terminal podomere unchanged. Seventh limbs developed into a cleaning limb consisting of three elongate podomeres. First podomere with a seta on the apical region. Terminal podomere capped in a fully developed cleaning pincer as seen in the adults. Caudal ramus bearing two equal claws (*Ga* and *Gp*, one short seta *Sa* and long seta *Sp*).

Instar A-1 (Figs. 9, 10)

Carapace length 800–890 μm; maximum height 520 μm. Antennule with two additional setae on

Fig. 8 *Heterocypris salina* instar A-2; L5, L6, L7, CR

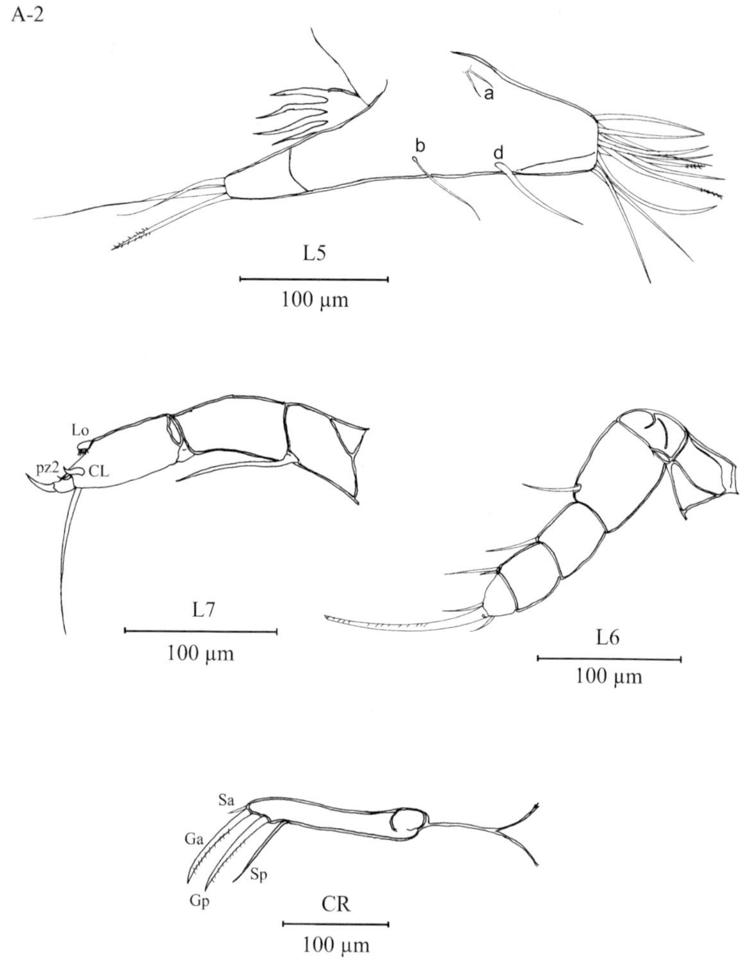

the ventro-apical and dorso-apical corners of the third podomere, which is now rectangular. One additional long seta of the fourth podomere. One medium-length seta appearing on the ventral corner of the fifth podomere; sixth and terminal podomeres unchanged. Antenna now with 5 natatory setae and their length longer than in instar A-2. Second podomere with additional seta *t3* (total of three) appearing, other podomeres unchanged except aesthetasc *Y*, which is now segmented at the tip. Mandibular coxa similar to instar A-2. Mandibula with an additional small seta (now total of three) on the dorsal edge of the second palp-segment. Third palp-segment with an additional apical seta on the inner side. Terminal podomere unchanged. Endopodite of maxillula with an extra seta on the outer edge and one additional sub-apical seta of the first podomere. Second podomere of endopodite and third endite unchanged. First and second endite with two additional setae (total of ten). The endopodite of fifth limbs with two segments; apical part bear one long, two medium-length setae. Endite unchanged in chaetotaxy, except four setae now being hirsute (two setae in instar A-2). Sixth limbs now with *d1* seta, other parts mostly unchanged. First podomere of seventh limbs with two additional setae, forming a group of three. Second podomere with one seta on the ventral edge. Third podomere with an additional seta in a medio-posterior position, but the apical chaetotaxy (pincer) similar to instar A-2. Caudal ramus more elongate than in instar A-2. Distal claw (*Ga*) longer than the proximal claw (*Gp*) and also *Sp* longer than in instar A-2.

Fig. 9 *Heterocypris salina* instar A-1; An1, An2, Md palp, Md coxa, Mx

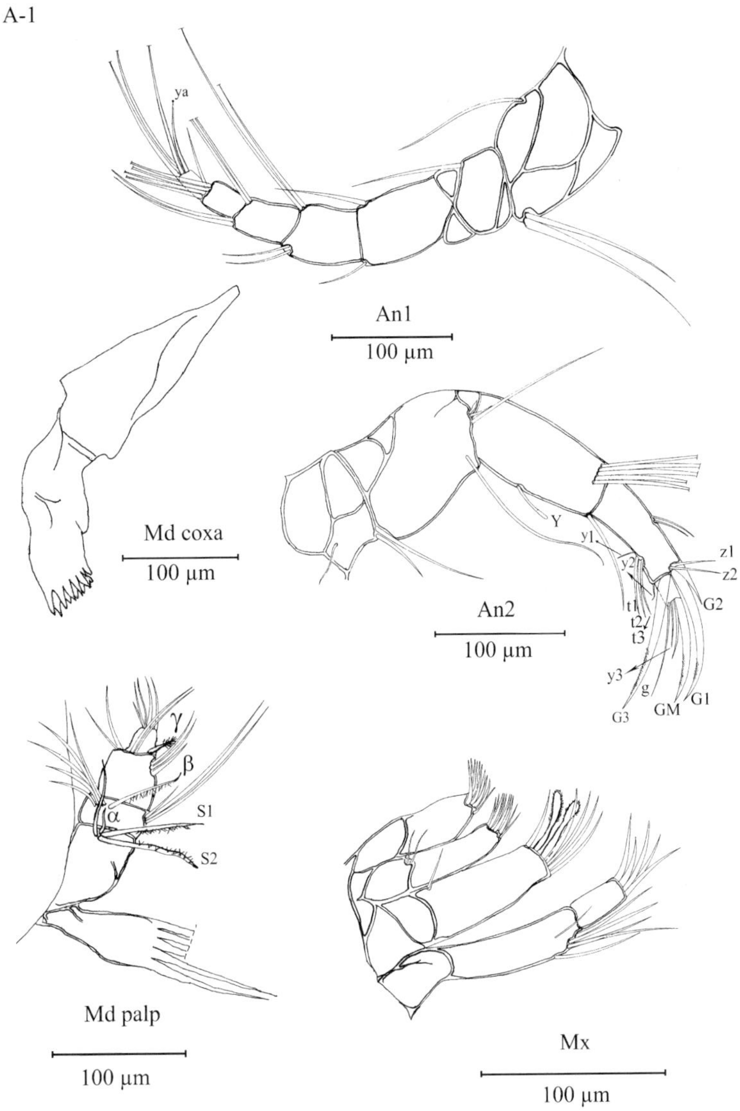

Adult (Figs. 11, 12)

Carapace length 980–1,000 μm; height 660–700 μm. Antennule similar to the previous instar with two exceptions: the two medium-length apical setae on the fifth podomere in A-1 are now longer and one additional medio-dorsal seta appears on the seventh podomere. Antenna larger than in instar A-1. Additional natatory seta on the inner face of the first podomere of the endopodite. Second podomere with additional seta *z3* (total of three) and *t4* (total of four) appearing and another large claw developed from the stout seta *G2*, joining the two other claws on this podomere. Terminal podomere unchanged. Mandibular coxa unchanged. Third segment of mandibular palp with an additional apical seta on the inner side (γ group). Other parts unchanged. First podomere of maxillula endopodite with the sub-apical seta situated further towards the apical edge. Third endite with three additional setae anterior to the *Zahnborsten*. First and second endite unchanged. Endopodites of fifth limbs have no segmentation. Endites of fifth limbs apically with one large group of 14 setae in this instar (11 in instar A-1), six of which are hirsute. Setae *b* and *d* are hirsute too. Chaetotaxy of sixth, seventh limbs and caudal ramus unchanged.

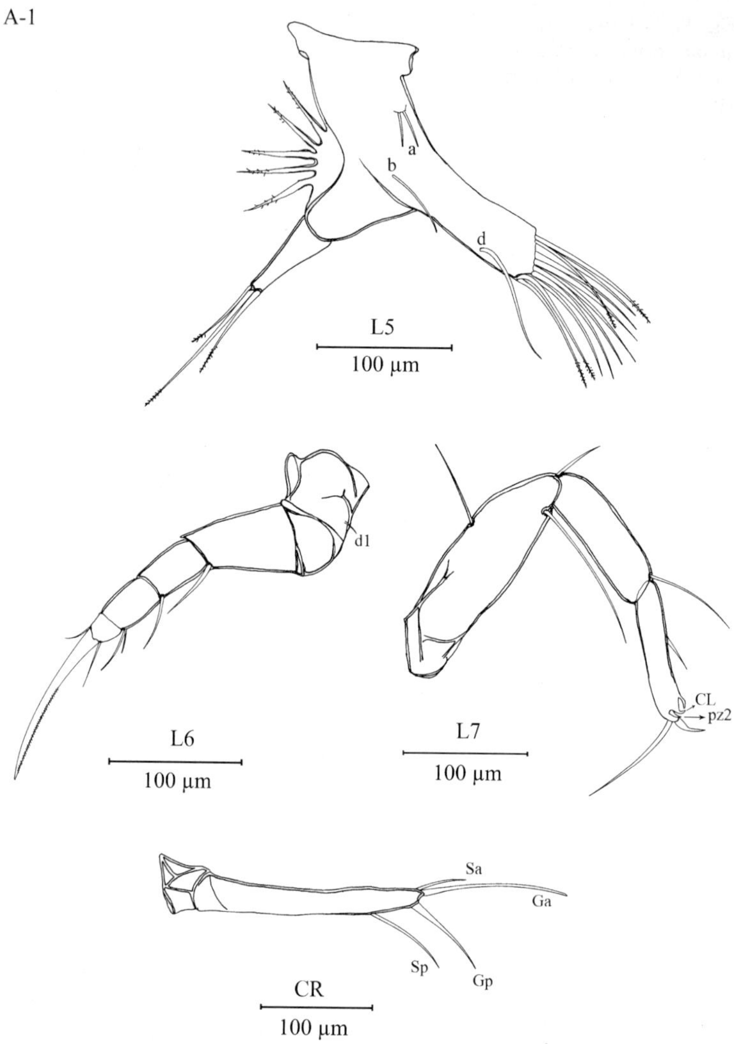

Fig. 10 *Heterocypris salina* instar A-1; L5, L6, L7, CR

Discussion

The timing of appearance and general morphology of appendages of *H. salina* during ontogeny are very similar to those of other Cyprididae species documented in previous studies. Some notable differences compared with previous studies are listed below:

(1) *H. salina* has the *Anlagen* of the antennal claw *G1* in instar A-8. Roessler (1983) figures a possible *G1 Anlagen* in the same instar for *H. bogotensis*, but it appears smaller than that of *H. salina* and may just be an extension of the cuticle. Schreiber (1922) and Smith & Martens (2000) do not figure such an *Anlagen* for the A-8 instar of *H. incongruens* and *E. virens* respectively, but Smith & Martens (2000) do show it in the next instar. Claus (1868) does not show it in either the first two instars of *Cyclocypris ovum*.

(2) The aesthetasc *Y* of the antenna of *H. salina* does not have any segmentation until instar A-1, when it becomes two-segmented. Roessler (1983) shows that this aesthetasc in *H. bogotensis* is three segmented from the A-7 instar onwards, whereas Smith & Martens (2000) illustrate a two-segmented *Y* aesthetasc in the A-6 instar and a three segmented *Y* in the A-5 instar onward for *E. virens*. Other studies do not illustrate this feature in

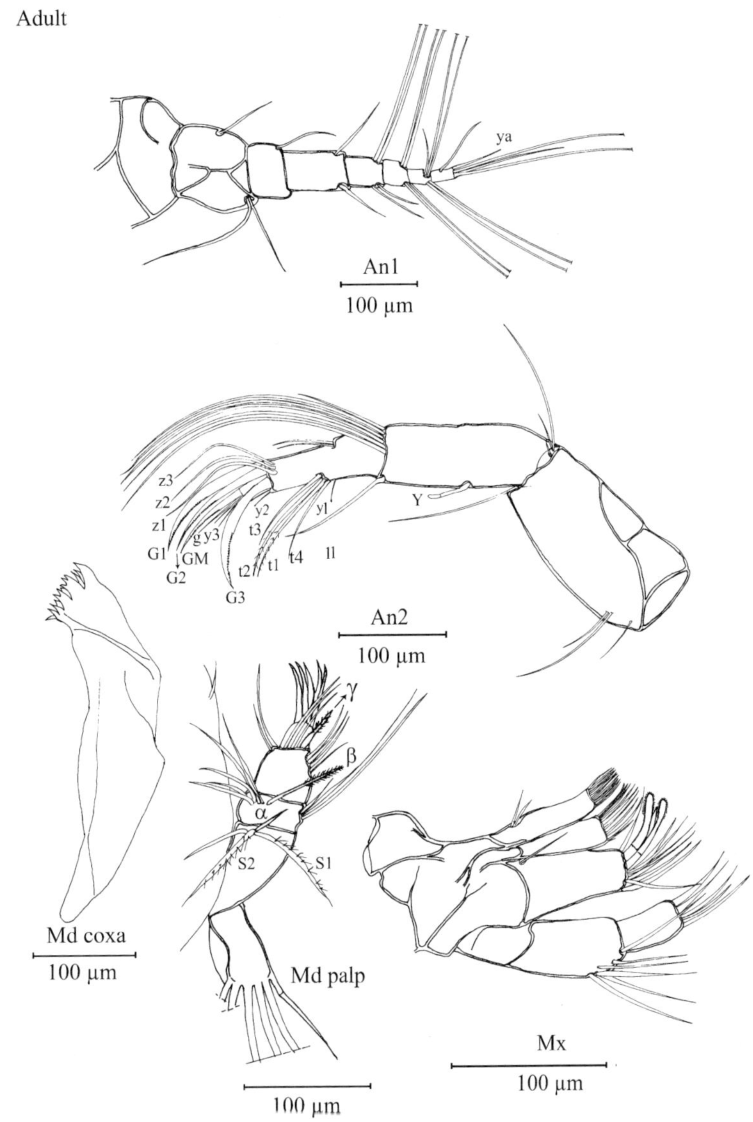

Fig. 11 *Heterocypris salina* Adult instar; An1, An2, Md palp, Md coxa, Mx

sufficient detail to accurately determine segmentation, but both Schreiber (1922) and Kesling (1951) figure it with at least two segments from the A-6 instar onwards for *H. incongruens* and *C. vidua* respectively.

(3) The mandibular palp of *H. salina* has four podomeres in the A-7 instar onwards, which concurs with both the studies of Claus (1868), Schreiber (1922), Kesling (1952), Fox (1964) and Roessler (1983). Smith & Martens (2000) show only three mandibular podomeres in the A-7 and A-6 instars of *E. virens*, but in both these instars there are setae that protrude from where the extra podomere boundary should be, indicating that it maybe just weakly expressed rather than missing altogether.

(4) In *H. salina* a single *Zahnborsten* appears on the third endite of the maxillula in instar A-6 for the first time, and the second *Zahnborsten* appears in the A-5 instar. Most other studies don't show the maxillula in detail, but both Schreiber (1922) and Smith & Martens (2000) show the third endite with one claw-like seta in the A-6

Fig. 12 *Heterocypris salina* Adult instar; L5, L6, L7, CR

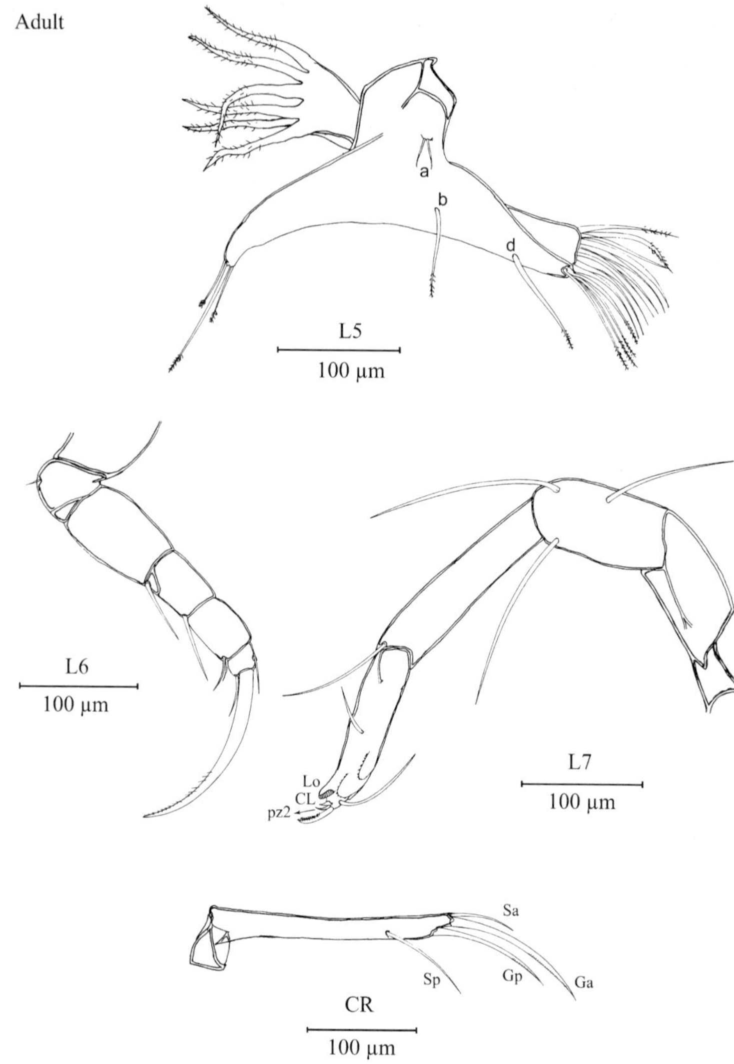

instar and two *Zahnborsten* in the A-5 instar for *H. incongruens* and *E. virens* respectively.

(5) The palp of the maxillula is segmented in the A-5 instar onwards in *H. salina*, which concurs with Schreiber (1922). Roessler (1983), however, figured the palp segmented from one stage earlier, the A-6 instar onwards for *H. bogotensis*. Both Fox (1964) and Ghetti (1970) both report that the palp is only indistinctly segmented in the A-5 instar and fully segmented in the A-4 instar. Smith & Martens (2000) noted that the palp is not segmented in the A-5 instar, but is segmented in the A-4 instar for *E. virens*.

(6) This study revealed that the A-4 instar has a caudal ramus that consists of the claws *Ga*, *Gp* and the seta *Sp*. Schreiber (1922) and Roessler (1983), however, both document a caudal ramus consisting of just the two claws in the A-4 instars of *Heterocypris* species. Fox (1964), Ghetti (1970) and Smith & Martens (2000) all illustrate similar caudal rami to those of Schreiber and Roessler for *Herpetocypris agilis*, *Stenocypris* species and *E. virens* respectively. Kesling (1951) noted that the caudal ramus of *Cypridopsis vidua* from the A-4 instar is reduced and resembles the reduced form of the adults. The early appearance of the *Sp* seta in *H. salina* therefore seems to be

Table 1 Summary tables of the ontogeny of six species

Instar	An1	An2	Md	Mx	L5	L6	L7	Cr
Heterocypris salina (Cypridoidae)								
1 (A-8)	•	•	○					
2 (A-7)	•	•	•	○				○
3 (A-6)	•	•	•	•				○
4 (A-5)	•	•	•	•	○			○
5 (A-4)	•	•	•	•	•	○		•
6 (A-3)	•	•	•	•	•	•	○	•
7 (A-2)	•	•	•	•	•	•	•	•
8 (A-1)	•	•	•	•	•	•	•	•
9 (A)	•	•	•	•	•	•	•	•
Eucypris virens (Cypridoidae)								
1 (A-8)	•	•	○					
2 (A-7)	•	•	•	○				○
3 (A-6)	•	•	•	•				○
4 (A-5)	•	•	•	•	○			○
5 (A-4)	•	•	•	•	•	○		•
6 (A-3)	•	•	•	•	•	•	○	•
7 (A-2)	•	•	•	•	•	•	•	•
8 (A-1)	•	•	•	•	•	•	•	•
9 (A)	•	•	•	•	•	•	•	•
Taken from Smith & Martens (2000)								
Loxoconcha japonica (Cytheroidea) and *Terrestricythere elisabethae* (Terrestricytheroidea)								
1 (A-8)	•	•	•					○
2 (A-7)	•	•	•	○				○
3 (A-6)	•	•	•	•				○
4 (A-5)	•	•	•	•	○			○
5 (A-4)	•	•	•	•	•	○		•
6 (A-3)	•	•	•	•	•	•	○	•
7 (A-2)	•	•	•	•	•	•	•	•
8 (A-1)	•	•	•	•	•	•	•	•
9 (A)	•	•	•	•	•	•	•	•
Taken from Smith & Kamiya (2003) and Horne et al. (2004)								
Uncinocythere occidentalis (Cytheroidea, Entocytheridae)								
1 (A-8)	First instar not recovered							
2 (A-7)	•	•	•					○
3 (A-6)	•	•	•	○				○
4 (A-5)	•	•	•	•	○			○
5 (A-4)	•	•	•	•	•	○		○
6 (A-3)	•	•	•	•	•	•	○	
7 (A-2)	•	•	•	•	•	•	•	
8 (A-1)	•	•	•	•	•	•	•	
9 (A)	•	•	•	•	•	•	•	
Taken from Smith & Kamiya (2005)								
Neonesidea oligodentata (Bairdioidea)								
1 (A-8)	First instar not recovered							
2 (A-7)	•	•	•	○				○
3 (A-6)	•	•	•	•				○
4 (A-5)	•	•	•	•	○			○
5 (A-4)	•	•	•	•	•	○		•
6 (A-3)	•	•	•	•	•	•	○	•
7 (A-2)	•	•	•	•	•	•	•	•
8 (A-1)	•	•	•	•	•	•	•	•
9 (A)	•	•	•	•	•	•	•	•
Taken from Smith & Kamiya (2002)								

○; Anlage of the limb, •; Limb resembling that of adult, An1 antennule, An2 antenna, Md mandible, Mx maxillula, L_5 fifth limb, L_6 sixth limb, L_7 seventh limb, Cr furca restricted to this species. From the A-3 instar onwards the caudal rami of all species are similar, with the exception of *C. vidua.*

(7) Smith & Martens (2000) mentions an antenna with the *Gm* claw in the A-3 instar of *E. virens*, whereas this is missing in the antenna of *H. salina*. Likewise, no *c* setae on the fifth limb and *d2* seta on the sixth limb were observed for *H. salina*, both of which are present in *E. virens*. These are differences that are observable in adult specimens, but it is noteworthy that the missing features in the adults of *H. salina* are missing in all juvenile stages as well, rather than just missing in the adult stage alone.

(8) The *z3* seta of the antenna of *H. salina* first appears in the adult stage, one stage later than that of *H. bogotensis, Sclerocypris* species, *E. virens* and *Pseudocandona serbani* (Roessler, 1983; Broodbakker & Danielopol, 1982; Martens, 1987; Smith & Martens, 2000).

(9) The endopodite (= palp) of the fifth limb of *H. salina* has three segments in instar A-3, and two segments in both the A-2 and A-1 instars. Fox (1964) reported that the palp was segmented in the A-3 instar, weakly segmented in the A-2 instar and with no segmentation in the A-1 instar for the cyprid species he studied. Schreiber (1922), Kesling (1951), Ghetti (1970), Roessler (1983) and Smith & Martens (2000) all reported that the A-3 instar had a three-segmented palp, but this segmentation was missing from the A-2 instar onwards. All studies reported a palp with no segmentation in the adult.

(10) This study documented an antennule with just four podomeres in the A-8 and A-7 instars of *H. salina*, in contrast to five documented by Claus (1868), Schreiber (1922), Kesling (1951), Ghetti (1970), Smith & Martens (2000) and Smith & Tsukagoshi (2005). From the A-6 instar onwards the development is similar.

Some of these features mentioned above appear either earlier or later in the development of *H. salina* compared with other species, even species in the same genus. For instance, the *G1* claw of the antennule, the *Sp* seta of the caudal ramus, and the *Zahnborsten* of the maxillula all appear earlier in *H. salina* compared with other species. Loss of segmentation in the fifth limb palp, appearance of segmentation of the antennal *Y* aesthetasc, and the appearance of the *z3* seta on the antenna however, all occur in later instars of *H. salina* compared with other species. This shows that even within a family, closely related species can show unexpected variations in development. The early appearance of the *Sp* seta on the caudal ramus in the A-4 instar, and the four-segmented antennule in the first two instars of *H. salina* are particularly noteworthy. Both of these features vary in the juveniles compared with those of other species, but the adult forms are similar to other species. Such a divergence in development in just one or two early instars, rather than all subsequent instars, is puzzling. The late appearance of the antennal *z3* seta in *H. salina* is also unexpected, as other Cyprididae (e.g., *H. bogotensis and E. virens*) and Candonidae (e.g., *Pseudocandona serbani*) species show a similar *z3* development to each other (Broodbakker & Danielopol, 1982; Roessler, 1983; Smith & Martens, 2000). This would indicate that the late appearance of this feature in *H. salina* is an apomorphy.

Within the genus, *H. salina* varies from the two other *Heterocypris* species previously studied, *H. incongruens* and *H. bogotensis*, in the following: the four-segmented antennules of the first two instars (five segmented in the other two species), the segmentation of the fifth limb palps in the A-2 and A-1 instars (not segmented in the other two species) and the appearance of the *Sp* seta of the caudal ramus in the A-4 instar (first appears in the A-3 instar in the other two species). (Other features are not clearly illustrated in all three studies to allow a comparison.) This suggests that *H. incongruens* and *H. bogotensis* are more closely related to each other than to *H. salina*. As some features mentioned above are present in very early instars (e.g., A-8 and A-7 instars for the antennule segmentation) they suggest that divergence of the lineage that led to *H. salina* occurred from an early point in the history of the genus.

If the variations in ontogeny of these features can be verified in other species/genera, then they maybe useful characters to use in phylogenetic analyses at the genus and family levels.

H. salina (Cypridoidae) has nine instars from the egg to the adult stage. This is the same number of instars as e.g., *E. virens* (Cypridoidea) (Smith & Martens 2000), *L. japonica* (Cytheroidea) (Smith & Kamiya, 2003) and *T. elisabethae* (Terrestricytheroidea) (Horne et al., 2004), but one more than *N. oligodentata* (Smith & Kamiya, 2002) and *U. occidentalis* (Entocytheridae) (Smith & Kamiya, 2005), which have only eight instar stages. Smith & Kamiya (2005) noted that this is probably due to the first instar in *N. oligodentata* and *U. occidentalis* moulting within the egg. Both the Terrestricytheroidea and Cytheroidea have the *Anlagen* of the caudal ramus in the A-8 instar, but it is missing in the A-8 instar of the Cypridoidea. From instar A-7 onwards the general appearance of the limbs of *H. salina* are similar to those of species from the superfamilies Cypridoidea, Cytheroidea, Bairdioidea and Terrestricytheroidea (Table 1), with the exception of the maxillula of the Entocytheridae (a family of the Cytheroidea), which first appears in instar A-6, one stage later than other groups.

Acknowledgements We are very grateful to Dr. Renate Matzke-Karasz an an anonymous reviewer for comments on earlier versions of this manuscript and are also grateful to Ferda Perçin and Hüseyin Akıncı for their helps in drawings.

References

Athersuch, J., D. J. Horne & J. E. Whittaker, 1989. Marine and Brackish Water Ostracods (Superfamilies Cypridacea and Cytheracea). In Kermack, D. M. & R. S. K. Barnes (eds), Synopses of the British Fauna (New Series), 43. Brill, Leiden.

Baltanás, A., M. Otero, L. Arqueros, G. Rossetti & V. Rossi, 2000. Ontogenetic changes in the carapace shape of non-marine ostracods *Eucypris virens* (Jurine). Hydrobiologia 419: 65–72.

Broodbakker, N. W. & D. L. Danielopol, 1982. The chaetotaxy of Cypridacea (Crustacea, Ostracoda) limbs: proposals for a descriptive model. Bijdragen tot de Dierkunde, 52: 103–120.

Claus, C., 1865. Zur näheren Kenntnis der Jugendformen von Cypris ovum. Zeitschrift für wissenschaftliche Zoologie 15: 391–398.

Claus, C., 1868. Beiträge zur Kenntnis der Ostracoden. 1: Entwicklungsgeschichte von *Cypris*. Schriften der Gesellschaft zur Beförderung der Gesammten Naturwissenschaften zu Marburg, 9: 151–166.

Fox, H. M., 1964. On the larval stages of Cyprids and on *Siphlocandona* (Crustacea. Ostracoda). Proceedings of the Zoological Society of London 142: 165–176.

Ghetti, P. F., 1970. The taxonomic significance of ostracod larval stages: with examples from the Burundi rice-fields. Bollettino di Zoologia 37: 103–119.

Hart, C. W. Jr., L. C. Hayek, J. Clark & W. H. Clark, 1985. The Life History and Ecology of the entocytherid ostracod *Uncinocythere occidentalis* (Kozloff and Whitmann) in Idaho. Smithsonian Contributions to Zoology 419: 1–22.

Horne, D. J. & R. J. Smith, 2004. First British record of *Potamocypris humilis* (Sars, 1924), a freshwater ostracod with a disjunct distribution in northern Europe and southern Africa. Bollettino della Società Paleontologica Italiana 43: 297–306.

Horne, D. J., R. J. Smith, J. E. Whittaker & J. W. Murray, 2004. The first British record and a new species of the superfamily Terrestricytheroidea (Crustacea, Ostracoda): morphology, ontogeny, lifestyle and phylogeny. Zoological Journal of the Linnean Society 142: 253–288.

Kesling, R. V., 1951. The morphology of ostracod molt stages. Illinois Biological Monographs 21, Urbana.

Martens, K., 1987. Homology and functional morphology of the sexual dimorphism in the antenna of *Sclerocypris* Sars, 1924 (Crustacea, Ostracoda, Megalocypridinae). Bijdragen tot de Dierkunde 57: 183–190.

Meisch, C., 2000. Freshwater Ostracoda of Western and Central Europe. In Schwoerbel, J. & P. Zwick (eds), Süßwasserfauna von Mitteleuropa 8/3. Spektrum Akademischer Verlag, Heidelberg-Berlin.

Roessler, E. W., 1983. Estudios taxonomicos, ontogeneticos, ecologicos y etologicos sobre los ostracodos de agua dulce en Colombia. IV. Desarrollo postembrionario de *Heterocyrpis bogotensis* Roessler (Ostracoda, Podocopa, Cyprididae). Caldasia, 13: 755–776.

Scheerer-Ostermeyer, E., 1940. Beitrag zur Entwicklungsgeschichte der Süßwasserostrakoden. Zoologische Jahrbücher, Abteilung für Anatomie und Ontogenie der Tiere, 66: 349–370.

Schreiber, E., 1922. Beiträge zur Kenntnis der Morphologie, Entwicklung und Lebensweise der Süßwasser-Ostracoden. Zoologische Jahrbücher, Abteilung für Anatomie und Ontogenie der Tiere 43: 485–539.

Smith, R. J. & K. Martens, 2000. The Ontogeny of cypridid Ostracoda *Eucypris virens* (Jurine, 1820) (Crustacea, Ostracoda). Hydrobiologia 419: 31–63.

Smith, R. J. & T. Kamiya, 2002. The Ontogeny of *Neonesidea oligodentata* (Bairdioidea, Ostracoda, Crustacea). Hydrobiologia 489: 245–275.

Smith, R. J. & T. Kamiya, 2003. The Ontogeny of *Loxoconcha japonica* Ishizaki, 1968 (Cytheroidea, Ostracoda, Crustacea). Hydrobiologia 490: 31–52.

Smith, R. J. & T. Kamiya, 2005. The Ontogeny of the entocytherid Ostracod *Uncinocythere occidentalis* (Kozloff & Whitman, 1954) Hart, 1962 (Crustacea). Hydrobiologia 538: 217–229.

Smith, R. J. & A. Tsukagoshi, 2005. The chaetotaxy, ontogeny and musculature of antennule of podocopid ostracods (Crustacea). Journal of Zoology, London 265: 157–177.

Weygoldt, P., 1960. Embryologische Untersuchungen an Ostrakoden: Die Entwicklung von *Cyprideis litoralis* (G. S. Brady) (Ostracoda, Podocopa, Cytheridae). Zoologische Jahrbücher, Abteilung für Anatomie und Ontogenie der Tiere, 78: 367–426.